MODERN COMMUNICATION CIRCUITS

McGraw-Hill Series in Electrical Engineering

Consulting Editor
Stephen W. Director, Carnegie-Mellon University

Circuits and Systems
Communications and Signal Processing
Control Theory
Electronics and Electronic Circuits
Power and Energy
Electromagnetics
Computer Engineering
Introductory
Radar and Antennas
VLSI

Previous Consulting Editors

Ronald M. Bracewell, Colin Cherry, James F. Gibbons, Willis W. Harman, Hubert Heffner, Edward W. Herold, John G. Linvill, Simon Ramo, Ronald A. Rohrer, Anthony E. Siegman, Charles Susskind, Frederick E. Terman, John G. Truxal, Ernst Weber, and John R. Whinnery

Communications and Signal Processing

Consulting Editor
Stephen W. Director, Carnegie-Mellon University

MODERN COMMUNICATION CIRCUITS

Jack Smith

Professor of Electrical Engineering
University of Florida

McGraw-Hill Book Company

New York   St. Louis   San Francisco   Auckland   Bogotá   Hamburg
London   Madrid   Mexico   Montreal   New Delhi
Panama   Paris   São Paulo   Singapore   Sydney   Tokyo   Toronto

MODERN COMMUNICATION CIRCUITS
INTERNATIONAL EDITION

Copyright © 1986
Exclusive rights by McGraw-Hill Book Co. — Singapore
for manufacture and export. This book cannot be re-exported from the
country to which it is consigned by McGraw-Hill.

2nd Printing 1987

This book was set in Times Roman by Scanway Graphics International Ltd.
The editors were Sanjeev Rao and David A. Damstra;
The production supervisor was Marietta Breitwieser.
The drawings were done by ANCO/Boston.

Library of Congress Cataloging in Publication Data

Smith, Jack date
  Modern communication circuits.

  (McGraw-Hill series in electrical engineering.
Communications and signal processing)
  Includes bibliographies and index.
  1.   Radio circuits.   I.   Title.   II.   Series.
TK6553.S5595   1986   621.3841 '2   84-26166
ISBN 0-07-058730-2
ISBN 0-07-058731-0 (solutions manual)

When ordering this title use ISBN 0–07–Y66544–3

Printed and Bound in Singapore by Fong & Sons Printers Pte Ltd

To My Mother
Laura Kornkven Smith

CONTENTS

Appendixes 503

Index 553

This book presents fundamental analysis and design techniques for modern communication circuits covering the frequency range up to approximately 100 MHz. The coverage reflects the practice of modern communication systems design to use integrated circuits, to minimize tuning operations by using broadband circuits, and to use more field-effect transistors in high-frequency circuits. Much of the information presented here appears in book form for the first time. Some of it is original, much of it has been gathered from a multifarious literature on communication electronics, and some has been adapted from early publications on vacuum-tube electronics. Practical approximations are emphasized rather than theoretical derivations based on complex circuit models. The approximations provide more insight into the design process than do lengthy derivations. Computer simulation is frequently used to establish the accuracy of the approximations.

The book is intended to be used as a textbook in senior-level and beginning-graduate-level courses in electrical engineering and as a reference book for practicing engineers. It is assumed that the reader has the basic background in linear transistor-circuit theory obtained from a junior-level electrical engineering course, but not necessarily in the subjects which are becoming restricted to specialized courses in high-frequency electronics—such as tuned-circuit analysis. Chapter 2 presents a review of the necessary electronic circuit background.

A characteristic of almost all realistic communication circuits is that they are too complex for a complete analysis and approximations must be made. The judicious use of approximations is also required in the design process. A goal of this book is to illustrate many of the approximations convenient and effective for the analysis and design of communication circuits. The circuit models are often incomplete, but more accurate and complex models are usually best treated with computer-aided analysis methods available to most electrical engineering students and electrical engineers. Computer-aided analysis is frequently used in the examples.

Most, if not all, communication circuits are available as integrated circuits, but there are several reasons why designers must still be familiar with the discrete-circuit techniques. First of all, there are many communication systems which cannot yet be realized in integrated-circuit form, because circuit flexibility is compromised and an inductor-capacitor tuned circuit cannot be fabricated on a chip. Second, it is the case that the performance of discrete communication circuits is superior. It is necessary for the communication circuit designer to consider the trade-offs between size, cost, output power, power consumption, noise, and distortion. Third, even when integrated circuits are used, a basic understanding of communication circuitry is required. Also, monolithic power amplifiers create many additional problems, which result in the complete amplifier occupying more space than would be required of a discrete power amplifier. This book incorporates the impact that integrated circuits are having on the design of communication systems. Two chapters are devoted to the phase-locked loop. The importance of this device in modern systems is largely due to its realization as an integrated circuit. The very recent digital phase-locked loops are discussed in Chap. 8.

It is not possible to completely cover a subject as broad as communication electronics in a single text. Much of the methodology presented here is applicable to communication systems in all frequency ranges, but distributed-parameter circuit analysis techniques, not covered in this book, are usually more accurate at frequencies above 100 MHz. Digital circuits are playing an increasingly important role in communication systems. Many applications of digital circuits are described in this text, but conventional logic circuits are not considered, as the subject is well covered in many texts.

Chapter 1 presents an introduction to communication circuits and discusses recent trends in receiver design. Chapter 2 reviews linear small-signal analysis of bipolar and field-effect transistor amplifiers. Operational amplifiers are also included, since there are now devices available with gain-bandwidth products sufficiently large that they can be used in high-frequency communication circuits. Chapter 3 defines the noise and distortion specifications used to describe communication systems. The fundamentals of low-noise amplifier design are also included in this chapter. The simple parallel and series tuned circuits still so important in modern communication systems are discussed in Chap. 4. Methods for analyzing the high-frequency performance of transistor amplifiers are given in Chap. 5, together with models for automatic gain control systems. The transformers which are so useful in modern communication systems are discussed in Chap. 6. The transmission-line transformer is still the most suitable lumped impedance-matching network for the realization of untuned broadband interstage and output-matching networks. Chapter 7 contains an in-depth treatment of the analysis and design of high-performance transistor oscillator circuits, including crystal and voltage-controlled oscillators. The phase-locked loop is one of the most versatile and widely applied circuits in communication systems. Its availability as an inexpensive integrated circuit implies that it will continue to find application in virtually all communication systems. Chapter 8 is devoted to applications of the phase-locked loop and Chap. 9 to its analysis. Integrated circuits have also resulted in the design of frequency synthesizers, which have altered the design of most new communication systems using frequency tuning. The frequency synthesizer has also allowed for the practical implementation of modern communication techniques such as frequency hopping. Chapter 10 provides a detailed treatment of frequency synthesizers. Methods for frequency shifting, modulation, and demodulation are discussed in Chap. 11, including methods for frequency- and phase-shift keying and other methods for handling digital signals. Chapter 12 discusses the design and analysis of power amplifiers and includes several curves useful for the design of class C power amplifiers.

The book contains enough material for a two-semester course, provided it is supplemented with articles from the current literature. Chapters 1, 3 through 7, and 10 or 12 have been used for a one-semester course at the University of Florida. Each chapter contains problems which emphasize particular points; several chapters include problems which extend the material covered. Component and integrated specification sheets contained in the Appendixes are used in some of the problems.

The writing of this book resulted from many discussions with my friend Dr. Ulrich Rohde, president of Communications Consulting Corporation, Upper Saddle River, New Jersey, and director of Radio Systems, RCA Corporation, Government Communications Systems, Camden, New Jersey. Much of the information on state-of-the-art circuits for receiver and frequency synthesizer design was obtained during my collaboration with him. The information and advice he provided were of tremendous help. The suggestions, corrections, and circuit designs of my students have also markedly improved the quality of this book.

Jack Smith

MODERN COMMUNICATION CIRCUITS

INTRODUCTION TO RADIO
COMMUNICATION SYSTEMS

1.1 INTRODUCTION

Communication systems transmit information from one place to another by means of electric energy. The frequency used for the information transmission varies from the very low frequencies used in direct telephone communication to optical frequencies, also used for telephone communication. This book describes the analysis and design of electronic circuits used in radio-frequency communication systems covering the frequency range up to approximately $100\,\mathrm{MHz}$ (10^8 cycles per second). The material is directly applicable to many other systems, including television and spectrum analyzers where the design of low-noise, high-frequency receivers is of paramount importance. For frequencies greater than $100\,\mathrm{MHz}$ different circuit models, particularly distributed-parameter circuits, are more accurate. Low-frequency circuitry is discussed, but the emphasis here is on the radio-frequency circuits.

The following chapters treat the analysis and design of fundamental circuits of communication receivers and transmitters. Integrated circuits have simplified the system design, but the communication system designer still needs to be familiar with many circuit techniques and the simplifying approximations which apply in this frequency range. The designer is often faced with a choice between using an integrated circuit (IC) or a discrete component version of the same circuit. The decision is based on many factors, including cost, size, power consumption, noise, and distortion. Chapter 3 presents quantitative criteria for evaluating a circuit's noise and distortion performance. The application of integrated circuits in a communication system requires a knowledge of electronic-circuit theory in order to properly interface the IC with the rest of the system. We will study the electronic circuits of the various subsystems, including oscillators, amplifiers, transformers, modulators, and demodulators, which make up a communication system. The mathematical analysis of the many modulation methods is not considered as it is well described in the many good texts on communications theory.

1.2 NETWORK THEORY

This section briefly reviews the concepts of network theory that are applied in the following chapters. The usual variables in an electronic circuit are the voltages and currents measured at various points in the circuit. The excitation and response can be described in the time domain, but determining the response in the time domain involves the solution of integrodifferential equations and rarely provides insight into the design process. For linear time-invariant systems it is usually easier to obtain the system response using the Laplace transform. The Laplace transform of the time variable $v(t)$ is

$$V(s) = \int_0^\infty v(t)e^{-st}\, dt \tag{1.1}$$

where s has the dimensions of frequency and is known as the *complex-frequency variable*. A linear system transfer function $H(s)$ is defined as

$$H(s) = \frac{R(s)}{V(s)} \tag{1.2}$$

where $R(s)$ is the Laplace transform of the response to an excitation $V(s)$. Linear circuit transfer functions can easily be obtained by interpreting an inductor as having a complex impedance sL and a capacitor as having a complex impedance $(sC)^{-1}$. The method is illustrated by the following example.

Example 1.1 Determine the transfer function $V_o(s)/V_i(s)$ of the circuit shown in Fig. 1.1.

SOLUTION In this circuit the inductor has been modeled as a complex impedance sL and the capacitor as a complex impedance $(sC)^{-1}$. By using the voltage-divider rule of circuit analysis we find that the transfer function

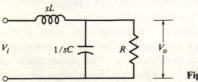

Figure 1.1 A low-pass filter.

$H(s)$ is

$$H(s) = \frac{V_o(s)}{V_i(s)} = \frac{R/(RsC+1)}{R/(RsC+1)+sL} = \frac{R}{s^2RLC+sL+R} \qquad (1.3)$$

For simplicity the excitation and response are often written as V_o and V_i when there is no possibility of confusion.

Another advantage of describing the system response by the linear transfer function $H(s)$ is that the frequency response of the network can be obtained by setting $s = j\omega$. That is, if the system is stable and the excitation is a sinusoid, the steady-state response (after the transients have decayed to a negligible value) will also be a sinusoid of the same frequency. If

$$v_i(t) = V \sin \omega t$$

the steady-state response is

$$r(t) = R \sin(\omega t + \phi)$$

where $\qquad |H(j\omega)| = \dfrac{V}{R} \qquad$ and $\qquad \arg H(j\omega) = \phi$

Example 1.2 Determine the frequency response of the network shown in Fig. 1.1 for $L = 0.5$ H, $C = 2$ F, and $R = 1\ \Omega$.

SOLUTION The frequency response is obtained by substituting $s = j\omega$ in the transfer function. In this case Eq. (1.3) becomes

$$H(j\omega) = [(j\omega)^2 + 0.5j\omega + 1]^{-1}$$

the magnitude of the response is the frequency-dependent function

$$|H(j\omega)| = \left\{ \left[(1-\omega^2)^2 + \left(\frac{\omega}{2}\right)^2 \right]^{1/2} \right\}^{-1}$$

and the phase shift is also frequency-dependent

$$\arg H(j\omega) = -\tan^{-1} \frac{0.5\omega}{1-\omega^2}$$

A linear transfer function without ideal delay elements will have the form

$$H(s) = \frac{A(s)}{B(s)} \qquad (1.4)$$

where $A(s)$ and $B(s)$ are polynomials in s. The zeros of $A(s)$ are referred to as *zeros of the transfer function*, and the zeros of $B(s)$ are referred to as *poles of the transfer function*. The poles are the values of s for which the magnitude of the transfer function is infinite. In order for the transfer function to be stable, all of the poles must lie in the left half of the s-plane (the real part of the pole must be negative). The stability problem is considered in detail in Chap. 9.

Example 1.3 Calculate the poles and zeros of the transfer function given in Example 1.2:

$$H(s) = (s^2 + 0.5s + 1)^{-1}$$

SOLUTION The transfer function has no finite zeros. Since the order of the denominator polynomial is two higher than that of the numerator, the transfer function has two zeros at infinity. The poles

$$s_1, s_2 = -0.25 \pm j\frac{(3.75)^{1/2}}{2}$$

are located in the left half plane. The real part of each pole is -0.25 and the imaginary parts are $\pm j(3.75)^{1/2}/2$.

1.3 MODULATION

In order for a signal to contain information, some feature of the signal must be varied in accordance with the information to be transmitted. Early radio communications conveyed information by the presence, or absence, of the signal. This method was soon surpassed by amplitude modulation of the radio wave by an audio signal. The amplitude-modulation process provides a means of transmitting· voice communications, and its development led to the rapid establishment of the radio broadcasting industry.

Angle modulation is another method of transmitting information widely used in high-frequency communication systems. An angle-modulated signal is described by the equation

$$S(t) = A \sin(\omega_o t + \phi)$$

The amplitude remains constant, and the angle ϕ is varied in response to the modulating signal. Both phase and frequency modulation can be used. One function of the receiver is to recover (demodulate) the original from the modulated signal. The circuitry for implementing the various types of modulation and demodulation are described in more detail in Chap. 12. Today digital modulation techniques are being more frequently employed, particularly in satellite and telephone communication systems. Digital modulation implies that a parameter of the signal is varied in response to a digital signal. Amplitude, phase, or frequency modulation can be varied in response to a digital signal. That is, digital modulation is an extension of one or more of the conventional amplitude- or angle-modulation methods.

Figure 1.2 illustrates a simplified block diagram of a digital single-channel-per-carrier (SPSC) satellite communications channel. Each voice channel is sampled and then converted to a digitally encoded signal that modulates a low-frequency (baseband) carrier signal. The modulated signal uses a different carrier frequency sufficiently separated so that the signals can be combined

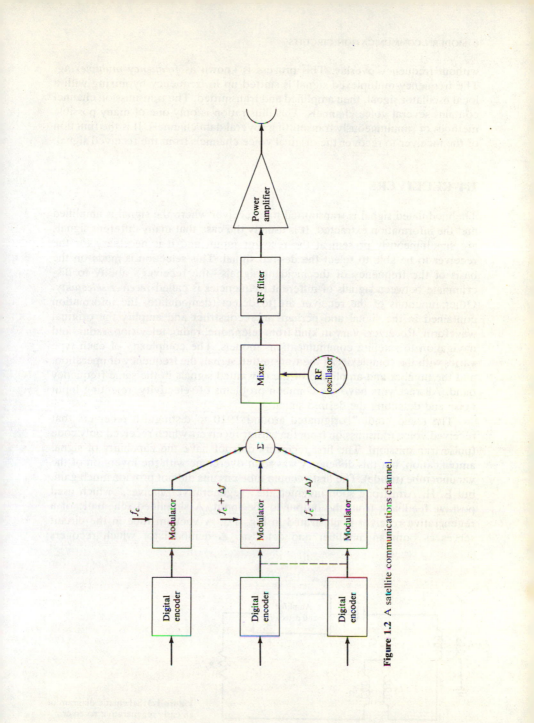

Figure 1.2 A satellite communications channel.

without frequency overlay. This process is known as *frequency multiplexing*. The frequency-multiplexed signal is shifted up in frequency by mixing with a local oscillator signal, then amplified and transmitted. The transmission channel contains several voice channels. This illustration is only one of many possible methods of simultaneously transmitting several data channels. It is the function of the receiver to recover the original voice channels from the received signal.

1.4 RECEIVERS

The modulated signal is transmitted to a receiver where the signal is amplified and the information extracted. It is usually the case that many different signals are simultaneously present at the receiver input, and it is necessary for the receiver to be able to select the desired signal. This selection is made on the basis of the frequency of the incident signals—the receiver's ability to discriminate between signals of different frequencies is called *receiver selectivity*. Other functions of the receiver are to detect (demodulate) the information contained in the signal and perhaps to reconstruct and amplify the original waveform. Receivers vary in kind from telephone, radio, television, radar, and navigation to satellite communications models. The complexity of each type varies with the complexity of the transmitted signal, the frequency of operation, and the number and amplitude of the unwanted signals in the same frequency band. All receivers have the common problems of selectivity, rejecting input noise and detecting the desired signal.

The name "radio" originated around 1910 to distinguish receivers that received voice transmission from the earlier receivers which received only code (pulse transmission). The first receivers did not have the capability of signal amplification, but this deficiency was soon overcome with the invention of the vacuum tube (triode). The first vacuum tube circuits did not provide much gain, but E. H. Armstrong soon invented the "regenerative receiver," which used positive feedback from the output to the input. A simplified schematic of a regenerative receiver is illustrated in Fig. 1.3. A vacuum tube in the circuit serves as both an amplifier and detector. A demodulator which recovers

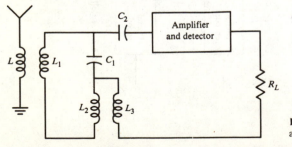

Figure 1.3 Schematic diagram of an early regenerative receiver.

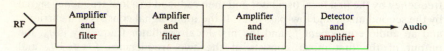

Figure 1.4 Block diagram of an early tuned-radio-frequency receiver.

(detects) the modulating signal is known as a *detector*. The ac output signal is fed back in phase with the input signal (regenerated), increasing the loop gain. The regenerative receiver was probably the first application of electronic feedback; it quickly led to the invention of the electronic oscillator, since the circuit was susceptible to oscillations. The electronic oscillator greatly improved transmitter design.

The regenerative receiver was soon replaced by the tuned-radio-frequency (TRF) receiver. A block diagram of a typical TRF receiver is illustrated in Fig. 1.4; it consists of three tuned RF amplifiers in cascade followed by a detector and power amplifier. This receiver was hard to operate because of the difficulty of tuning all the RF amplifiers to the same frequency. The TRF receiver became obsolete with the invention of the superheterodyne receiver by E. H. Armstrong. The "superhet" eliminates the need for tuning all the RF amplifiers to the frequency of the input signal by shifting the input signal frequency to that of the fixed frequency of the receiver filter. It is possible to build fixed-frequency filters and amplifiers that are far superior to variable-frequency filters. The superheterodyne principle is still used in virtually all receivers. It consists of multiplying, or beating (heterodyne is from the Greek *heteros*, "other," and *dynamis*, "force"), the input signal with a signal generated in the local oscillator. If a sine wave of frequency ω_c is multiplied by a sine wave of frequency ω_L, the resultant signal consists of the sine waves of frequencies $\omega_c \pm \omega_L$. That is,

$$\sin \omega_c t \sin \omega_L t = \frac{\cos (\omega_c - \omega_L)t - \cos (\omega_c + \omega_L)t}{2}$$

A simplified block diagram of a superheterodyne receiver is illustrated in Fig. 1.5. In this type of receiver the incoming signal is converted to an intermediate

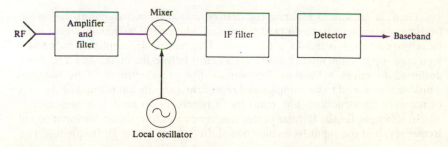

Figure 1.5 A superheterodyne receiver.

frequency by the first local oscillator and then reduced to a low-frequency signal by the second mixer and low-pass filter. If the input signal consists of a carrier f_c and an audio component f_a and the first local oscillator frequency is f_o, the output of the first mixer consists of the two frequencies $f_c + f_a + f_o$ and $f_c + f_a - f_o$. The local oscillator frequency f_o is selected so that one of these frequencies is equal to the center frequency of the intermediate-frequency (IF) filter (f_{IF}). Since f_a is usually much less than f_c, for all practical purposes

$$f_o = |f_{IF} - f_c| \quad \text{or} \quad f_o = f_{IF} + f_c$$

in order that the mixer-output frequency be at the center of the IF filter bandwidth.

One advantage of this form of detection is that the same high-quality filter can be used for all input frequencies. The frequency selection is obtained by varying the local oscillator frequency f_o. The IF filter output $f_{IF} + f_a$ is then reduced to f_a in the second mixer, which mixes the IF output with the second oscillator frequency (which is fixed at f_{IF}). A problem with this form of detection occurs when there are a large number of signals of different frequencies present at the input. Consider, for example, the receiver designed to select the difference frequency at the output of the first mixer. That is,

$$f_{IF} = |f_o - f_c| \quad \text{or} \quad f_{IF} = f_o + f_c$$

There exists another signal frequency f_{IM}, referred to as the *image frequency*, which when mixed with the local oscillator frequency f_o will produce a signal at the IF frequency. If $f_{IF} = |f_o - f_c|$, then $f_{IM} + f_o = f_{IF}$, or

$$f_{IM} = f_{IF} - f_o = 2f_{IF} + f_c$$

Example 1.4 Consider a receiver with the IF filter centered at 455 kHz. If it is desired to receive a 1-MHz input signal, the local oscillator is tuned to 1.455 MHz. Then an image frequency

$$f_{IM} = f_{IF} - f_o \quad \text{or} \quad f_{IM} = f_o + 2f_{IF} = 1.91 \text{ MHz}$$

if present at the input would also pass through the IF filter.

There is no way to separate the desired signal from a signal at the image frequency after they have entered the mixer. The image frequency signal must be removed before it arrives at the mixer. This can be accomplished by adding an image suppression filter (called a *preselector*) before the mixer. For a receiver designed to cover a band of frequencies the preselector must be tunable. Tunable filters tend to be complex and represent a significant portion of the cost of receiver construction. The majority of receivers do include a preselector.

In Example 1.4 the IF filter center frequency was lower than the input signal frequency, and the input frequency was shifted down to the IF frequency. The

IF center frequency can also be selected above the input signal frequency (up conversion). The following example illustrates the advantages of up conversion.

Example 1.5 Consider again the receiver that is designed to cover the frequency band 1 to 30 MHz, but which uses an IF filter centered at 40 MHz. For an input signal frequency f_s of 1 MHz there are two local oscillator frequencies f_o (41 and 39 MHz) that will result in a mixer product at 40 MHz. If the local oscillator frequency is 41 MHz, the image frequency will be 81 MHz; and if the local oscillator frequency is 39 MHz, the image frequency will be 79 MHz. Table 1.1 lists the local oscillator frequency and corresponding image frequency for several input frequencies spanning the frequency band to be covered. Either set of local oscillator frequencies could be selected, but normally the first set, f_o, would be used, since the ratio of the highest to the lowest frequency, 70/41, is lower than that used for f_o', 39/10. In the design and construction of variable-frequency oscillators, the ratio of highest to lowest frequencies is an important factor (the lower the ratio, the simpler the design).

One important feature of the up-conversion technique is that all image frequencies lie above the frequency band to be covered. This implies that all image frequencies can be suppressed by adding a low-pass filter to the input (30-MHz bandwidth); a tunable bandpass filter is not needed with up conversion. Until recently it was not possible to use up conversion in this frequency range because high-quality bandpass filters were not readily available in the 30- to 50-MHz region; however, recent improvements in the manufacturing technology now provide for high-quality crystal filters in this frequency range.

Another advantage of up conversion is that the oscillator-tuning ratio f_{max}/f_{min} is less than that for down conversion. If a down-conversion receiver with a 455-kHz IF is used to cover the same frequency band, the local oscillator frequency will need to vary from $f_{min} = 1.455$ MHz to $f_{max} = 30.455$ MHz, a tuning range of 20.93 : 1.

Table 1.1 Local oscillator frequency and corresponding image frequency, MHz

f_s	f_o	f_{IM}	f_o'	f_{IM}'
1	41	81	39	79
2	42	82	38	78
10	50	90	30	70
30	70	110	10	50

A Modern Communications Receiver

A block diagram of the high-frequency section of a modern radio receiver is essentially the same as that of the superheterodyne receiver illustrated in Fig. 1.5, but the circuits differ from the earlier models. One difference is that up conversion is used so that the input filter can remain a relatively simple low-pass filter that need not be tuned. Whether or not an amplifier is required before the mixer depends on the particular application and the receiver specifications. It will be shown in Chap. 3 that the absence of this amplifier can actually improve receiver performance in many applications.

Modern receivers differ from older receivers in many ways. A main difference is the frequency synthesizer used to generate the frequencies needed from the variable-frequency oscillator. The frequency synthesizer is capable of generating a large number of relatively precise frequencies from a single reference frequency. Although it does contain at least two oscillators, it has less frequency drift and noise than the older, conventional variable-frequency oscillators. Today the output frequency of frequency synthesizers can be precisely controlled using digital circuitry. This makes possible the microprocessor control of radio receivers and spectrum analyzers. Frequency synthesizers are described in detail in Chap. 10. Since most synthesizers now incorporate a phase-locked loop, a thorough understanding of phase-locked loop characteristics is necessary for frequency synthesizer design. This information is presented in Chaps. 8 and 9. This discussion of receivers has considered only the reception of amplitude-modulated signals. Receivers for FM reception are similar; the following section describes the main components of the FM receiver.

An Integrated-Circuit FM Receiver

Integrated-circuit technology has made it possible to design a radio with a minimum of external components. Figure 1.6 contains a simplified block diagram of the Siemans 5469, a high-frequency integrated circuit with a high

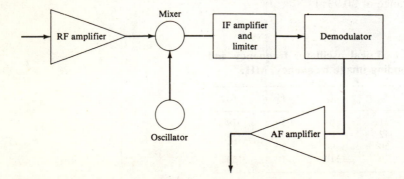

Figure 1.6 Block diagram of an integrated-circuit FM receiver.

level of system integration for FM communication. This bipolar device contains a complete FM receiver for input frequencies up to 50 MHz. The device includes (1) an RF input amplifier, (2) an oscillator, (3) a mixer, (4) an intermediate-frequency (IF) amplifier and limiter, (5) a coincidence demodulator, and (6) audio-frequency (AF) amplifiers with mute and volume control.

A complete FM system can be constructed with this one integrated circuit plus a crystal, a minimum of two inductors, and several resistors and capacitors used for frequency trimming and gain adjustment. The external resistors and capacitors provide the circuit with additional flexibility. Although the resistors and capacitors could be included in the IC, an inductor-capacitor tuned circuit cannot be fabricated in an IC chip, so a complete, self-contained IC receiver or transceiver (combined transceiver and receiver) is not possible with the current IC technology.

1.5 TRANSMITTERS

The transmitter modulates the information to be communicated onto a carrier, amplifies the waveform to the desired power level, and delivers it to the transmitting antenna. The elements of a transmitter are illustrated in Fig. 1.7. The transmitter includes a radio-frequency oscillator that is modulated by the message signal. The modulated signal is then multiplied in frequency up to the desired transmitting frequency and amplified to the desired power level in the power amplifier. The first radio transmitters worked by charging two electrodes separated by a gap. When the charge became sufficiently large a spark was created across the gap and electric energy was radiated. However, the spark-gap transmitter was slow, and it was difficult to accurately control the waveform and frequency of oscillation. The spark-gap transmitter became obsolete with the development of the electronic oscillator and vacuum tubes that could handle large amounts of power.

The transmitter topology illustrated in Fig. 1.7 is only one of many types. The modulation can actually take place in the power amplifier. Transmitter

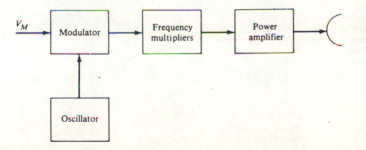

Figure 1.7 Block diagram of a transmitter.

topology depends on the type of modulation used and the necessary power level. Narrowband transmitters usually employ pulse, amplitude, or frequency modulation. Wideband transmitters use single-sided or multimode modulation and are used for long-range military, marine, aircraft, and amateur communications. Many transmitters and receiver circuits are similar; both require low-noise amplifiers and oscillators. Receivers are designed for the minimum detectable signal, while output power is of prime importance in the transmission of signals. The RF power amplifier is described in Chap. 11.

PROBLEMS

1.1 Determine the transfer function $V_o(s)/I(s)$ of the circuit shown in Fig. P1.1.

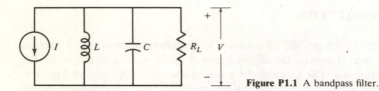

Figure P1.1 A bandpass filter.

1.2 Calculate the poles and zeros of the circuit in Fig. P1.1 for $L = 1$ H, $C = 15$ F, and $R = 1\ \Omega$.

1.3 A receiver is to be designed to cover the frequencies 20 to 40 MHz using an IF filter centered at 10 MHz. Specify two different local oscillator frequencies for each input frequency and determine the corresponding image frequency for each input frequency.

1.4 Select the local oscillator frequencies and specify the preselector frequency response for an up-conversion receiver covering the 2- to 30-MHz frequency range. The center frequency of the IF filter is 50 MHz.

1.5 Figure P1.5 illustrates a direct-conversion receiver in which the input signal is converted directly to an audio signal. Specify the local oscillator frequencies and the corresponding image frequencies for a direct-conversion receiver covering the 2- to 30-MHz range.

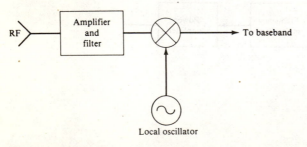

Figure P1.5 A direct-conversion receiver.

1.6 The double-conversion receiver in Fig. P1.6 employs two IF filters. Specify the required first-stage local oscillator frequencies f_1 and f_2 for a receiver covering the 8- to 10-MHz range. The center frequency of the first IF filter is 70 MHz and that of the second is 10 MHz.

ADDITIONAL READING

Carlson, A. B., *Communication Systems*, 3rd ed., McGraw-Hill, New York, 1986.

Franks, L. E., *Signal Theory*, rev. ed., Dowden & Culver, Stroudsburg, 1981.

Haykin, S., *Communication Systems*, 2nd ed., Wiley, New York, 1983.

Stremler, F. G., *Introduction to Communication Systems*, 3rd ed., Addison-Wesley, Reading, 1990.

Ziemer, R. E., and W. H. Tranter, *Principles of Communications*, 3rd ed., Houghton Mifflin, Boston, 1990.

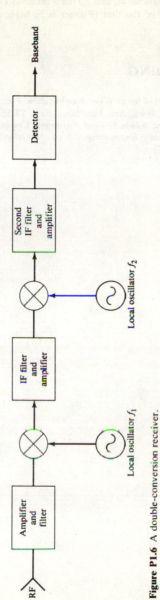

Figure P1.6 A double-conversion receiver.

1.6 The double-conversion receiver in Fig. P1.6 employs two IF filters. Specify the required local oscillator frequencies (f_1 and f_2) for a receiver covering the 2- to 30-MHz range. The center frequency of the first IF filter is 50 MHz and that of the second is 10 MHz.

ADDITIONAL READING

Armstrong, E. H.: A New System of Short-Wave Amplification, *Proc. Radio Eng.*, **9**:3–27 (1919).
Fessenden, R. H.: Wireless Telephony, *Am. Inst. Elect. Eng.*, **27**:553–629 (1908).
Lessing, L.: *Man of High Fidelity: Edwin Howard Armstrong*, Lippincott, Philadelphia, 1956.
Terman, F. E.: *Electronic and Radio Engineering*, 4th ed., McGraw-Hill, New York, 1955.

CHAPTER TWO

SMALL-SIGNAL AMPLIFIERS

2.1 INTRODUCTION

Most amplifiers used in communications circuits can be considered small-signal amplifiers. These are amplifiers in which the input and output signals are sufficiently small that the amplifier performance is described with linear equations. In this chapter we shall first discuss the low-frequency small-signal models for bipolar and field-effect transistor (FET) amplifiers. The hybrid-π model is used to describe the bipolar transistor since it is the easiest to analyze. It will be shown that the same small-signal model can be used for the FET. (It will be shown in Chap. 5 that this model can be readily extended for the evaluaion of high-frequency amplifier performance.) The importance of push-pull and operational amplifiers will be outlined here as will the importance of the versatile differential amplifier used in the majority of linear integrated circuits. Operational amplifier circuits will be discussed, and we will look at recent advances in integrated-circuit fabrication that have extended the gain-band-width product (a term defined in this chapter) of the operational amplifier to the extent that they now find many applications in high-frequency circuits.

2.2 BIPOLAR TRANSISTOR AMPLIFIERS

Equivalent Circuits

In the hybrid-π low-frequency equivalent circuit for the bipolar transistor (Fig. 2.1) terminal b' represents the base junction and b, the base terminal. The resistor r_b' connected between these two terminals is usually considered a

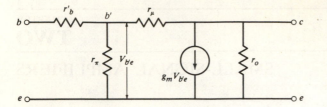

Figure 2.1 A small-signal, midfrequency equivalent-circuit model of a bipolar transistor.

constant in the range of 10 to 50 Ω. The resistor r_π is the base-emitter junction resistance and is usually much larger than r'_b. A useful estimate of r_π is given by the expression[2.1]

$$r_\pi = \frac{kT}{q}\frac{\beta}{I_C} = \frac{0.026\beta}{I_C} \tag{2.1}$$

where β = transistor base-to-collector current gain
$\quad I_C$ = dc bias current in collector
$\quad q$ = charge on an electron
$\quad k$ = Boltzmann's constant
$\quad T$ = the temperature
At room temperature $T = 290$ K, and $kT/q = 0.026$ V. The base-emitter resistance r_π is inversely proportional to the collector bias current.

> **Example 2.1** A 2N3904 transistor has a current gain β of 100 and is biased so that the quiescent collector current is 10^{-3} A. What is the transistor base-emitter resistance?
>
> SOLUTION From Eq. (2.1)
>
> $$r_\pi = \frac{0.026(100)}{10^{-3}} = 2600\ \Omega$$

The collector-to-emitter resistance r_o and the collector-to-base resistance r_u are also inversely proportional to the dc collector current. A typical value for r_o is 50 kΩ. r_u is a large resistance, on the order of megohms, used to model basewidth modulation effects in the transistor. r_u will be assumed to be an open circuit in all cases considered in this text. (It is of considerable importance in high-voltage-gain amplifiers with very large values of load impedance, but these applications are not normally found in high-frequency circuits.)

The transistor transconductance g_m is determined from the formula

$$g_m r_\pi = \beta \tag{2.2}$$

or

$$g_m = \frac{qI_C}{kT} \approx 40 I_C \tag{2.3}$$

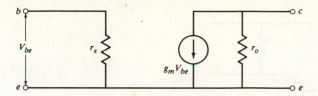

Figure 2.2 A simplified small-signal, midfrequency equivalent-circuit model of a bipolar transistor.

The transconductance g_m is directly proportional to the dc collector current.

Although the circuit appears rather complicated, under most conditions the resistor r_b' can also be neglected. The equivalent circuit is then as shown in Fig. 2.2. The model now consists of three independent parameters g_m, r_π, and r_o. g_m and r_π are determined by the dc collector current and the current gain β of the transistor.

The Common-Emitter Amplifier

In order to analyze the small-signal behavior of a linear amplifier such as the common-emitter amplifier illustrated in Fig. 2.3a, the transistor is replaced by the small-signal model and the complete equivalent circuit becomes as shown in Fig. 2.3b. If the coupling capacitor is assumed to be a short circuit, the base voltage V is given by

$$V = \frac{R}{R + R_s} V_i \tag{2.4}$$

where R is the equivalent resistance R_1, R_2, and r_π connected in parallel, or

$$R = \frac{R_1 R_2 r_\pi}{R_1 R_2 + r_\pi R_1 + r_\pi R_2} \tag{2.5}$$

The output voltage is given as

$$V_o = \frac{-g_m R_L r_o V}{r_o + R_L} = \frac{-g_m R_L r_o}{r_o + R_L} \frac{R V_i}{R + R_s} \tag{2.6}$$

and the midfrequency voltage gain is given as

$$A_v = \frac{V_o}{V_i} = \frac{-g_m R_L r_o}{r_o + R_L} \frac{R}{R + R_s} \tag{2.7}$$

The phase shift of the midfrequency voltage gain of the common-emitter amplifier is 180°. The input impedance of the amplifier is by definition

$$Z_i = \frac{\Delta V_i}{\Delta I_i} \tag{2.8}$$

An increment of voltage is applied, and the change in input current is measured

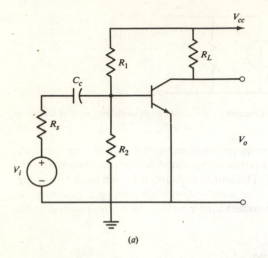

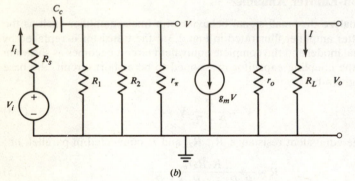

Figure 2.3 (*a*) A common-emitter amplifier; (*b*) small-signal, midfrequency equivalent circuit of the amplifier of Fig. 2.3*a*.

(assuming that all other independent sources remain constant) as shown in Fig. 2.4. In this figure the source resistance is not included in determining the input impedance of the amplifier. For this circuit

$$Z_i = R_i = \frac{R_1 R_2 r_\pi}{R_1 R_2 + R_1 r_\pi + R_2 r_\pi} \tag{2.9}$$

Since r_π depends on the dc bias current, the input impedance will depend on it also. The current gain is the load current I_L divided by the input current, or

$$A_i = \frac{I_L}{I_i} = \frac{V_o/R_L}{V_i/(R_i + R_s)} = \frac{R_s + R_i}{R_L} A_v \tag{2.10}$$

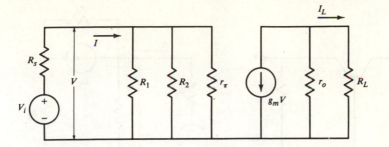

Figure 2.4 The input impedance of the amplifier circuit is defined as V/I.

The current gain from the base to the output is $-\beta$, provided that $r_o \gg R_L$. The midfrequency current gain also has a 180° phase shift.

Amplifier output impedance is determined by applying a voltage across the output terminals and measuring the change in the output current (with all other independent sources held constant). This is the Thévenin equivalent impedance seen by the load impedance:

$$Z_o = \frac{\Delta V_o}{\Delta I_o} \tag{2.11}$$

The load resistance is usually excluded from the definition, so

$$Z_o = r_o \tag{2.12}$$

for the common-emitter amplifier.

The Common-Base Amplifier

The same small-signal equivalent circuit can also be used for the common-base and common-collector amplifiers. Figure 2.5a contains a common-base amplifier, and Fig. 2.5b contains the midfrequency small-signal circuit. The direction of the dependent current source is now from emitter to collector, since the dependent voltage V has been taken from emitter to base rather than from base to emitter. Since the sum of the currents leaving the emitter junction is zero,

$$\frac{V - V_i}{R_s} + \frac{V}{r_\pi} + \frac{V - V_o}{r_o} + g_m V = 0 \tag{2.13}$$

Also

$$\frac{V - V_o}{r_o} + g_m V = \frac{V_o}{R_L} \tag{2.14}$$

If the dependent voltage V is eliminated from these two equations, we obtain an expression for the voltage gain of the common-base amplifier. This equation is

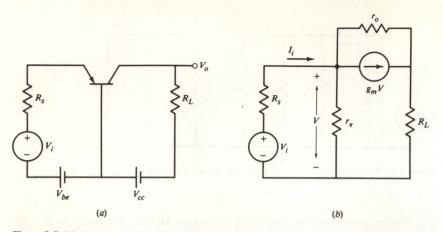

Figure 2.5 (a) A common-base amplifier; (b) a small-signal, midfrequency equivalent circuit of the common-base amplifier.

rather complicated, and usually little accuracy is sacrificed by assuming that r_o is large compared with R_s, r_π, and R_L. The voltage gain is then

$$A_v = \frac{V_o}{V_i} = \frac{g_m R_L r_\pi}{R_s + r_\pi + g_m R_s r_\pi} = \frac{\beta R_L}{r_\pi + R_s(1 + \beta)} \qquad (2.15)$$

(Note that there is no phase inversion in the voltage gain of the common-base amplifier.) If the source impedance is small, $r_\pi \gg R_s(1 + \beta)$, the magnitude of the voltage gain will be the same as that of the common-emitter amplifier.

The input impedance of the common-base amplifier is determined using the simplified circuit of Fig. 2.6. Since

$$I_i = \frac{V_i}{r_\pi} + g_m V_i \qquad (2.16)$$

then

$$Z_i = \frac{r_\pi}{1 + g_m r_\pi} = \frac{r_\pi}{1 + \beta} \approx \frac{1}{g_m} \qquad (2.17)$$

The input impedance of the common-base amplifier is smaller than that of the

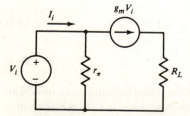

Figure 2.6 The input impedance of this common-base amplifier equivalent circuit is V_i/I_i.

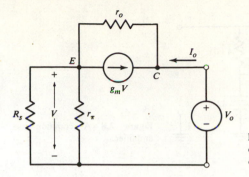

Figure 2.7 A simplified circuit model for calculating the output impedance of the common-base amplifier.

common-emitter amplifier. Since it is inversely proportional to g_m, it is inversely proportional to the dc collector current. This property is found useful in setting the amplifier impedance to a desired level for impedance matching. Also, since

$$I_L = g_m V_i = g_m I_i Z_i \tag{2.18}$$

the current gain is

$$A_i = \frac{g_m r_\pi}{1 + g_m r_\pi} \approx \frac{\beta}{1 + \beta} \tag{2.19}$$

which is slightly less than 1. The output impedance can be determined from the two node equations (see Fig. 2.7):

$$-I_o = g_m V + \frac{V - V_o}{r_o} \tag{2.20}$$

and

$$\frac{V}{R} + \frac{V - V_o}{r_o} + g_m V = 0 \tag{2.21}$$

where

$$R = \frac{r_\pi R_s}{r_\pi + R_s} \tag{2.22}$$

Therefore, the output impedance is

$$Z_o = r_o + R(1 + g_m r_o) \tag{2.23}$$

which is larger than that of the common-emitter amplifier.

Example 2.2 Calculate the midband voltage gain, the current gain, and the input impedance of the common-base amplifier shown in Fig. 2.8. The collector bias current is 10^{-3} A and the transistor $\beta = 100$.

SOLUTION The small-signal, midfrequency model is shown in Fig. 2.9. The collector-emitter resistance is assumed to be infinite. Since $I_C = 1$ mA,

$$g_m = 40 I_C = 40 \times 10^{-3} \text{ S}$$

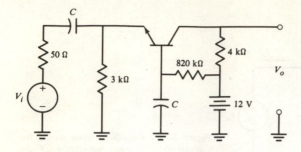

Figure 2.8 A common-base amplifier.

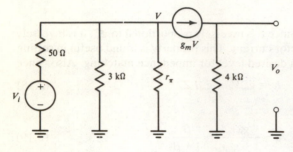

Figure 2.9 A small-signal equivalent circuit of the amplifier illustrated in Fig. 2.8.

and

$$r_\pi = \frac{\beta}{g_m} = 2500 \; \Omega$$

Therefore, from Eq. (2.15), the voltage gain is given by

$$A_v = \frac{100(4 \times 10^3)}{2.5 \times 10^3 + 50(101)} = 52.98$$

and the current gain is [using Eq. (2.19)]

$$A_i = \frac{100}{101} = 0.99$$

The input impedance is calculated using Eq. (2.17):

$$Z_i \approx \frac{1}{g_m} = 25 \; \Omega$$

The common-base amplifier has a voltage gain, but the current gain is less than unity. This amplifier is used in applications in which it is desired to build a noninverting amplifier or an amplifier with a low-input or a high-output impedance. It does have much better high-frequency response than the common-emitter amplifier and is often used in high-frequency circuits. An application of this amplifier is described in the section on multistage amplifiers, and additional examples are provided in Chap. 5.

The Emitter-Follower

The common-collector amplifier, better known as an emitter-follower, has various applications. It will be shown that this amplifier has a noninverting voltage gain of less than 1, and a current gain approximately equal to the β of the transistor used. Although it has a voltage gain less than 1, it can be combined with another amplifier stage, such as a common-emitter stage, to realize a greater combined voltage gain than could be achieved from the use of an emitter-follower stage alone. This is particularly useful when a low impedance load is used.

An emitter-follower amplifier is illustrated in Fig. 2.10, and the small-signal, midfrequency equivalent circuit is given in Fig. 2.11. It is normally the case that the base biasing resistor R_b is much larger than the source resistance R_s and can therefore be neglected. If this is the case, the equivalent circuit is as shown in Fig. 2.12. The voltage gain is determined from three equations:

$$V_i = I_i(R_s + r_\pi) + V_o \tag{2.24}$$

$$V = I_i r_\pi \tag{2.25}$$

and

$$V_o = (I_i + g_m V)\frac{r_o R_L}{r_o + R_L} \tag{2.26}$$

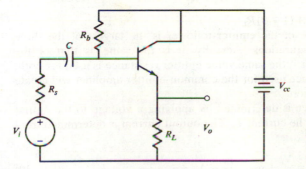

Figure 2.10 An emitter-follower (common-collector amplifier).

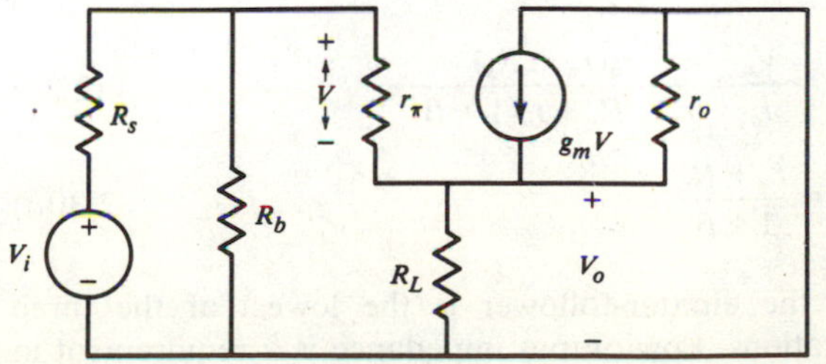

Figure 2.11 A small-signal equivalent circuit of the emitter-follower illustrated in Fig. 2.10.

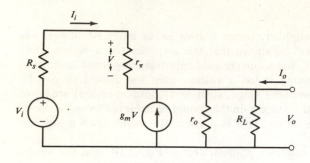

Figure 2.12 A simplified small-signal equivalent circuit of the emitter-follower.

If V and I_i are eliminated, the voltage gain is found to be

$$A_v = \frac{V_o}{V_i} = \frac{(1+\beta)[r_o R_L/(r_o + R_L)]}{R_s + r_\pi + (1+\beta)[r_o R_L/(r_o + R_L)]} \tag{2.27}$$

The voltage gain is noninverting (positive) and less than 1. As R_L increases, the gain approaches 1. The emitter-follower input impedance can be found from Eqs. (2.24) and (2.27) by eliminating V_o and setting $R_s = 0$. The input impedance is

$$Z_i = r_\pi + \frac{(1+\beta)r_o R_L}{r_o + R_L} \tag{2.28}$$

If r_o is very large $Z_i \approx r_\pi + (1+\beta)R_L$.

The input impedance of the emitter-follower is the largest of the three transistor amplifier configurations. Actually, it is the same as that of the common-emitter amplifier if the same value emitter resistance is used for both, but the additional resistance used for the common-emitter amplifier will create additional power dissipation.

The output impedance is determined by applying a voltage to the output terminals and measuring the current I_o. The output current is determined from the equation

$$I_o = \frac{V_o}{r_o} + g_m \frac{r_\pi}{r_\pi + R_s} V_o + \frac{V_o}{R_s + r_\pi} \tag{2.29}$$

The output impedance is

$$Z_o = \frac{V_o}{I_o} = \frac{r_o(r_\pi + R_s)}{r_\pi + R_s + r_o(1+\beta)} \tag{2.30}$$

Then

$$Z_o \approx \frac{r_\pi + R_s}{1+\beta} \tag{2.30a}$$

The output impedance of the emitter-follower is the lowest of the three transistor amplifier configurations. Low output impedance is a requirement in many amplifier applications. The emitter-follower is used when a low output impedance is needed.

The emitter-follower current gain is

$$A_i = \frac{I_L}{I_i} = \frac{V_o/R_L}{I_i} = \frac{V_o}{V_i}\frac{R_s + Z_i}{R_L} \tag{2.31}$$

and using Eqs. (2.7) and (2.28) the current gain is found to be

$$A_i = \frac{(1 + g_m r_\pi)r_o}{r_o + R_L} \approx 1 + \beta \tag{2.32}$$

Although the emitter-follower has a voltage gain less than 1 it has a large current gain. It is also frequently used as a power amplifier for low-impedance loads.

Example 2.3 Calculate the power gain of the common-collector amplifier shown in Fig. 2.13 a. The transistor is a 2N5901 with $\beta = 40$. The transistor output impedance is large and can be neglected. The dc collector current is 40 mA.

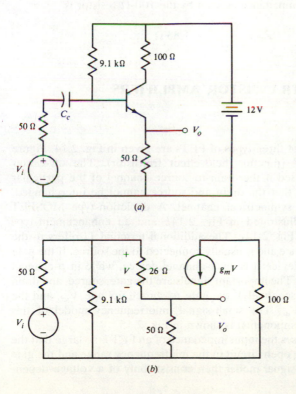

(a)

(b)

Figure 2.13 (a) An emitter-follower; (b) a simplified equivalent circuit of the emitter-follower.

SOLUTION The small-signal, midfrequency equivalent circuit is shown in Fig. 2.13b (r_o has been neglected). Since the 100-Ω resistor, which is used for biasing, is in series with the dependent current source, it has no effect on the output signal. The collector bias current $I_C = 40 \times 10^{-3}$ and r_π is [Eq. (2.1)]

$$r_\pi \approx \frac{26(40)}{40} = 26 \ \Omega$$

The voltage gain is [using Eq. (2.27) and neglecting the 9.1-kΩ bias resistor]

$$A_v = \frac{\beta R_L}{R_s + r_\pi + (1 + \beta) R_L} \doteq \frac{40(50)}{50 + 26 + 41(50)} = 0.94$$

The current gain is [Eq. (2.31)]

$$A_i \approx \beta + 1 = 41$$

The input and load impedances are real so the voltages and currents are in phase and the power gain is

$$A_p = A_i A_v = 38.54$$

The amplifier output impedance as seen by the 100-Ω resistor is

$$Z_o = \frac{26 + 50}{41} = 1.85 \ \Omega$$

2.3 FIELD-EFFECT TRANSISTOR AMPLIFIERS

Equivalent Circuits

Symbolic representations of three types of FETs are given in Fig. 2.14. Figure 2.14a illustrates a JFET (junction field-effect transistor). The off-center location of the arrow is used if the drain-to-source channel of the particular device is asymmetrical so that the drain and source cannot be interchanged. Many JFETs do have a symmetrical channel. A depletion-type MOSFET (insulated-gate FET) is illustrated in Fig. 2.14b and an enhancement-type MOSFET is illustrated in Fig. 2.14c. The additional terminal U refers to the substrate (body) of the device and is usually connected to the source. If the gate arrow points toward the device it is an n-channel device, while in p-channel FETs the arrow points out. The important signals are the gate, source, and drain currents (I_g, I_s, and I_d, respectively), the gate-to-source voltage V_{gs}, and the drain-to-source voltage V_{ds}. The small-signal, midfrequency model of this device (ignoring the dc components) is shown in Fig. 2.15.

For all practical purposes the input impedance of an FET is so large that the input can be considered an open circuit in the midfrequency range and the gate current $I_g = 0$. The small-signal model then consists only of a voltage-depen-

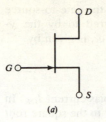

(a)

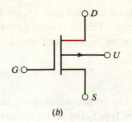

(b)

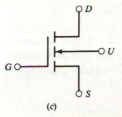

(c)

Figure 2.14 Circuit models for field-effect transistors: (*a*) a JFET; (*b*) a depletion-type MOSFET; (*c*) an enhancement-type MOSFET.

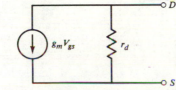

Figure 2.15 A small-signal, low-frequency equivalent circuit for a field-effect transistor.

dent current source whose value is proportional to the difference between the gate and source small-signal voltages and a drain-to-source resistance r_d. The same model is used for both junction and insulated-gate FETs. Only the relation between g_m and the dc biasing varies for the different devices. By definition, the transistor's transconductance relates the change in drain current to the change in gate-to-source voltage. That is,

$$g_m = \frac{dI_d}{dV_{gs}} \tag{2.33}$$

For JFETs it is readily shown that

$$g_m = g_{mo}\left(\frac{I_D}{I_{DSS}}\right)^{1/2} \tag{2.34}$$

where g_{mo} is the transconductance when gate-to-source bias voltage is zero.[2.2]

I_D is the dc drain current and I_{DSS} is the drain current when the gate-to-source voltage is zero. The transconductance g_m is often referred to by the y-parameter symbol y_{fs} on data sheets. For a MOS transistor g_m is given by

$$g_m = g_{mR} \left(\frac{I_D}{I_{DR}} \right)^{1/2} \tag{2.35}$$

where g_{mR} is the transconductance at some specified drain bias current I_{DR}. In both transistor types, the drain current varies proportionally to the square root of the dc bias current.

The Common-Source Amplifier

The common-source (or source-follower) amplifier is similar to the common-emitter amplifier. The midfrequency equivalent circuit of the common-source amplifier shown in Fig. 2.16a is given in Fig. 2.16b. Normally $R_g \gg R$, so

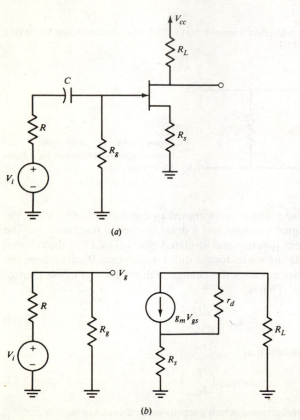

(a)

(b)

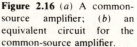

Figure 2.16 (a) A common-source amplifier; (b) an equivalent circuit for the common-source amplifier.

$V_i \approx V_g$. The dependent current source $g_m V_{gs}$ depends on both the gate and the source voltages. The gate voltage is known ($V_g = V_i$), but the source voltage must be determined from the following equations:

$$\frac{V_o - V_s}{r_d} + \frac{V_o}{R_L} + g_m V_{gs} = 0 \qquad (2.36)$$

and

$$\frac{V_s}{R_s} + \frac{V_s - V_o}{r_d} = g_m V_{gs} \qquad (2.37)$$

The source voltage, in terms of the output voltage, is

$$V_s = \frac{-V_o R_s}{R_L} \qquad (2.38)$$

and the voltage gain is

$$\frac{V_o}{V_i} = \frac{-g_m R_L r_d}{R_L + r_d + R_s(1 + g_m r_d)} \qquad (2.39)$$

If the transistor output resistance r_d is much larger than R_s and R_L, as it normally will be, the voltage-gain expression simplifies to

$$\frac{V_o}{V_i} = \frac{-g_m R_L}{1 + g_m R_s} \qquad (2.40)$$

and if $g_m R_s \gg 1$,

$$\frac{V_o}{V_i} = \frac{-R_L}{R_s} \qquad (2.41)$$

The common-source amplifier with a source resistance can be used to design an inverting amplifier whose gain is independent of the transistor (provided $g_m R_s \gg 1$). This is a particularly valuable design procedure since it eliminates the costly process of having to carefully select a transistor.

Since the input impedance of the amplifier is very large, the current gain is also very large (for practical purposes it is infinite). If $R_s = 0$, the output impedance is given by

$$Z_o = r_d \qquad (2.42)$$

It is left as an exercise to calculate the output impedance for a nonzero r_d.

Example 2.4 Figure 2.17a contains an FET amplifier and Fig. 2.17b is the small-signal, midfrequency equivalent-circuit model. The coupling and bypass capacitors are assumed to act as short circuits in this frequency range. The manufacturer's data sheet for the 2N5486 gives the following minimum parameter values for $V_{DS} = 15$ V: $I_{DSS} = 8$ mA, $g_{mo} = 4 \times 10^{-3}$ S, and $r_d = 13$ kΩ. R_s is selected so that the amplifier is biased with $I_D = 2.0$ mA. What is the midband voltage gain of the amplifier?

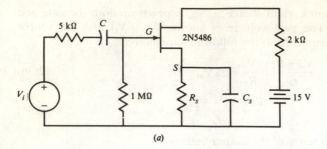

(a)

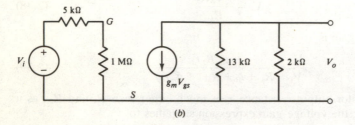

(b)

Figure 2.17 (a) A common-source amplifier; (b) an equivalent circuit for the amplifier shown in Fig. 2.17a.

SOLUTION From the equivalent circuit we see that

$$V_g = V_i \frac{10^6}{10^6 + 5 \times 10^3} = V_i$$

Since the source is grounded, $V_s = 0$ and

$$V_{gs} = V_g - V_s = V_i$$

The equivalent load resistance is

$$R'_L = \frac{2\ \text{k}\Omega \times 13\ \text{k}\Omega}{2\ \text{k}\Omega + 13\ \text{k}\Omega} = 1.73\ \text{k}\Omega$$

and the output voltage is

$$V_o = -g_m V_{gs} R'_L = -g_m V_i R'_L$$

To determine the voltage gain the transconductance must first be determined. From Eq. (2.34) the transconductance is

$$g_m = 4 \times 10^{-3} (\tfrac{2}{8})^{1/2} = 2 \times 10^{-3}\ \text{S}$$

Therefore, the voltage gain is

$$\frac{V_o}{V_i} = -g_m R'_L = -3.46$$

The Source-Follower

The same equivalent circuit is used for the FET whether it is used in the common-source, common-gate, or common-drain configuration. Figure 2.18a illustrates a common-drain or source-follower circuit, and the small-signal, midfrequency equivalent circuit is given in Fig. 2.18b. If, as will usually be the case, $R_b \gg R_s$, then

$$V_g = V_i$$

and
$$V_{gs} = V_i - V_o \qquad (2.43)$$

Also,
$$V_o = g_m V_{gs} \frac{r_d R_L}{r_d + R_L} \qquad (2.44)$$

so that
$$A_v = \frac{V_o}{V_i} = \frac{g_m[r_d R_L/(r_d + R_L)]}{1 + g_m[r_d R_L/(r_d + R_L)]} \qquad (2.45)$$

which is a noninverting voltage gain less than 1. Another important parameter of this amplifier is the output impedance Z_o. By definition, from Eq. (2.11),

$$Z_o = \frac{\Delta V_o}{\Delta I_o}$$

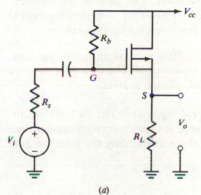

(a)

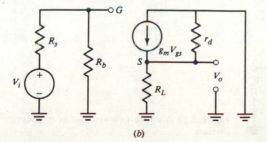

(b)

Figure 2.18 (a) A MOSFET source-follower; (b) a low-frequency equivalent circuit of the amplifier shown in Fig. 2.18a.

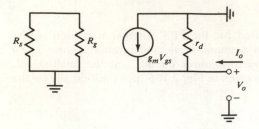

Figure 2.19 A simplified equivalent circuit for measuring the output impedance of a source-follower.

with all other independent voltage and current sources held constant. For this amplifier, the output impedance can be determined by calculating the response I_o to a voltage V_o as illustrated in Fig. 2.19. Since $V_g = 0$, $V_{gs} = -V_o$, and

$$I_o = g_m V_o + \frac{V_o}{r_d} \tag{2.46}$$

The output impedance is then

$$Z_o = \frac{r_d}{1 + g_m r_d} \tag{2.47}$$

The output impedance of the source-follower, like that of the emitter-follower, is much smaller than those of the other two FET amplifier configurations. This is the principal reason for using this amplifier configuration.

Example 2.5 The circuit shown in Fig. 2.20 is used as an active antenna. Calculate the voltage gain for a 1-MHz input signal. The transconductance $g_m = 60 \times 10^{-3}$ S. The antenna source impedance is 50 Ω.

SOLUTION The midfrequency small-signal equivalent circuit for this amplifier is shown in Fig. 2.21. Since the source resistance is much smaller than the parallel combination of the two bias resistors shunting the gate, the gate

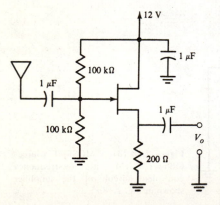

Figure 2.20 A source-follower used as an active antenna.

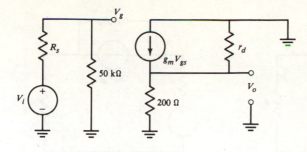

Figure 2.21 A small-signal equivalent circuit of the source-follower shown in Fig. 2.20.

voltage is

$$V_g = V_i$$

and

$$V_{gs} = V_i - V_o$$

A typical value for r_d is 2.5 kΩ, in which case the output impedance is

$$\frac{r_d}{1 + g_m r_d} = \frac{2.5 \times 10^3}{1 + 60(2.5)} = 16.6 \ \Omega$$

If the 200-Ω load is considered part of the amplifier, then the output impedance of the active antenna is

$$Z_o = \frac{200(16.6)}{216.6} = 15.3 \ \Omega$$

The voltage gain is found from Eq. (2.45) to be

$$A_v = \frac{(60 \times 10^{-3})(185)}{1 + (60 \times 10^{-3})(185)} = 0.92$$

The Common-Gate Amplifier

The common-gate amplifier is often used in high-frequency applications. It will be shown in Chap. 5 that its bandwidth is much greater than that of the common-source amplifier. Another reason for its use is that the circuit configuration offers low input impedance, which is convenient for matching to transmission lines and other low-impedance sources. The basic circuit configuration is shown in Fig. 2.22a, and the small-signal, midfrequency equivalent circuit is given in Fig. 2.22b. Since the gate is grounded,

$$V_{gs} = - V_s \tag{2.48}$$

and the circuit can be redrawn as in Fig. 2.22c. If the currents are summed at the output node so that

$$\frac{V_o}{R_L} + \frac{V_o - V_s}{r_d} - g_m V_s = 0 \tag{2.49}$$

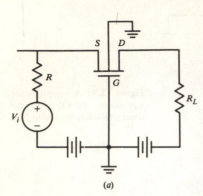

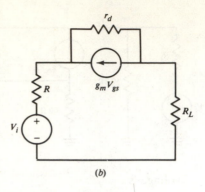

(b)

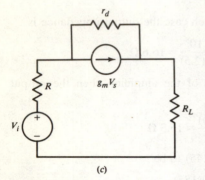

(c)

Figure 2.22 (a) A common-gate amplifier; (b) a simplified equivalent circuit for the common-gate amplifier; (c) another simplified circuit for the same amplifier.

we find that

$$\frac{V_o}{V_s} = \frac{R_L(1 + g_m r_d)}{r_d + R_L} \qquad (2.50)$$

and

$$\frac{V_o}{V_s} \approx g_m R_L \qquad (2.51)$$

for large $r_d \gg R_L$.

The voltage gain is noninverting, and for load resistances much less than the dynamic drain resistance the magnitude of the voltage gain is approximately the same as that of the common-source amplifier. The current gain is close to 1.

Input Impedance

The input impedance at the source can be found by solving for the source current:

$$I_s = g_m V_s + \frac{V_s - V_o}{r_d} \tag{2.52}$$

After Eq. (2.46) is used to eliminate V_o we obtain

$$Z_i = \frac{V_s}{I_s} = \frac{r_d + R_L}{1 + g_m r_d} \tag{2.53}$$

If $r_d \gg R_L$,

$$Z_i \approx (g_m)^{-1} \tag{2.54}$$

The common-gate amplifier has a low input impedance which is inversely proportional to the square root of the dc drain current.

Voltage Gain

Since

$$V_s = \frac{V_i Z_i}{Z_i + R} \tag{2.55}$$

the voltage gain is determined by combining Eqs. (2.50) and (2.53)

$$A_v = \frac{V_o}{V_i} = \frac{R_L}{R + (r_d + R_L)/(1 + g_m r_d)} \tag{2.56}$$

Output Impedance

The output impedance can be found by applying a voltage to the output terminals and determining the output current, as shown in Fig. 2.23:

$$I_o = \frac{V_o - V_s}{r_d} - g_m V_s \tag{2.57}$$

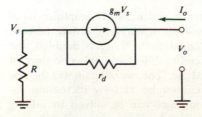

Figure 2.23 A circuit for determining the output impedance of the common-gate amplifier.

but I_o must also be the current through the source resistance R, so

$$I_o = \frac{V_s}{R} \tag{2.58}$$

Combining these two equations we find that

$$Z_o = \frac{V_o}{I_o} = r_d + (1 + g_m r_d)R \tag{2.59}$$

The common-gate amplifier has the largest output impedance of the three FET amplifier configurations.

2.4 MULTISTAGE AMPLIFIERS

It is often necessary to use more than one amplifier stage for impedance matching or to obtain additional amplification. Power transistors possess a smaller gain-bandwidth product than low-power transistors; hence, the power-amplification stage is often operated near unity voltage gain in order to maximize the bandwidth. The voltage amplification is carried out in the stages preceding the power-amplification stage. The following example illustrates the use of an additional stage, with a voltage gain less than unity, to increase the overall voltage gain.

Example 2.6 Design an amplifier using 2N3904 transistors to realize a voltage-gain magnitude of at least 10. The source resistance is 500 Ω, and the load resistance is 50 Ω.

SOLUTION The small value of the load resistance creates a design problem. If the transistor input impedance is large, the voltage gain will be

$$A_v \approx - g_m R_L$$

Since $R_L = 50$, a voltage gain of 10 can be realized with a common-emitter amplifier, provided $g_m \geq 0.2\,\mathrm{S}$. The transistor input resistance will be (assuming $\beta = 100$)

$$r_\pi = \frac{\beta}{g_m} \leq 500\ \Omega$$

Therefore, with a 500-Ω source resistance and a 500-Ω amplifier input resistance, only one-half of the signal voltage will appear across the base and the overall amplification will be equal to 5. If g_m is doubled, the transistor input impedance will be reduced to 250 Ω and only one-third of the applied signal will appear across the base. The overall gain will then be about $\frac{20}{3}$, but the gain specification cannot be met by increasing I_C. However, the problem of low load resistance can be solved by adding another stage, which simply increases the load impedance seen by the first stage. The input impedance of the second stage serves as the load impedance for the first stage. Since the input impedance of an emitter-follower from Eq. (2.28) is

$$Z_i \approx r_\pi + (1 + \beta) R_L$$

it is a logical choice for the second stage. For this transistor, β is at least 100 so the input impedance will be more than 5 kΩ. It is not a problem to realize a voltage gain of 10 with this large (5 kΩ) a load resistance. A complete two-stage amplifier is shown in Fig. 2.24. The resistors R_{L_1}, R_{b_1}, and R_{b_2} are selected to complete the biasing. Typical values for use with a 12-V supply are

$$R_{L_1} = 2 \text{ k}\Omega \qquad R_{b_1} = 500 \text{ k}\Omega \qquad \text{and} \qquad R_{b_2} = 4.5 \text{ k}\Omega$$

With these values $I_{C_1} \approx 2$ mA and $I_{C_2} \approx 120$ mA. The load resistance seen by the first stage will be the 2-kΩ resistor in parallel with R_{b_2}, and the input impedance of the second stage (approximately 5 kΩ). That is,

$$R'_{L_1} = 1.05 \text{ k}\Omega$$

The voltage gain of the second stage will be close to unity, and the voltage gain of the first stage will be

$$A_{v_1} \approx -g_{m_1} R'_L \frac{r_{\pi_1}}{r_{\pi_1} + R_s}$$

Since

$$g_{m_1} \approx 40 I_{C_1} = 0.08 \text{ S} \qquad \text{and} \qquad r_{\pi_1} = \frac{\beta}{g_{m_1}} = 1250 \ \Omega$$

$$A_{v_1} \approx \frac{-84(1250)}{1250 + 500} = -60$$

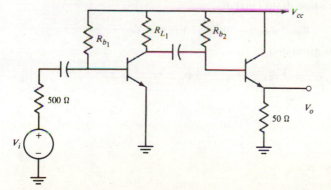

Figure 2.24 A two-stage amplifier consisting of a common-emitter amplifier followed by an emitter-follower.

If less gain is desired, the gain can readily be reduced by adding a potentiometer in series with the source resistance.

2.5 THE DUAL-GATE FET

The FET cascode circuit has so many applications in high-frequency amplifiers that two FETs are often fabricated as a single transistor with two gates. The source of the one transistor is continuous with the drain of the other so the device has one source, one drain, and two gates; it is referred to as a *dual-gate FET*. The schematic representation of a dual-gate MOSFET is shown in Fig. 2.25. The dual-gate MOSFET offers low noise and high gain in radio-frequency applications. It is a versatile device which can be used as a mixer or automatic gain control (AGC) amplifier, as will be described in Chap. 5. The construction is such that the capacitance from gate 2 to the drain is very small (approximately 1 percent of a single-gate MOSFET). This accounts for the excellent high-frequency performance of the device.

A linear amplifier circuit is shown in Fig. 2.26*a*, and the small-signal equivalent circuit is shown in Fig. 2.26*b* (where the dynamic drain resistances have been ignored). Drain 1 and source 2 are shown only as aids in understanding the circuit. Here gate 2 and the source are grounded for medium-frequency signals.

$$V_{D_1} = -g_{m_1} V_{gs_1} R_{L_1}$$

The load resistance of the first stage is the input resistance of the second stage, and the second stage is a common-gate amplifier. Therefore

$$V_{D_1} = -g_{m_1} V_i \frac{1}{g_{m_2}}$$

since the input impedance of a common-gate amplifier is g_m^{-1}. Since both transistors have the same drain current,

$$g_{m_1} = g_{m_2} \quad \text{and} \quad V_{D_1} = -V_i$$

Also, since gate 2 is grounded,

$$V_{gs_2} = -V_{s_2} = -V_{D_1} = V_i$$

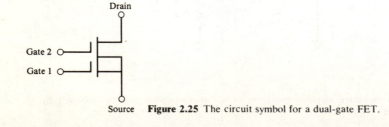

Source   **Figure 2.25** The circuit symbol for a dual-gate FET.

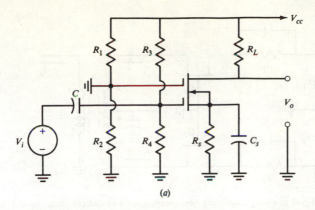

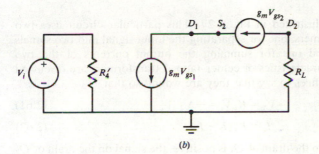

Figure 2.26 (a) An amplifier utilizing a dual-gate FET; (b) a simplified equivalent circuit of the amplifier shown in Fig. 2.26a.

and the output voltage is

$$V_o = -g_m V_{gs_2} R_L = -g_m R_L V_i \qquad (2.60)$$

The small-signal gain of the dual-gate MOSFET is equivalent to that of a single-gate MOSFET. The advantage of the dual-gate device for small-signal operation is that the much smaller gate-to-drain capacitance provides a much larger bandwidth than does the standard MOSFET. Also, the transconductance g_m, and hence the voltage gain, can be controlled by the bias voltage applied to gate 2.

2.6 PUSH-PULL AMPLIFIERS

Transistors all exhibit a nonlinear characteristic that causes distortion of the input signal even at small-signal levels. Much of the distortion can be eliminated by an amplifier configuration known as the push-pull amplifier. An example of

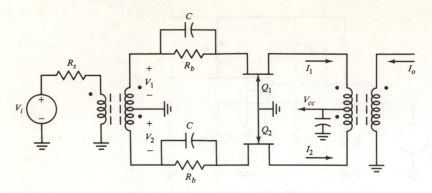

Figure 2.27 A balanced push-pull amplifier employing two common-gate amplifiers.

such an amplifier is illustrated in Fig. 2.27. This particular circuit uses two center-tapped transformers, one for separating the input signal into two signals 180° out of phase and one for summing the output currents of the two transistors. Circuit characteristics of center-tapped transformers are discussed in Chap. 6. Here it suffices to say that they are wound so that

$$V_1 = K_v V_i = -V_2 \qquad (2.61)$$

and
$$I_o = K_I(I_1 - I_2) \qquad (2.62)$$

Thus when the signal on the drain of Q_1 is positive, the signal on the drain of Q_2 is of equal magnitude but of opposite sign. The two drain signals are 180° out of phase. If the input signal is sinusoidal

$$V_i = A \cos \omega t$$

then the output currents of the two transistors are also periodic and they can be expressed in a Fourier series:

$$I_1 = B_0 + B_1 \cos \omega t + B_2 \cos 2\omega t + B_3 \cos 3\omega t + \cdots \qquad (2.63)$$

The higher-frequency components are created by the transistor nonlinearities. If the two transistors and associated components are identical, then I_2 and I_1 are identical except that I_2 lags I_1 by 180°. That is,

$$I_2(\omega t) = I_1(\omega t + \pi)$$

Therefore,

$$I_2 = B_0 - B_1 \cos \omega t + B_2 \cos 2\omega t - B_3 \cos 3\omega t + \cdots \qquad (2.64)$$

and the output current is

$$I_o = 2K_I(B_1 \cos \omega t + B_3 \cos 3\omega t \cdots) \qquad (2.65)$$

The output does not contain any even harmonics since the even harmonics of the two transistors have been cancelled by the push-pull arrangement. This is particularly beneficial with FET amplifiers because they have a square-law characteristic that generates a relatively large second harmonic. Push-pull amplifiers are also used in power amplifiers. Several applications of their use in power amplifiers are provided in Chap. 11.

2.7 THE DIFFERENTIAL AMPLIFIER

A differential amplifier provides an output proportional to the difference between the signals present at the two input terminals; neither input need be grounded. Ideally, the output will be zero in response to a signal common to the two input terminals. A differential amplifier can be applied where it is desired to measure the voltage difference between two points, neither of which is grounded. Recent advances in transistor fabrication now make it possible to manufacture two transistors with closely matched characteristics on the same wafer. This has made the differential amplifier one of the most versatile building blocks at the designer's disposal, allowing for the relatively easy design of low-noise, low-drift, dc coupled amplifiers. Today the majority of analog integrated circuits use a differential-amplifier input stage.

Consider the differential amplifier illustrated in Fig. 2.28. The positive terminal is referred to as the *noninverting terminal* since any voltage at this terminal will be amplified and appear at the output without phase inversion, whereas any voltage on the negative (inverting) terminal will be amplified and appear at the output with 180° phase shift (at least at low frequencies). The output may be either single-ended (one terminal grounded) or floating (neither output terminal grounded). Any combination of input signals can be decomposed into differential and common input signals. It will be shown that this decomposition greatly simplifies the analysis of differential amplifiers. The differential input voltage is defined as the difference between the two input signals; that is,

$$e_d = V_1 - V_2 \qquad (2.66)$$

and the common input voltage is defined as

$$e_c = \frac{V_1 + V_2}{2} \qquad (2.67)$$

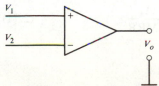

Figure 2.28 The circuit symbol for a differential amplifier.

Any two input signals, then, consist of a component in which the signals on the two input terminals are equal in magnitude and 180° out of phase and a component in which the two input terminal signals are equal in magnitude and in phase.

Example 2.7 Determine the differential and common input voltages of the two signals

$$V_1 = 5 + 3 \sin \omega t \quad \text{and} \quad V_2 = 3 - \sin \omega t$$

SOLUTION The differential signal is

$$e_d = V_1 - V_2 = 2 + 4 \sin \omega t$$

and the common signal is

$$e_c = \frac{V_1 + V_2}{2} = 4 + \sin \omega t$$

The amplification of a differential amplifier is characterized by its differential gain:

$$A_d = \frac{V_o}{e_d} \tag{2.68}$$

The common-mode gain is defined as

$$A_c = \frac{V_o}{V_1}\bigg|_{V_1 = V_2} = \frac{V_o}{e_c} \tag{2.69}$$

In an ideal differential amplifier the common-mode gain is zero, but an actual differential amplifier always has a finite common-mode gain.

The Common-Mode Rejection Ratio

The ratio of differential gain to common-mode gain is known as the *common-mode rejection ratio* (*CMRR*):

$$\text{CMRR} = \frac{A_d}{A_c} \tag{2.70}$$

The actual amplifier output to any input signal is

$$V_o = A_d e_d + A_c e_c = A_d e_d + \frac{A_d}{\text{CMRR}} e_c \tag{2.71}$$

The term

$$\frac{A_d}{\text{CMRR}} e_c$$

is an error term caused by the finite CMRR. It is not possible to determine whether the output signal is due to a differential signal or a common signal. The larger the CMRR the smaller the error term and the greater will be the amplifier accuracy.

Example 2.8 Consider the problem of measuring the voltage across the 5-Ω resistor of the circuit shown in Fig. 2.29. This is a common problem in biomedical engineering and telemetry applications where the large impedances in series with the small resistors might represent the impedance of transducers or probe interfaces.

SOLUTION The voltage on the positive terminal is

$$V_1 = \frac{(5 + 10^6) V_i}{5 + 2 \times 10^6}$$

and the voltage on the inverting terminal is

$$V_2 = \frac{10^6 V_i}{5 + 2 \times 10^6}$$

So the differential voltage is

$$e_d = \frac{V_1 - V_2}{2} = \frac{5 V_i}{2 \times 10^6}$$

and the common-mode voltage is

$$e_c = \frac{V_1 + V_2}{2} = \frac{(5 + 2 \times 10^6) V_i}{2(5 + 2 \times 10^6)} = \frac{V_i}{2}$$

If the differential amplifier has a differential gain of 2×10^3 and a common-mode rejection ratio of 1 million (120 dB), the output voltage will be

$$V_o = \left[\frac{5}{2 \times 10^6} (2 \times 10^3) \right] V_i + \left(\frac{V_i}{2} \frac{10^3}{10^6} \right)$$

$$= (5 \times 10^{-3} + 1 \times 10^{-3}) V_i$$

The second term represents a 20 percent error due to the finite CMRR. To

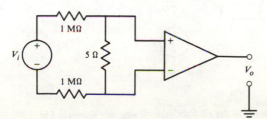

Figure 2.29 An application of a differential amplifier.

realize a 1 percent measurement accuracy for this problem requires a differential amplifier with a minimum CMRR of 2×10^7 (146 dB).

The FET Differential Amplifier

Consider the differential amplifier connection shown in Fig. 2.30. We will assume that the transistors are identical and at the same temperature. The small-signal, midfrequency equivalent circuit is then as shown in Fig. 2.31. Although this circuit can be readily analyzed for arbitrary input voltages, we will first calculate the differential gain and then the common-mode gain, as this greatly simplifies the analysis.

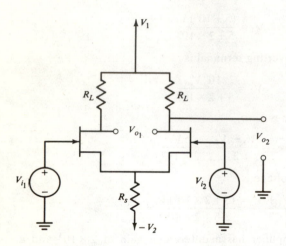

Figure 2.30 A differential amplifier realized with two field-effect transistors.

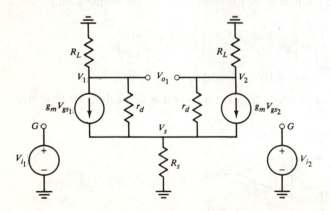

Figure 2.31 A small-signal equivalent circuit of the FET differential amplifier shown in Fig. 2.30.

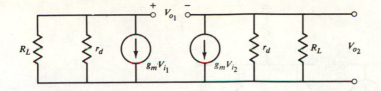

Figure 2.32 The small-signal equivalent circuit for determining the differential gain of the FET differential amplifier.

Assume first that $V_{i_1} = -V_{i_2}$. If V_{i_1} increases by ΔV, V_{i_2} will decrease by the same amount, so the changes in the two dependent current sources in Fig. 2.31 will be equal in magnitude but opposite in sign. In this case it is readily shown that the voltage at the source terminal does not change, and so the current through R_s does not change. Therefore, the small-signal equivalent circuit for differential inputs can be simplified to that shown in Fig. 2.32. The output voltage is

$$V_{o_1} = \frac{-g_m r_d R_L (V_{i_1} - V_{i_2})}{r_d + R_L} = \frac{-e_d g_m r_d R_L}{r_d + R_L} \tag{2.72}$$

and
$$V_{o_2} = \frac{e_d}{2} \frac{g_m r_d R_L}{r_d + R_L} \tag{2.73}$$

The single-ended output differential gain is one-half of the differential output differential gain. Also with the single-ended output as shown, input terminal 1 would be the noninverting input terminal.

To calculate the common-mode signal we will assume (with no loss of generality) that the two input signals are the same, that is,

$$V_{i_1} = V_{i_2}$$

and the single-ended output voltage can be determined from an analysis of the circuit shown in Fig. 2.33 a. Because of the circuit symmetry the output can be derived from the simplified circuit shown in Fig. 2.33 b. Here

$$V_{gs} = V_i - V_s$$

The sum of the currents at the source node is zero, so

$$\frac{V_s}{2R_s} + \frac{V_s - V_{o_2}}{r_d} = g_m V_{gs}$$

and the sum of the currents at the output node is zero, so

$$\frac{V_{o_2}}{R_L} + \frac{V_{o_2} - V_s}{r_d} + g_m V_{gs} = 0$$

If V_s is eliminated from these equations, we obtain the common-mode gain

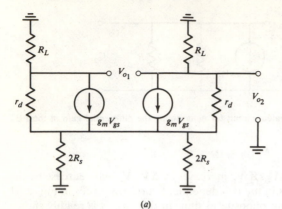

(a)

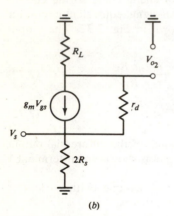

(b)

Figure 2.33 (a) A simplified circuit for calculating the common-mode gain of the FET differential amplifier shown in Fig. 2.30; (b) one-half of the symmetric equivalent circuit shown in Fig. 2.33a.

expression

$$A_c = \frac{V_{o_2}}{V_i} = \frac{V_{o_2}}{e_d} = \frac{g_m r_d R_L}{(1 + g_m r_d) 2 R_s + r_d + R_L}$$

$$\approx \frac{g_m R_L}{1 + 2 g_m R_s} \quad \text{for } r_d \geq R_L$$

(2.74)

The common-mode gain should ideally be zero. We see that the larger the source resistance, the smaller will be the common-mode gain. Also, the differential gain does not depend on R_s. Since an ideal current source has infinite impedance, the source resistance can be replaced by a current source (which also provides a path for the bias current). A current-source-biased differential amplifier is shown in Fig. 2.34.

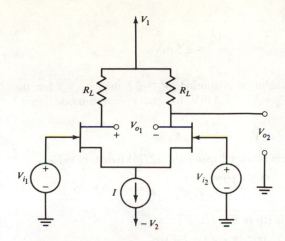

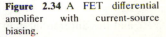

Figure 2.34 A FET differential amplifier with current-source biasing.

Example 2.9 Calculate the output voltage of the circuit shown in Fig. 2.35 in response to an input signal $V_i = 5 \times 10^{-3}$ V. The equivalent output impedance of the current source is 150 kΩ. Assume that the SU2366 dual JFET is used. For the SU2366 dual JFET, $g_m = 1.5 \times 10^{-3}$ S and $r_d = 500$ kΩ.

SOLUTION Although it is possible to directly calculate the response to the single 5-mV input, it is far easier to consider the circuit to be that of a differential amplifier and calculate the response to the differential and common-mode inputs. The differential input is

$$e_d = V_{i_1} - V_{i_2} = 5 \text{ mV}$$

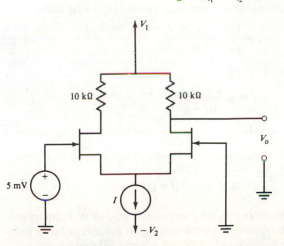

Figure 2.35 The amplifier discussed in Example 2.9.

and the common input is

$$e_c = \frac{V_{i_1} + V_{i_2}}{2} = 2.5 \text{ mV}$$

(Note that this is equivalent to assuming $V_{i_1} = 2.5 \text{ mV} = -V_{i_2}$ for the differential input and $V_{i_1} = V_{i_2} = 2.5 \text{ mV}$ for the common-mode input.) Then by superposition

$$V_o = e_c A_c + e_d A_d$$

The common-mode gain is obtained using Eq. (2.74) (since $r_d \gg R_L$):

$$A_c = -\frac{1.5 \times 10^{-3} \times 10^4}{1 + 2(1.5 \times 10^{-3})(150 \times 10^3)} = -0.033$$

and the differential gain is [from Eq. (2.73)]

$$A_d = \frac{g_m R_L}{2} = \frac{1.5 \times 10^{-3} \times 10^4}{2} = 7.5$$

so that the output to the applied signal is

$$V_o = -0.033(2.5) + 7.5(5) = 37.42 \text{ mV}$$

The BJT Differential Amplifier

A bipolar transistor differential amplifier is shown in Fig. 2.36a and the small-signal equivalent circuit is shown in Fig. 2.36b (with r_o and r_b' neglected). The circuit can be analyzed with the method used for the FET differential amplifier. For differential inputs the emitter is at ground potential and

$$A_d = \frac{V_{o_2}}{e_d} = \frac{g_m R_L r_\pi}{2(R_s + r_\pi)} \tag{2.75}$$

The differential output gain is

$$\frac{V_{o_1}}{e_d} = \frac{-g_m R_L r_\pi}{R_s + r_\pi} \tag{2.76}$$

For common-mode input signals the circuit is equivalent to two identical circuits in parallel, and

$$A_c = \frac{V_{o_2}}{e_c} = \frac{-g_m R_L r_\pi}{R_s + r_\pi + 2(\beta + 1)R_E} \tag{2.77}$$

The common-mode gain can be reduced by increasing R_E, so BJT differential amplifiers usually use emitter-current source biasing to achieve a large resistance to the emitters of the two transistors of the differential amplifier.

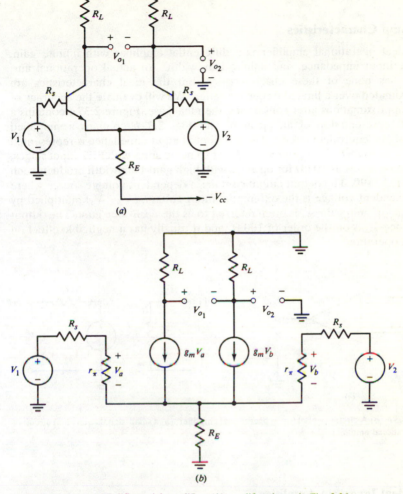

Figure 2.36 (a) A BJT differential amplifier; (b) amplifier shown in Fig. 2.36a.

2.8 THE OPERATIONAL AMPLIFIER

An operational amplifier (op amp) is a direct-coupled differential amplifier with high gain and large input impedance. The device is used with large amounts of negative feedback to obtain the desired "gain-versus-frequency" characteristic. It has been used mainly in low-frequency circuits, but integrated-circuit op amps are now available with gain-bandwidth products in excess of 1.7×10^9 Hz

(1.7 GHz). Op amps are currently being used in communication circuits as wideband or video amplifiers.

Op-Amp Characteristics

The ideal operational amplifier is a differential amplifier with infinite gain, infinite input impedance, and infinite bandwidth. An actual operational amplifier has none of these characteristics, but the ideal characteristics are approximated over a limited frequency range. We will evaluate the accuracy of these approximations after considering the ideal case. Figure 2.37a contains a circuit representation of an op amp, and Fig. 2.37b contains a simplified small-signal equivalent circuit. The differential input impedance is represented by Z_i, which can be greater than 10^{12} Ω for some op amps with FET input stages, but can be as low as 10 kΩ for op amps with high gain-bandwidth products such as the HA5190. The output circuit contains a dependent voltage source where the dependent voltage is the differential input voltage $(V_2 - V_1)$ multiplied by A_a. The op-amp gain A_a is often referred to as the *open-loop gain*. The output resistance Z_o is on the order of 100 Ω, and it usually has a negligible effect on circuit operation.

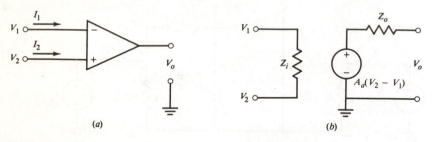

Figure 2.37 (a) A circuit symbol for an operational amplifier; (b) a small-signal equivalent circuit of the operational amplifier.

The Ideal Inverting Amplifier

Figure 2.38 illustrates the most commonly used operational amplifier configuration, the inverting amplifier. Here the noninverting input terminal is grounded, and the output voltage is fed back to the inverting input terminal through the impedance Z_f. For the ideal operational amplifier the circuit analysis is simple. It is based on the fact that no current will flow into the input terminals, and for any ideal op-amp circuit the voltage on the positive terminal is equal to the voltage on the negative input terminal:

$$V_1 = V_2$$

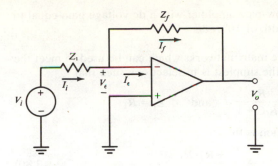

Figure 2.38 An inverting op-amp circuit.

That the two voltages are equal follows from the fact that the output voltage is

$$V_o = A_a(V_2 - V_1) \tag{2.78}$$

In the ideal op amp, A_a is infinite; therefore, the only way for the output voltage to be finite is for V_1 to equal V_2. The two voltages are made equal through the feedback network connected around the op amp. It is, of course, possible that the feedback is such that the amplifier will be unstable, but it is assumed here that the circuit configuration is such that the amplifier is stable. In the inverting amplifier shown in Fig. 2.38, if the amplifier is ideal the voltage at the inverting terminal V_ϵ must be zero, since the positive terminal voltage $V_2 = 0$ and $V_1 = V_2$ in the ideal op-amp circuit. Also, in the ideal op amp the input current $I_\epsilon = 0$, so

$$\frac{V_i}{Z_1} = I_i = I_f = \frac{-V_o}{Z_f}$$

That is, the transfer function of the inverting-amplifier configuration, using an ideal op amp, is

$$A_v = \frac{V_o}{V_i} = \frac{-Z_f}{Z_1} \tag{2.79}$$

The closed-loop transfer function is independent of the operational amplifier and depends only on the two impedances used in the feedback network.

Example 2.10 Design an amplifier with a voltage gain equal to -50.

SOLUTION If an ideal operational amplifier is used, the circuit configuration of Fig. 2.38 can be used, provided that $Z_f/Z_1 = 50$. Here one of the impedances can be arbitrarily chosen. For example, if $Z_1 = 1 \text{ k}\Omega$, then $Z_f = 50 \text{ k}\Omega$ will give the desired voltage gain.

The inverting amplifier configuration can also be used to realize a frequency-dependent transfer function. This concept is illustrated in the following example.

Example 2.11 Design a low-pass amplifier with a dc voltage gain equal to -50 and a -3-dB bandwidth of 10^6 rad/s.

SOLUTION Again there are many networks which can be used to meet the specifications, but one of the simplest is to select one such that

$$Z_f = \frac{R_f}{sR_fC + 1} \quad \text{and} \quad Z_1 = R_1$$

The voltage transfer function is then

$$A_v = \frac{-R_f/R_1}{sCR_f + 1} \tag{2.80}$$

which is a low-pass transfer function with a low-frequency gain equal to $-R_f/R_1$ and a -3-dB frequency:

$$\omega_L = \frac{1}{R_fC} \tag{2.81}$$

Therefore, we arbitrarily select $R_1 = 1\,\text{k}\Omega$. Then $R_f = 50\,\text{k}\Omega$ and $C = (10^6 \times 5 \times 10^4)^{-1} = 20\,\text{pF}$. The complete circuit is shown in Fig. 2.39.

An almost limitless number of different frequency-dependent transfer functions can be constructed using the inverting configuration.

Since the inverting terminal is also at ground potential, the input impedance of the amplifier shown in Fig. 2.38 is

$$Z_i = \frac{V_i}{I_i} = Z_1 \tag{2.82}$$

The impedance is easily controlled by the selection of Z_1. The output impedance, assuming an ideal op amp, is zero.

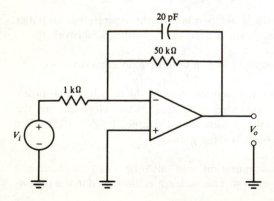

Figure 2.39 The low-pass amplifier discussed in Example 2.11.

The Nonideal Inverting Amplifier: The Effect of Finite Loop Gain

The gain of an actual operational amplifier is never infinite and, in fact, is only large over a limited frequency range. The effects of finite loop gain on the transfer function can be determined by analyzing Fig. 2.40, which is the small-signal equivalent circuit for the operational amplifier with finite open-loop gain A_a. Since the amplifier input impedance is assumed to be infinite,

$$\frac{V_i - V_\epsilon}{Z_1} = \frac{V_\epsilon - V_o}{Z_2}$$

and if the op-amp output impedance is neglected,

$$V_o = -A_a V_\epsilon$$

If V_ϵ is eliminated from these two equations, the transfer function is found to be

$$A_v = \frac{V_o}{V_i} = \frac{-Z_f/Z_1}{1 + (Z_1 + Z_f)/A_a Z_1} \tag{2.83}$$

If the operational amplifier gain is known, the deviation of the ideal transfer function from its actual value can be calculated.

> **Example 2.12** An inverting amplifier with a gain of 100 is designed by using an operational amplifier with an open-loop gain $A_a = 10^4$. The feedback resistor $R_f = 10^4$, and the input resistance $R_1 = 100$. What is the actual loop gain?
>
> SOLUTION The actual gain is calculated using Eq. (2.83):
>
> $$A_v = \frac{-100}{1 + (100 + 1)/10^4} = \frac{-100}{1.01} = -99.0$$
>
> The finite op-amp gain has resulted in a 1 percent deviation in the closed-loop gain from the ideal value.

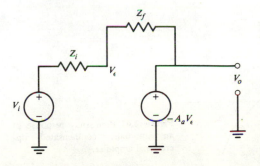

Figure 2.40 An equivalent circuit of the inverting amplifier with finite op-amp gain (A_a).

The Gain-Bandwidth Product

In addition to being finite, the amplifier gain A_a is also frequency-dependent. The majority of operational amplifiers are internally compensated so that the frequency dependence is of the form

$$A_a = \frac{A_o}{1 + j\omega/\omega_o} \tag{2.84}$$

where A_o is the low-frequency gain and ω_o is the -3-dB frequency. This particular form of frequency dependence is selected because it enhances the stability characteristics of the closed-loop system. The magnitude of A_a as a function of frequency is plotted in Fig. 2.41. The gain decreases at a rate of -6 dB per octave at high frequencies. At the frequency ω_o the gain is 0.707 (-3 dB) of the dc value. This frequency is the -3-dB bandwidth of the amplifier. The frequency ω_T is defined as the frequency where the magnitude of the gain is unity. That is,

$$A_a = 1 = \frac{A_o}{|1 + j\omega_T/\omega_o|} \approx \frac{A_o\omega_o}{\omega_T}$$

or $\qquad\qquad \omega_T = A_o\omega_o$

$$\tag{2.85}$$

The frequency ω_T is equal to the low-frequency gain A_o times the open-loop bandwidth ω_o. For this reason it is known as the *gain-bandwidth product*. The gain-bandwidth product is usually contained in the op-amp specification sheets.

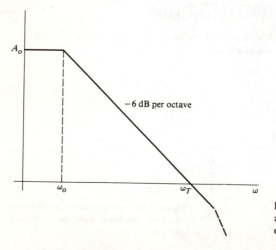

Figure 2.41 Frequency response of an internally compensated operational amplifier.

Example 2.13 The HA5190 operational amplifier has a gain-bandwidth product of 150 MHz and a low-frequency voltage gain of 90 dB. What is the transfer function for the open-loop voltage gain?

SOLUTION Since the low-frequency gain is 90 dB,

$$20 \log A_o = 90 \qquad \text{and} \qquad A_o = 3.16 \times 10^4$$

Since

$$\omega_o A_o = \omega_T = 2\pi(150 \times 10^6)$$

the bandwidth is

$$\omega_o = 2\pi(4.74 \times 10^3) \text{ rad/s}$$

and the transfer function is

$$A_v = \frac{3.16 \times 10^4}{1 + j\omega/[2\pi(4.7 \times 10^3)]}$$

The open-loop gain of an op amp declines with increasing frequency so that the deviation of the actual transfer function from the ideal value will increase with frequency. At any frequency the actual value can be calculated using Eq. (2.83) with the value of A_a given by Eq. (2.84).

The Effect of Finite Input Impedance

For the case where the input impedance is finite the equivalent circuit is as shown in Fig. 2.42a. This circuit is most easily analyzed by replacing the input circuit and Z_i by the Thévenin equivalent, as shown in Fig. 2.42b. Here

$$V_{th} = \frac{V_i Z_i}{Z_i + Z_1} \tag{2.86}$$

and

$$Z_{th} = \frac{Z_i Z_1}{Z_1 + Z_i} \tag{2.87}$$

A useful rule of thumb is that the effect of finite op-amp input impedance can be neglected, provided $Z_i > 100Z_1$. In either case the output voltage will be

$$V_o = \frac{-Z_f/Z_{th} V_{th}}{1 + (Z_{th} + Z_f)/A_v Z_{th}} \tag{2.88}$$

where V_{th} and Z_{th} are given by Eqs. (2.86) and (2.87), respectively.

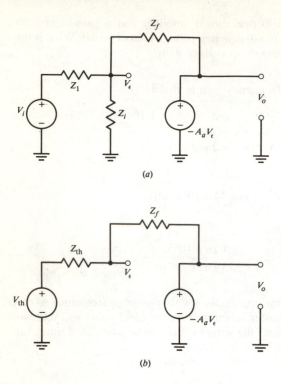

(a)

(b)

Figure 2.42 (a) A small-signal equivalent circuit of an inverting amplifier with finite op-amp input impedance Z_i and op-amp gain A_a; (b) a Thévenin equivalent circuit of the amplifier shown in Fig. 2.42a.

The Effect of Finite Input and Output Impedance

If neither Z_i nor Z_o is neglected, the small-signal equivalent circuit is as shown in Fig. 2.43a. The circuit seen by the load Z_L can be replaced by its Thévenin equivalent circuit as shown in Fig. 2.43b. To determine the Thévenin equivalent impedance (which is the output impedance of the closed-loop amplifier), a voltage V_o is applied across the output terminals and the current I_o is determined. The current I_o is given by

$$I_o = \frac{V_o + A_a V_\epsilon}{Z_o} + \frac{V_o - V_\epsilon}{Z_f}$$

because

$$V_\epsilon = \frac{V_o Z_1'}{Z_1' + Z_f}$$

where

$$Z_1' = \frac{Z_i Z_1}{Z_i + Z_1}$$

The Thévenin equivalent impedance (the output impedance) is

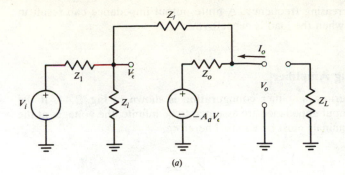

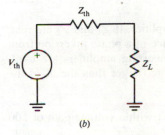

Figure 2.43 (a) An inverting amplifier equivalent circuit illustrating the op-amp input impedance Z_i, output impedance Z_o, and gain A_a; (b) a Thévenin's equivalent circuit of the amplifier shown in Fig. 2.43a.

$$Z_{th} = \frac{Z_o}{1 + (Z_f/Z_1' + A_a)/(2 + Z_f/Z_1')} \tag{2.89}$$

The feedback circuit has the effect of reducing the output impedance from its open-circuit value (Z_o), and the larger the op-amp gain A_a, the smaller the output impedance. The Thévenin equivalent voltage V_{th} is the output voltage without the load Z_L connected.

After eliminating V_ϵ, the transfer function is found to be

$$\frac{V_o}{V_{th}} = \frac{-(Z_f - Z_o/A_a)}{Z_{th} + (Z_o + Z_f + Z_{th})/A_a} \tag{2.90}$$

This equation describes the inverting amplifier configuration with finite Z_o, Z_i, and A_a. Communication circuits frequently employ small Z_i and Z_f, and the effects of nonzero Z_o must be considered. Note that the significance of the output impedance becomes more important as the open-loop gain A_a decreases, as it does with increasing frequency. A finite output impedance can result in stability problems when the load is capacitive.[2,3]

The Noninverting Amplifier

The basic noninverting amplifier configuration is shown in Fig. 2.44. If the op-amp gain and input impedance are assumed to be infinite, the voltages at the plus and minus terminals must be equal. Therefore,

$$V_+ = V_i = V_- = \frac{V_o Z_1}{Z_1 + Z_2}$$

or

$$\frac{V_o}{V_i} = 1 + \frac{Z_2}{Z_1} \qquad (2.91)$$

This is the basic equation for the noninverting amplifier. If Z_1 and Z_2 are real, the gain is positive and greater than 1. The absence of a phase inversion from the input to the output is one reason that noninverting amplifiers are used. Another reason is that the input impedance is large (larger than the Z_i of the device).

Example 2.14 Design a noninverting amplifier with a voltage gain of 100 using an ideal op amp.

SOLUTION For the ideal op amp used in the amplifier configuration of Fig. 2.43 a, the voltage gain is given by

$$A_v = \frac{V_o}{V_i} = 1 + \frac{Z_2}{Z_1}$$

Either Z_2 or Z_1 can be chosen arbitrarily. For example, if $Z_1 = 1 \text{ k}\Omega$, then $Z_2 = 99 \text{ k}\Omega$ will realize a voltage gain equal to 100.

The effects of finite voltage gain and input impedance are treated just as they were for the inverting amplifier. The calculations are left as an exercise.

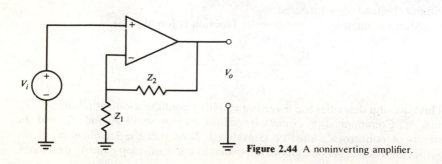

Figure 2.44 A noninverting amplifier.

PROBLEMS

2.1 What is the approximate midfrequency voltage gain V_o/V_i of the circuit shown in Fig. P2.1? The transistor β is 110, and the transistor output impedance is 50 kΩ. What is the midfrequency power gain?

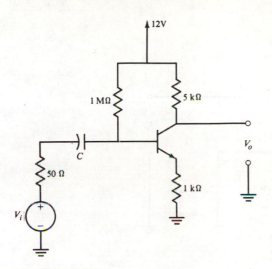

Figure P2.1 An emitter-loaded common-emitter amplifier.

2.2 What is the midfrequency voltage gain and input impedance Z_i of the amplifier shown in Fig. P2.2? The transistor β is 110, and the transistor output impedance is 100 kΩ.

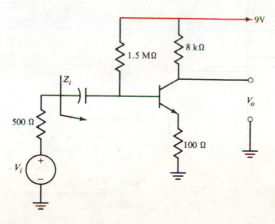

Figure P2.2 An emitter-loaded common-emitter amplifier.

2.3 Design an amplifier, using one of the transistors specified in the appendixes, with a midfrequency voltage gain of at least 26 dB. The source resistance is 600 Ω, and the load resistance is to be 5 kΩ. Show the complete circuit, including the biasing network.

2.4 Calculate the approximate midfrequency voltage gain of the amplifier shown in Fig. P2.4. The g_m of the transistor is 4×10^{-3} S and r_d is 10 kΩ.

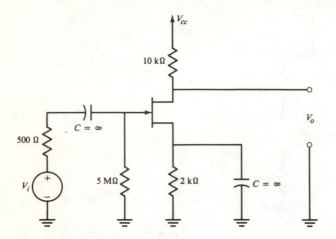

Figure P2.4 A common-source amplifier.

2.5 Calculate the midfrequency voltage gain of the amplifier shown in Fig. P2.5. The transistor g_m is 10^{-3} S, and the transistor output impedance r_d is 100 kΩ.

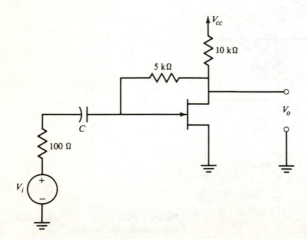

Figure P2.5 A common-source amplifier with current feedback.

2.6 What is the midfrequency voltage gain and input impedance of the amplifier shown in Fig. P2.6?

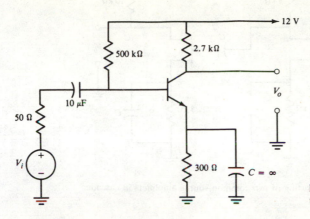

Figure P2.6 A common-emitter amplifier.

2.7 Calculate the midfrequency voltage, power, and current gains of the amplifier shown in Fig. P2.7. The transistor g_m is 3×10^{-3} S, and the transistor output impedance is 20 kΩ.

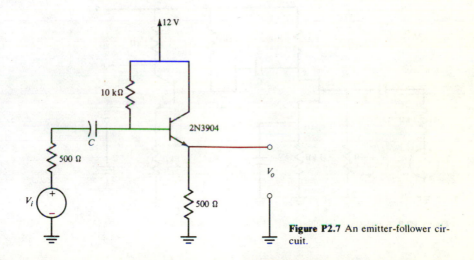

Figure P2.7 An emitter-follower circuit.

2.8 For the two-stage amplifier shown in Fig. P2.8 calculate the midfrequency voltage and current gains. The g_m of both transistors is 3×10^{-3} S and r_d is 20 kΩ.

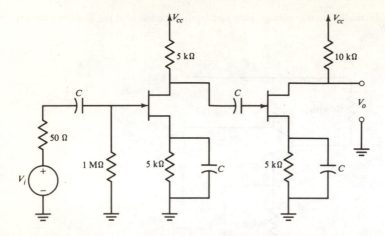

Figure P2.8 An amplifier consisting of two common-source amplifiers in cascade.

2.9 Determine the approximate midfrequency voltage gain of the two-stage amplifier shown in Fig. P2.9. (All capacitors can be assumed to be short circuits.) The β of the BJT is 100, and g_m is 3×10^{-3} S for the FET. The transistor output impedances are large enough to be neglected.

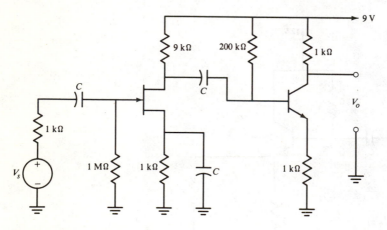

Figure P2.9 A two-stage amplifier.

2.10 An amplifier configuration often referred to as a Darlington amplifier is shown in Fig. P2.10. Calculate the voltage gain and input resistance of this circuit. The transistors have identical β's and their output impedances can be neglected, but the g_m of the two devices will not be the same.

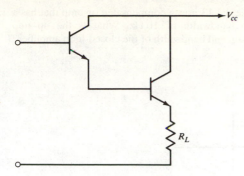

Figure P2.10 A Darlington amplifier.

2.11 The amplifier circuit shown in Fig. P2.11 employs an operational amplifier with ideal characteristics except that its open-loop gain A_a equals 20,000. How much does the actual closed-loop gain differ from the ideal value?

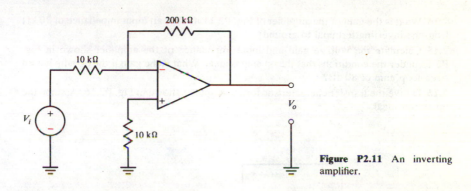

Figure P2.11 An inverting amplifier.

2.12 Determine the gain of the amplifier shown in Fig. P2.12. The operational amplifier has ideal characteristics.

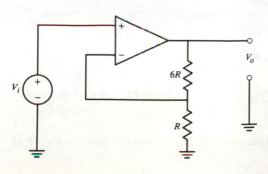

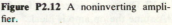

Figure P2.12 A noninverting amplifier.

2.13 The inverting amplifier shown in Fig. P2.13 uses a compensated op amp that has a dc gain of 140 dB and an open-loop bandwidth of 10 Hz. What is the op-amp gain-bandwidth product? What are the gain and bandwidth of the closed-loop amplifier?

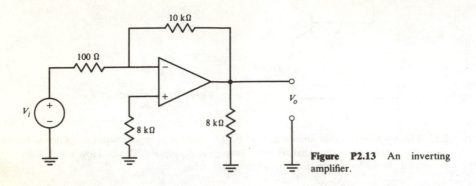

Figure P2.13 An inverting amplifier.

2.14 What is the gain of the amplifier of Fig. P2.11 if it has an input impedance of 20 kΩ from the inverting terminal to ground?

2.15 Calculate the voltage gain and input impedance of the amplifier shown in Fig. P2.13 under the conditions that the op amp is ideal. What is the gain if the op amp has an open-loop gain of 80 dB?

2.16 Derive the transfer characteristic for the amplifier shown in Fig. P2.16. Assume the op amp is ideal.

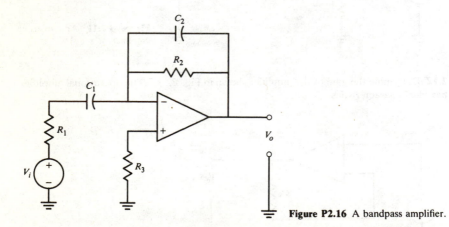

Figure P2.16 A bandpass amplifier.

2.17 Derive an expression for I_o in terms of V_i for the amplifier shown in Fig. P2.17. Assume the op amp is ideal.

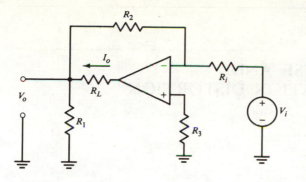

Figure P2.17 An inverting amplifier.

2.18 Calculate the gain of the cascode circuit shown in Fig. P2.18. Assume the dc collector current is 1 mA.

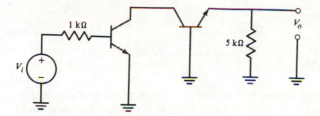

Figure P2.18 A cascode amplifier.

2.19 The HA5190 wideband operational amplifier is used in the circuit shown in Fig. P2.11. What is the low-frequency voltage gain? See Appendix 4 for amplifier specifications.

2.20 What is the bandwidth of the amplifier illustrated in Fig. P2.13 if the Burr–Brown 3554 is used as the operational amplifier? See Appendix 5 for the 3554 specifications.

REFERENCES

2.1. Searle, C. L. et al.: *Elementary Circuit Properties of Transistors*, Wiley, New York, 1964.
2.2. Evans, A. D. (ed.): *Designing with Field Effect Transistors*, McGraw-Hill, New York, 1981.
2.3. Roberge, J. K.: *Operational Amplifiers: Theory and Practice*, Wiley, New York, 1975.

THREE

NETWORK NOISE AND INTERMODULATION DISTORTION

3.1 INTRODUCTION

One of the most important factors to consider in evaluating the performance of a communications network is the network's ability to process low-amplitude signals. Every system creates noise, which limits its ability to process weak signals; the principal noise sources are (1) random noise generated in the resistors and transistors, (2) mixer noise created by the nonideal properties of the mixers, (3) the undesired cross-coupling of signals between two sections of the receiver, and (4) power-supply noise. Except for random noise, all of these sources of noise can be eliminated, at least in theory, by proper design and construction.

Random noise is inherent in all resistors and transistors. It is a critical factor in the performance of communication receivers since it determines the minimum signal level that can be detected. A measure of receiver performance, referred to as the *noise figure*, has long been used to quantitatively describe the noise generated in a network. This chapter discusses the nature of random noise and how it affects receiver performance and design of low-noise amplifiers.

At one time the noise figure was considered sufficient for characterizing a receiver's performance, but today large-amplitude, unwanted signals are often at the receiver input. It is often necessary to be able to detect low-amplitude signals that are adjacent in frequency to large-amplitude, unwanted signals, and the noise figure is not adequate to completely describe a receiver's performance. For this reason, the term "dynamic range" is introduced in this chapter to more completely describe receiver performance.

3.2 NOISE

All signals, whether at the network input or output, are contaminated by noise which degrades the system performance. The "noisiness" of a signal is usually specified in terms of the signal-to-noise ratio S/N, which is, in general, a function of frequency. The signal-to-noise ratio can be defined as

$$\frac{S(f)}{N(f)} = \frac{\text{rms signal voltage}}{\text{rms noise voltage}}$$

or

$$= \frac{\text{peak signal power}}{\text{average noise power}}$$

or

$$= \frac{\text{average signal power}}{\text{average noise power}}$$

Unless otherwise stated, the last definition will be used here for the signal-to-noise ratio, that is,

$$\frac{S}{N} = \frac{\text{average signal power}}{\text{average noise power}}$$

In order to define the average noise power, the origins and characteristics of network noise will first be discussed. An appropriate definition of noise is that it is "any unwanted input." Noise consists of both nonrandom, or periodic, components and random components. Types of nonrandom, or periodic, noise include power-supply noise and noise due to the unwanted cross-coupling of large signals such as that from a local oscillator. Note that the oscillator signal is considered to be a noise if it occurs at a point in the system where it is not desired. Noise of human origin is usually the dominant factor in receiver noise. Most of this type of noise is deterministic and can (at least theoretically) be eliminated through proper circuit design, layout, and shielding.

Random noise, by its very nature, cannot be eliminated. It places a lower theoretical limit, much like the uncertainty principle in physics, on the receiver noise level. An understanding of the properties of random noise allows one to control, by system design, its effect on receiver performance.

Random noise is described in terms of its statistical properties. At any instant noise amplitude cannot be predicted exactly, but it can be expressed in terms of a probability density function. For system design and evaluation it suffices to describe the noise in terms of its mean-square or root-mean-square values. The mean-square noise voltages (currents) are often referred to as the *noise power* since they are inversely proportional to the power dissipated in a resistor. The mean-square power is normally frequency-dependent and is usually expressed as a power spectral density function (unit of power per hertz). The total noise power P is

$$P = \int_{f_1}^{f_2} p(f)\, df$$

In the following sections lowercase letters denote spectral density functions (volts squared per hertz, etc.) and uppercase letters denote mean-square values (volts squared, etc.).

Random noise can be subdivided into noise occurring external to the receiver, such as atmospheric and interstellar noise over which the receiver designer has no control, and random noise occurring in the receiver. The most common form of random noise originating in the receiver is thermal noise.

Thermal Noise

J. B. Johnson discovered *thermal noise*—the minute currents caused by the thermal motion of the conduction electrons in a resistor that constitute a random noise. H. Nyquist was able to demonstrate, using statistical thermodynamics, that the thermal noise generated in an impedance $Z(f)$ in a frequency interval Δf is given by (Fig. 3.1)

$$E_n^2 = 4kTR(f)\,\Delta f \qquad (3.1)$$

where E_n = rms value of thermal noise voltage
$\quad R(f)$ = resistive component of impedance $Z(f)$, in Ω
$\quad T$ = absolute temperature, K
$\quad k$ = Boltzmann's constant, 1.38×10^{-23} W $\cdot$ s/K

Since the real part of the impedance R will in general be a function of frequency, the thermal noise voltage will also be frequency-dependent. However, for a resistor R, Eq. (3.1) states that the thermal noise voltage squared will be proportional to the frequency interval Δf. This implies that if the interval is infinite the noise power contributed by the resistor is also infinite. Actually, Eq. (3.1) must be modified at very high frequencies (above 100 MHz), but it is sufficiently accurate for the purposes of this text.

For a linear network the total mean-square voltage density appearing across any two terminals is given by

$$E_n^2 = 4kTR(f)\,df \qquad (3.2)$$

where $R(f)$ is the real part of the impedance $Z(f)$. $Z(f)$ is the impedance looking into the two terminals between which E_n is measured. The total mean-square noise is then obtained by integrating this expression over all frequencies.

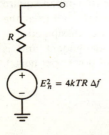

$$E_n^2 = 4kTR\,\Delta f$$

Figure 3.1 A resistor together with the mean-square thermal noise voltage of the resistance.

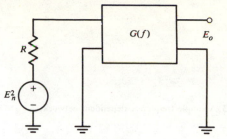

Figure 3.2 A resistor connected to a linear network with a frequency-dependent transfer function $G(f)$.

An alternative but equivalent interpretation is to find the noise due to each resistor. If a resistor is connected to a frequency-dependent network as shown in Fig. 3.2, then the total noise at the output will be given by

$$E_o^2 = \int_0^\infty 4kTRG(f)\, df \qquad (3.3)$$

where $G(f)$ is the magnitude squared of the frequency-dependent transfer function between the input and the output voltages, or

$$G(f) = \left| \frac{E_o(f)}{E_n(f)} \right|^2 \qquad (3.4)$$

Since in this case R is not a function of frequency, Eq. (3.3) is equivalent to

$$E_o^2 = 4kTR \int_0^\infty G(f)\, df \qquad (3.5)$$

The integral of the magnitude squared of the transfer function (normalized for unity gain) is referred to as the *noise bandwidth B_n* of the system. The noise bandwidth differs from the system's 3-dB bandwidth in that it is the area under the curve $G(f)$. A system can have a narrow 3-db bandwidth and yet have a large noise bandwidth.

Example 3.1 The impedance of the parallel combination of a resistor R and capacitor C shown in Fig. 3.3 is given by

$$Z(\omega) = \frac{R}{1 + j\omega RC} \qquad (3.6)$$

The real part of $Z(\omega) = R(\omega)$ is given by

$$R(\omega) = \frac{R}{1 + \omega^2 R^2 C^2} \qquad (3.7)$$

so the total output noise power generated by the impedance will be

$$E_o^2 = 4kTR \int_0^\infty \frac{df}{1 + \omega^2 R^2 C^2} = \frac{kT}{C} \qquad (3.8)$$

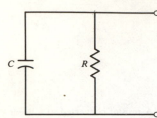

Figure 3.3 A simple frequency-dependent network.

which is independent of the size of the resistor and depends only on the capacitance C and temperature T.

Example 3.2 The circuit of Example 3.1 could also be interpreted as shown in Fig. 3.4. Here the noise is represented as the equivalent thermal noise of the resistor R. The transfer function is given by

$$G(f) = \left|\frac{E_o}{E_n}\right|^2 = (1 + \omega^2 R^2 C^2)^{-1} \tag{3.9}$$

Thus, using Eq. (3.4), the output noise voltage squared is found to be

$$E_o^2 = 4kTR \int_0^\infty \frac{df}{1 + \omega^2 R^2 C^2} = \frac{kT}{C} \tag{3.10}$$

which is the same result as obtained in Example 3.1. Since

$$E_o^2 = 4kTRB_n$$

the circuit noise bandwidth $B_n = 1/4RC$ Hz. The 3-dB bandwidth is $2\pi/RC$ Hz.

In this example the same result was obtained by interpreting the noise to originate in the real part of the impedance $Z(f)$ or by interpreting the noise to originate in the resistor. It makes no difference whether the thermal noise voltage is interpreted to arise in the real part $R(f)$ of the impedance $Z(f)$ or whether the resistive elements are taken as the ultimate source of the thermal noise.

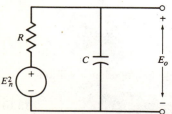

Figure 3.4 A simple frequency-dependent network including the thermal noise source.

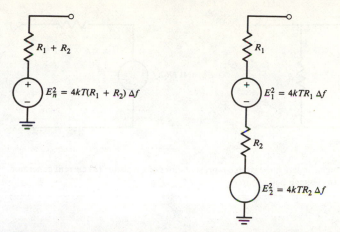

Figure 3.5 Equivalent mean-square noise voltage representation of two resistors in series.

Resistors in series. Equation (3.1) states that the noise voltage is the rms value of a randomly varying signal. If two resistors are added in series, as shown in Fig. 3.5, it is the noise voltages squared, not the noise voltages, which are added.

$$E_n^2 = E_1^2 + E_2^2 = 4kT(R_1 + R_2)\,\Delta f \tag{3.11}$$

Similarly, for two resistors in parallel the equivalent noise voltage is

$$E_n^2 = 4kT\,\frac{R_1 R_2}{R_1 + R_2}\,\Delta f \tag{3.12}$$

The noise sources described here all refer to rms quantities, and the noise source has no polarity associated with it. In order to keep the notation simple, the noise sources are expressed in terms of the square of the voltage or current, and the values referred to are always the mean-square values.

Current source representation. Equation (3.1) states that the thermal noise can be represented by a voltage source in series with a noiseless resistor. Norton's theorem shows that the voltage noise source illustrated in Fig. 3.6b can also be represented by a current generator in parallel with a noiseless resistor as shown in Fig. 3.6a.

Excess resistor noise. The thermal-noise-power density generated in resistors does not vary with frequency, but many resistors also generate a frequency-dependent noise referred to as *excess noise*. The excess noise power has a $1/f$ spectrum; the excess noise voltage is inversely proportional to the square root of the frequency. Noise that exhibits a $1/f$-power spectral characteristic is often referred to as *pink noise*.

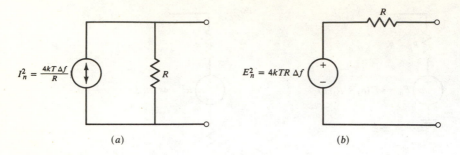

Figure 3.6 Equivalent mean-square noise source representations of a resistor: (*a*) current generator with noiseless resistor; (*b*) voltage noise source.

The amount of excess noise generated in a resistor depends upon the resistor's composition. Carbon resistors generate the largest amount of excess noise, whereas the amount generated in wire-wound resistors is usually negligible. However, the inductance inherent in wire-wound resistors restricts them to low-frequency applications. Metal film resistors are usually the best choice for high-frequency communications circuits, where low noise and constant resistance are required.

Active Device Noise

Besides the thermal noise of resistors, the other sources of random noise of importance in network design are the active devices—integrated circuits, diodes, and transistors. The two main types of device noise are $1/f$, or flicker, noise and shot noise. *Flicker noise* is a low-frequency phenomenon in which the noise power density follows a $1/f^\alpha$ curve, the value of α is close to unity.

An electric current composed of discrete charge carriers flows through an active device. The discreteness of the charge-carrier fluctuations are present in the current crossing a barrier where the charge carriers pass independently of one another. Examples of such barriers are the semiconductor *pn* junction in which the passage takes place by diffusion and the cathode of a vacuum tube where electron emission occurs as a result of thermal motion. The current fluctuations represent a noise component referred to as *shot noise*, which can be represented by an appropriate current source in parallel with the dynamic resistance of the barrier across which the noise originates. The spectral density of this shot noise is given by

$$i_{n_o}^2 = qkI_o \qquad \text{A}^2/\text{Hz} \tag{3.13}$$

where q is the charge on an electron, I_o is the direct current, and k is a constant that varies from device to device and also depends on how the junction is biased. In a junction transistor k is equal to 2. Figure 3.7 illustrates the

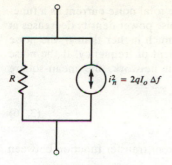

Figure 3.7 Network including the shot-noise current source of a *pn* junction (lowercase i^2 denotes current spectral density).

shot-noise equivalent circuit for a forward-biased *pn* junction. Lowercase letters are used to denote the noise density as mean square per hertz. Shot noise, like thermal noise, has a uniform power spectral density, and the total noise current squared is proportional to the bandwidth. That is,

$$I_n^2 = i_{n_o}^2 \, \Delta f \qquad (3.14)$$

The current source represented in Fig. 3.7 denotes that no direction is associated with the current source since it is a mean-square value. If the additional $1/f$ noise is included, the total mean-square current noise density can be written as

$$i_n^2(f) = i_{n_o}^2 \left(1 + \frac{f_L}{f}\right) \qquad A^2/Hz \qquad (3.15)$$

where f_L is the frequency where the shot-noise current is equal to the $1/f$-noise current. f_L varies from device to device and is usually determined empirically.

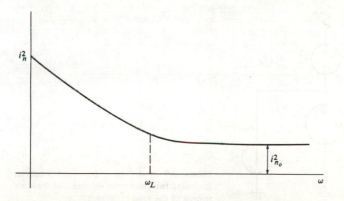

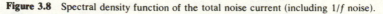

Figure 3.8 Spectral density function of the total noise current (including $1/f$ noise).

Figure 3.8 illustrates the power density of the total noise current as a function of frequency. At frequencies below f_L the noise power density decreases at a rate of 6 dB per octave, while at frequencies much higher than f_L the noise power is equal to the shot noise and is independent of frequency. If the noise current is connected to a frequency-dependent network, the mean-square current at the output will be

$$I_o^2 = \int_0^\infty A_i(f) i_n^2 \, df \qquad (3.16)$$

where $A_i(f)$ is the magnitude squared of the current transfer function between input and output.

Noise in Transistor Amplifiers[3.1]

The previous discussion has shown that any amplifier must generate noise, which consists of thermal noise generated in the resistors plus the shot and $1/f$ noise generated in the active devices. An equivalent circuit of a transistor amplifier, which identifies the shot-noise sources, is shown in Fig. 3.9. i_{n_2} represents the shot-noise current density due to the bias current on the output of the device, and i_{n_1} is the shot-noise current density due to the input bias current. The other noise source is due to the load resistor R_L.

If the transistor output impedance is much larger than R_L, the output noise voltage due to i_{n_2} will be

$$(e_o)_{n_2}^2 = i_{n_2}^2 R_L^2$$

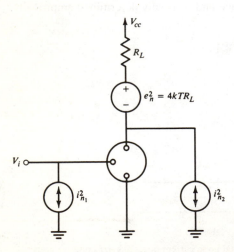

Figure 3.9 Model of a transistor amplifier including two shot-noise sources and the noise source of the load resistance.

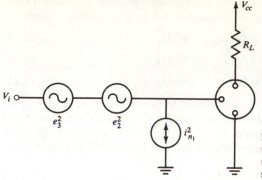

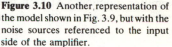

Figure 3.10 Another representation of the model shown in Fig. 3.9, but with the noise sources referenced to the input side of the amplifier.

It is convenient to refer all of the noise sources to the input. The amplifier voltage gain is approximately

$$\left|\frac{V_o}{V_i}\right| = g_m R_L$$

so the output noise current source can be replaced by an equivalent input noise voltage source.

$$e_2^2 = \frac{i_{n_2}^2 R_L^2}{g_m^2 R_L^2} = \frac{i_{n_2}^2}{g_m^2} \tag{3.17}$$

The noise source e_2^2 can be interpreted as due to thermal noise of the transconductance g_m. Likewise, the thermal noise of the load can also be represented as a noise e_3 in series with the input, where

$$e_3^2 = \frac{4kTR_L}{g_m^2 R_L^2} \tag{3.18}$$

Normally $e_3 \ll e_2$ and can be neglected. The amplifier, with the noise sources referred to the input, can be designated as shown in Fig. 3.10. The amplifier is considered noiseless, and the noise is represented by the noise voltage and current sources. The model as represented has been simplified by assuming that the voltage gain and amplifier transadmittance are independent of frequency. In this case the total mean-square noise voltage will be proportional to frequency, but in the more general case frequency-dependent transfer functions must be used, and the total mean-square noise voltage can only be obtained by integrating the instantaneous values over the frequency region of interest. The model also does not include any thermal noise present in the amplifier.

Transistor noise can originate from one of two sources.

BJT noise. The principal noise sources in a bipolar transistor are the two shot-noise sources and the thermal noise created in the base spreading resistor r_b':

$$e_n^2 = 4kTr_b' \qquad \text{V}^2/\text{Hz} \tag{3.19}$$

$$i_{n_1}^2 = 2qI_B \qquad \text{A}^2/\text{Hz} \tag{3.20}$$

$$i_{n_2}^2 = 2qI_C \qquad \text{A}^2/\text{Hz} \tag{3.21}$$

The noise current source $i_{n_1}^2$ is connected between the base and emitter junctions, and the other current source is connected between the collector and emitter junctions. At high frequencies (above f_β of the transistor) the noise currents increase with graduating frequencies, and the complete expression is

$$i_{n_c}^2 = i_n^2 \left(1 + \beta \frac{f^2}{f_T^2}\right) \tag{3.22}$$

Section 3.4 describes a more detailed model of the noise sources in a BJT amplifier.

FET noise. The FET noise sources (excluding excess noise) are given by the following expressions:

$$e_n^2 = \frac{2.8kT}{g_m} \qquad \text{V}^2/\text{Hz} \tag{3.23}$$

and $$i_n^2 = 2qI_g \qquad \text{A}^2/\text{Hz} \tag{3.24}$$

where g_m is the mutual conductance and I_g is the gate leakage current. The noise sources of MOSFETs and JFETs are the same, but I_g is negligible for MOSFETs. The shot noise increases with frequency at very high frequencies; the total noise current is

$$i_n^2 = 2qI_g + \frac{2.8kT}{g_m} \omega^2 C_{gs'}^2 \qquad \text{A/Hz} \tag{3.25}$$

where $C_{gs'}$ is approximately two-thirds of the transistor gate-to-source capacitance.

The transistor amplifier noise model will subsequently be used for the design of low-noise amplifiers, but first one of the parameters most often used to characterize the "noisiness" of a system, the noise figure, will be described.

3.3 NOISE FIGURE, NOISE FACTOR, AND SENSITIVITY

Although signal-to-noise ratio is the best measure of system input and output signal quality, a quantitative measure is also needed of how much noise is added by the circuit, whether the circuit is a passive filter, an amplifier, or an entire receiver. The noise factor F has become a standard figure of merit of the amount of noise added in a circuit. According to the *IEEE Standards*, "The

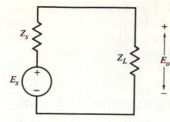

Figure 3.11 A simple circuit model for determining the maximum power transfer to the load impedance.

noise factor, at a specified input frequency, is defined as the ratio of (1) the total noise power per unit bandwidth available at the output port when the noise temperature of the input termination is standard (290 K) to (2) that portion of (1) engendered at the input frequency by the input termination."[3.2]

Available power refers to the maximum power that can be delivered from a generator with a source impedance R_g to a load impedance R_L. For the network of Fig. 3.11 it is easily shown that the load receives maximum power when the load is matched to the source, that is, when

$$Z_L = Z_s^*$$

where Z^* is the complex conjugate of the impedance Z. Under matched conditions the load power will be

$$P_o = \frac{E_g^2}{4 R_s} \tag{3.26}$$

This is the maximum available power from the source E_s. Therefore the available noise power from a resistor R is equal to $kT \, \Delta f$. That is, the available noise power is independent of the size of the resistor!

The IEEE definition for noise factor F can be stated as

$$F = \frac{\text{available output noise power}}{\text{available output noise due to the source}} \tag{3.27}$$

If N_i is used to denote the noise power available from the source and N is the total noise,

$$F = \frac{N_o}{(N_i)_o} \tag{3.28}$$

The o subscripts indicate that the noise powers are specified at the output. It must be kept in mind that the symbols refer to the available noise powers. The noise factor depends upon the noise generated in a device and on its input termination, but not upon the output termination. Since the output noise power in a linear system is the sum of the noise due to the source plus the noise N_a added in the system, the noise factor can be written as

$$F = \frac{N_i + N_a}{N_i} \tag{3.29}$$

The definition for available noise power used in Eqs. (3.26) to (3.29) must specify whether the noise is the total noise or the noise per unit bandwidth. Noise per unit bandwidth will be used unless otherwise noted. In Eq. (3.29) the values for noise can refer to the values at the input or output, and it is important only that one be consistent. We will use as standard notation

N_i = available input noise per unit bandwidth
N_a = available noise power added per unit bandwidth (referred to the input)

Since the output power S_o is the available input signal power S_i times the power gain $G(f)$, Eq. (3.29) can be written as

$$F = \frac{N_i + N_a}{N_i} \frac{G(f)S_i}{S_o} = \frac{S_i}{N_i} \frac{N_o}{S_o} \qquad (3.30)$$

[The noise at the output N_o is the noise at the input $(N_i + N_a)$ multiplied by the power gain $G(f)$.] Therefore the noise factor can also be written as

$$F = \frac{\text{input signal-to-noise ratio}}{\text{output signal-to-noise ratio}}$$

The noise factor F is a measure of the degradation of the signal-to-noise ratio due to the noise added in the system. Since the maximum available noise power is $E^2/4R_g$, the maximum available noise power per unit bandwidth from a source resistance is

$$N_i = \frac{4kTR_g}{4R_g} = kT \qquad (3.31)$$

independent of the size of the source resistance. Hence the noise factor of a receiver is

$$F = 1 + \frac{N_a}{kT} \qquad (3.32)$$

This is the noise factor measured in a unit bandwidth at a particular frequency and is often referred to as the *spot-noise factor*. The source impedance does not appear in this expression for the noise factor, but it will subsequently be shown that the noise added depends on the source impedance, and hence so does the noise factor.

Note that in an ideal receiver no noise is added ($N_a = 0$), so the receiver has a unity noise factor. Since the noise factor is always greater than 1, the output signal-to-noise ratio is always less than the input signal-to-noise ratio. That this does not agree with experience is a result of the definition of noise factor. A receiver will usually improve the signal-to-noise ratio through filtering of the input noise. Since the noise factor definition uses the same bandwidth in defining the two signal-to-noise ratios, the noise factor does not reflect the filtering quality of the receiver, and it is only one parameter to be considered in completely describing receiver performance.

Average Noise Factor

The noise performance of a communications system normally needs to be described over a range of frequencies. One method is to determine the spot-noise factor at several frequencies. Another method found useful in noise measurements is to specify the average noise factor. The average noise factor is defined as the ratio of (1) the total noise power delivered into the output termination by the transducer when the noise temperature of the input termination is standard (290 K) at all frequencies to (2) that portion of (1) engendered by the input termination.

The average noise factor $\bar{F}$ is

$$\bar{F} = \frac{\displaystyle\int F(f)\,G(f)\,df}{\displaystyle\int G(f)\,df} \tag{3.33}$$

where $G(f)$ is the system power (transducer) gain and $F(f)$ is the frequency-dependent noise factor. For a heterodyne system, the noise created by the input includes only that portion of the noise from the input termination that appears in the output via the principal frequency transformation of the system and does not include spurious contributions such as those from an image-frequency transformation.

Noise Figure

The noise factor is often expressed in decibels. In this case it is called the *noise figure* (NF) and is defined as

$$\text{NF} = 10 \log_{10} F \tag{3.34}$$

Since the minimum value of $F = 1$, the noise figure NF of an ideal noiseless network is 0 dB.

Noise Factor of Cascaded Networks

If the noise factor and power gain of individual networks are known, the noise factor of cascaded networks is readily determined. First consider the series combination of two networks with noise factors and power gains F_1, G_1 and F_2, G_2, respectively. If the available input noise power N_i is equal to kT, the noise added by network 1 is

$$F_1 N_i - N_i = N_{a_1} = kT(F_1 - 1) \tag{3.35}$$

Likewise, the noise added by network 2 is

$$N_{a_2} = kT(F_2 - 1)$$

and the noise added in network 2, referred to the input, is

$$\frac{N_{a_2}}{G_1} = \frac{kT(F_2 - 1)}{G_1} \tag{3.36}$$

The overall noise factor is thus

$$F = \frac{\text{available input noise power} + \text{noise added}}{\text{available input noise power}}$$

$$= \frac{kT + (F_1 - 1)kT + (F_2 - 1)kT/G_1}{kT} \tag{3.37}$$

$$= F_1 + \frac{F_2 - 1}{G_1} \tag{3.38}$$

Equation (3.38) states that if the power gain of the first stage is large, the overall noise factor will be essentially that of the first stage. In other cases, the noise factor of the second stage, and even of succeeding stages, will be an important factor in the overall noise factor. Equation (3.34) is readily extended to n stages. For an n-stage system

$$F = F_1 + \frac{F_2 - 1}{G_1} + \frac{F_3 - 1}{G_1 G_2} + \cdots + \frac{F_n - 1}{G_1 G_2 \cdots G_{n-1}} \tag{3.39}$$

Example 3.3 For the system shown in Fig. 3.12, the first stage has a noise figure of 2 dB and a gain of 12 dB; the second stage has a noise figure of 6 dB and a power gain of 10 dB. What is the overall noise figure?

SOLUTION Equation (3.38) expresses the noise factor F in terms of the noise factors for each stage. Thus the noise figures must first be converted to noise factor values:

$$F_1 = 1.59 \qquad F_2 = 4$$

The corresponding gain values are

$$G_1 = 15.9 \qquad G_2 = 10$$

The overall noise factor is

$$F = 1.59 + \frac{4 - 1}{15.9} = 1.779$$

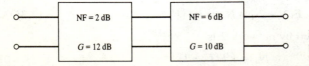

Figure 3.12 Numerical example of two cascaded, noisy networks.

and the noise figure of the two-stage system is

$$NF = 10 \log 1.779 = 2.5 \text{ dB}$$

Example 3.4 If G_1 and G_2 of Example 3.3 are independent of frequency, what will be the total output noise power of the cascaded system in a 3-kHz bandwidth? The operating temperature is 290 K.

SOLUTION Since $N_i + N_a = FkTB$,

$$kTB = 1.37 \times 10^{-23} \times 290 \times 3 \times 10^3$$
$$= 1.192 \times 10^{-17} \text{ W}$$
$$N_i + N_a = 1.779 \, kTB = 2.12 \times 10^{-17} \text{ W}$$

and the output noise

$$N_o = G_1 G_2 (N_i + N_a) = 159 \times 2.12 \times 10^{-17} = 337 \times 10^{-17} \text{ W}$$

Noise Temperature

The noise factor will normally lie between 1 and 10. For situations in which an expanded scale is needed, the system noise factor is usually expressed in terms of noise temperature. The noise factor is given by

$$F = 1 + \frac{N_a}{N_i} = 1 + \frac{N_a}{kT} \tag{3.40}$$

where T is the reference noise temperature. The noise added can be interpreted as the available noise from a resistor whose temperature is T_r. That is,

$$F = 1 + \frac{T_r}{T} \tag{3.41}$$

or

$$T_r = (F - 1)T \tag{3.42}$$

T_r is referred to as the system noise temperature.

Example 3.5 What is the variation in noise temperature as the noise factor varies from 1 to 1.6 (NF varies from 0 to 2 dB)? Assume the reference temperature is 290 K.

SOLUTION When the noise factor is 1, the noise temperature is 0. When the noise factor is 1.6,

$$T_r = (1.6 - 1)290 = 174 \text{ K}$$

Thus the change in the noise temperature is much greater than the change in the noise factor. This is a principal reason for using noise temperature to describe system noise.

Sensitivity

The available input-signal level S_i for a given output signal-to-noise ratio $(S/N)_o$ is referred to as the *system sensitivity*, or *noise floor*. The input voltage level corresponding to S_i is called the *minimum detectable signal*. Although the signal-to-noise ratio will depend on the system frequency response, we will assume for simplicity that the frequency response can be represented by the ideal characteristic shown in Fig. 3.13. Although this frequency response can never be realized in an actual receiver, it is closely approximated in many communication systems, especially those which include a narrow bandpass filter. When the frequency characteristic is ideal, Eq. (3.5) for the total available noise power from a resistor can be written as

$$\frac{E_n^2}{4R} = kTB \tag{3.43}$$

where B is the bandwidth. Therefore Eq. (3.30) can be rewritten as

$$S_i = F(kTB) \left(\frac{S}{N}\right)_o \tag{3.44}$$

where N_o is now the total noise power at the output.

Example 3.6 What minimum input signal will give an output signal-to-noise ratio of 0 dB in a system that has an input impedance equal to 50 Ω, a noise figure (NF) of 8 dB, and a bandwidth of 2.1 kHz?

SOLUTION For a 0-dB output signal-to-noise ratio and a 290 K operating temperature, Eq. (3.44) can be written

$$10 \log S_i = \text{NF} - 144 + 10 \log B$$

where S_i is in milliwatts and B is in kilohertz. For a bandwidth of 2.1 kHz,

$$S_i = -133 \text{ dBm} \qquad 133 \text{ dB below 1-mW level}$$

S_i is the available input power and is related to the input signal voltage by

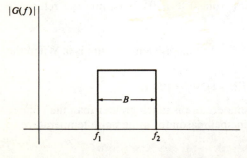

Figure 3.13 Frequency response of the magnitude of an ideal bandpass filter.

Eq. (3.26). Thus

$$S_i = \frac{E_i^2}{4R_s} = 5.02 \times 10^{-17} \text{ W}$$

Since $R_s = 50 \ \Omega$,

$$E_i = 0.10 \ \mu\text{V}$$

That is, for these specifications the noise floor for an output signal-to-noise ratio of 1 is 0.10 μV.

Example 3.7 What is the minimum detectable signal or noise floor of the system in the previous example for an output signal-to-noise ratio of 10 dB?

SOLUTION In this case Eq. (3.40) becomes

$$10 \log S_i = \text{NF} - 134 + 10 \log B = -123 \text{ dBm}$$

$$S_i = 5 \times 10^{-13} \times 10^{-3} \text{ W}$$

and the minimum detectable signal is

$$E_i = 0.32 \ \mu\text{V}$$

Sensitivity is always specified for a given signal-to-noise ratio. Although the required output signal-to-noise ratio may not be the same as that used in the sensitivity specification, sensitivity does provide an objective measure for comparing receiver performance. The required signal-to-noise ratio at the receiver output will depend on the function of the receiver and on whether or not additional signal processing (such as correlation detection) is performed. An output signal-to-noise ratio between 0 and 10 dB is adequate for normal listening.

Receiver noise figure is a measure of how much noise is added by the system. A low noise figure is often desirable, but there are situations in which this is of little importance. This is particularly true when the input noise is much greater than the noise added by the system. Numerical examples will illustrate this point.

Example 3.8 Consider a communications receiver with a 50-Ω input impedance, a B of 3 kHz and a 4-dB noise figure. The noise floor of this receiver for an output signal-to-noise ratio of 10 dB is found to be, using Eqs. (3.26) and (3.44),

$$S_i = -125 \text{ dBm} = 3 \times 10^{-16} \text{ W}$$

$$E_i = 0.245 \ \mu\text{V}$$

An input signal of 0.245 μV will produce a 10-dB output signal-to-noise ratio. Now consider the performance of this receiver when it is connected to an antenna with a noise figure of 20 dB. Expressing the noise at the antenna in terms of the noise figure has become accepted practice as it facilitates the numerical analysis. Antenna noise factor is defined to be

$$F_{\text{ant}} = \frac{\text{noise}_{\text{ant}} + \text{thermal noise}}{\text{thermal noise}} \tag{3.45}$$

or $\text{noise}_{\text{ant}} = (F_a - 1) \times$ thermal noise. Antenna noise refers to the total noise picked up by the antenna, primarily from external sources. The antenna noise factor in this example is 100. Hence the antenna noise is seen to be

$$N_{\text{ant}} = 99 \times \text{thermal noise} = 99\,kTB$$

The total input noise is the antenna noise plus the source noise, of $100\,kTB$. The output noise (referred to the input) is

$$N_o = N_{\text{ant}} + N_i + N_a = (F_a - 1)kTB + F_r kTB$$
$$= N_{\text{ant}} + F_r kTB$$

where F_r refers to the receiver noise factor. The output signal-to-noise ratio is thus

$$\frac{S_i}{N_i F} = \frac{S_o}{N_o}$$

$$\left(\frac{S}{N}\right)_o = \frac{S_i}{(F_a + F_r - 1)kTB} \tag{3.46}$$

In this example the antenna noise figure is 20 dB which corresponds to a noise factor of 100. Since the receiver noise factor is 2.5 (NF = 4 dB), the input signal required for a 10-dB output signal-to-noise ratio is

$$S_i = \left(\frac{S}{N}\right)_o (100 + 2.5 - 1)kTB$$
$$= 10 \times 101.5 \times 397 \times 10^{-23} \times 3 \times 10^3$$
$$= 1.203 \times 10^{-14} \text{ W}$$

Thus the minimum detectable signal for a 10-dB output signal-to-noise ratio is

$$E_i = 1.56 \ \mu\text{V}$$

This is much larger than the 0.245 μV required if there were no antenna noise.

Example 3.9 What would be the minimum detectable signal level in the previous example if a receiver with a noise figure of 10 dB is substituted?

SOLUTION Since the receiver noise factor is 10, Eq. (3.46) becomes for this system

$$S_i = \left(\frac{S}{N}\right)_o \times (100 + 10 - 1)kTB$$

$$= 1.29 \times 10^{-14} \text{ W}$$

and the minimum detectable signal is

$$E_i = 1.6 \ \mu\text{V}$$

A 6-dB reduction in the receiver noise figure results in only a 0.3-dB reduction in the output signal-to-noise ratio because the noise added by the receiver is much less than the antenna noise.

Examples 3.8 and 3.9 illustrate that if the input noise is large, very little is gained by reducing the system noise figure below some acceptable level. For communications receivers operating below 30 MHz, 8 to 10 dB is usually taken as an acceptable receiver noise figure because of the large antenna noise figure. However, as the frequency is increased above 30 MHz, receivers with lower noise figures are desirable because the antenna noise is much less at the higher frequencies. When the input noise is large, not only the receiver bandwidth but the actual passband must be selected for optimum performance by considering the frequency characteristics of the antenna noise.

Noise comparisons of two receivers must be used with care since the network with the lowest noise figure does not necessarily have the lowest output signal-to-noise ratio. The following section proves this important point.

3.4 DESIGN OF LOW-NOISE NETWORKS

Network Noise Representation

Any linear noisy network (or amplifier) can be represented by a noiseless network plus two noise generators e_n and i_n, as shown in Fig. 3.14. Two noise sources are required to represent the network noise because noise may exist at the output with the input terminals short- or open-circuited. In general e_n and i_n will be frequency-dependent. In the noise model for the common-emitter amplifier described in Sec. 3.2 the noise current i_n was due to the input bias current shot noise, and the noise voltage e_n was due to the output bias current shot noise plus the thermal noise due to the load resistor. The equivalent short-circuit input rms noise voltage e_n is the noise voltage that would appear to originate at the input of the noiseless network if the input terminals were short-circuited. The rms noise current i_n represents the remainder of the network noise. To determine i_n, a resistor R_s is shunted across the input terminals and the output noise is measured.

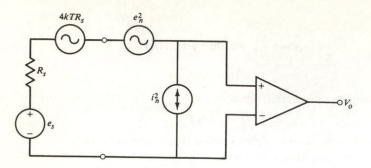

Figure 3.14 An amplifier noise source model.

Equivalent Noise Representation of a BJT Amplifier

The equivalent input noise representation will now be illustrated by considering a common-emitter amplifier. The small-signal, midfrequency circuit model, including noise sources, is illustrated in Fig. 3.15. The values for the shot-noise current sources are given by Eqs. (3.20) and (3.21). The equivalent input noise voltage is determined by shunting the source and determining the output voltage. If $R_s = 0$, the output noise voltage squared is

$$(e_n)^2_0 = 4kTr'_b \left(\frac{r_\pi \hat{g}_m R_L}{r_\pi + r_b} \right)^2 + i^2_{n_2} R^2_L + 4kTR_L \qquad \text{V}^2/\text{Hz} \qquad (3.47)$$

since the voltage transfer function with $R_s = 0$ is given by

$$\frac{V_o}{V_i} = \frac{-g_m R_L r_\pi}{r_\pi + r'_b}$$

The output noise voltage squared, referred to the input, is

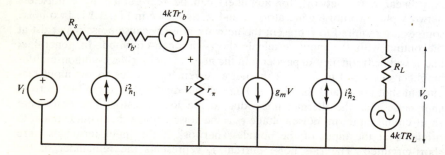

Figure 3.15 A small-signal equivalent circuit, including noise sources of a common-emitter amplifier.

$$e_n^2 = 4kTr_b' + i_{n_2}^2 \left(\frac{r_\pi + r_b'}{g_m r_\pi} \right)^2 + \frac{4kT}{R_L} \left(\frac{r_\pi + r_b'}{g_m r_\pi} \right)^2$$

$$= 4kTr_b' + \left(\frac{r_\pi + r_b'}{\beta} \right)^2 \left(i_{n_2}^2 + \frac{4kT}{R_L} \right) \qquad \text{V}^2/\text{Hz} \qquad (3.48)$$

This is the equivalent noise-voltage generator. The noise-current generator is determined by calculating the output noise with R_s connected across the input, and then determining the component due to i_n. The total output noise voltage (excluding the thermal noise from R_s) is

$$V_o^2 = \left(\frac{g_m r_\pi R_L}{r_\pi + r_b' + R_s} \right)^2 \left[e_n^2 + \left(i_{n_2} \frac{r_\pi + r_b'}{\beta} \right)^2 + (i_{n_1} R_s)^2 + 4kTR_L \right] \qquad \text{V}^2/\text{Hz}$$

$$(3.49)$$

The total output noise voltage consists of a component due to e_n and a component due to i_n. The component due to e_n is found by setting R_s equal to zero:

$$V_o|_{i_{n_1}} = \frac{i_{n_1} R_s r_\pi g_m R_L}{R_s + r_b' + R_L}$$

so the equivalent input noise current

$$i_n = i_{n_1} \qquad \text{A}^2/\text{Hz} \qquad (3.50)$$

is the shot-noise current due to the base bias current. Note that the input noise voltage e_n depends on the load resistance as well as on the transistor parameters.

Network Noise Factors

In this section we will consider the more general problem in which the source and network input resistances are not necessarily matched.

The output noise power referred to the input in a unit bandwidth of the amplifier illustrated in Fig. 3.14 is then (assuming i_n and e_n are uncorrelated)

$$N = e_n^2 + i_n^2 R_s^2 + 4kTR_s \qquad (3.51)$$

The noise voltage e_n is determined by measuring the output voltage with the input terminals short-circuited. The output noise is then measured with a specified source resistance. From this measurement i_n can be determined, provided the power gain is known. The noise added by the network is

$$N_a = e_n^2 + i_n^2 R_s^2 \qquad (3.52)$$

The network noise factor in any unit bandwidth can be defined as

$$F = \frac{e_n^2 + i_n^2 R_s^2 + 4kTR_s}{4kTR_s} \qquad (3.53)$$

Example 3.10 Plots of the input noise voltage and input noise current for

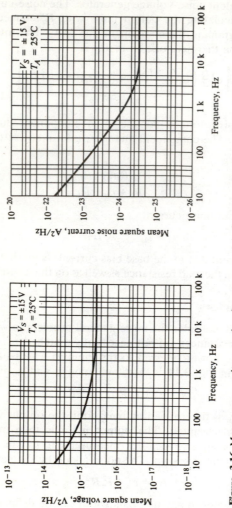

Figure 3.16 Mean-square voltage and noise sources of a 741 operational amplifier. (*Courtesy of Fairchild Semiconductor.*)

the Fairchild 741 operational amplifier are presented in Fig. 3.16. At 1 kHz the input noise voltage is approximately

$$e_n^2 \gtrsim 8 \times 10^{-16} \ V^2/Hz$$

and the input noise current is approximately

$$i_n^2 = 9 \times 10^{-25} \ A^2/Hz$$

If a 10-kHz source impedance is used, the amplifier noise factor will be

$$F = 1 + \frac{8 \times 10^{-16} + 0.9 \times 10^{-16}}{4 \times 1.37 \times 10^{-23} \times 290 \times 10^4}$$

$$= 6.6$$

Equation (3.53) indicates that a network's noise factor depends upon the source resistance. The value of source resistance that minimizes the noise factor can be formed by differentiating Eq. (3.53) with respect to R_s:

$$\frac{dF}{dR_s} = \frac{1}{4kT} \left(i_n^2 - \frac{e_n^2}{r_s^2} \right) = 0$$

or

$$R_s = \frac{e_n}{i_n} \tag{3.54}$$

The value of source resistance that minimizes the noise factor is equal to the input noise voltage divided by the input noise current. If i_n is large relative to e_n, a low value of source resistance is called for; if i_n is small (such as in a FET input amplifier), a large value of R_s is needed to minimize the noise factor. Since e_n and i_n are, in general, frequency-dependent, the source resistance required for minimization will be also; R_s is usually chosen to minimize the spot-noise factor at a specified frequency.

Example 3.11 What will be the minimum noise figure of the 741 operational amplifier of the preceding example at 1 kHz?

SOLUTION The value of source resistance that minimizes the noise figure is found from Eq. (3.54) (and Fig. 3.16):

$$(R_s)_{opt} = \left(\frac{8 \times 10^{-16}}{9 \times 10^{-25}} \right)^{1/2} = 30 \times 10^3 \ \Omega$$

The minimum noise factor is found using this value of source resistance in Eq. (3.53):

$$F = 1 + \frac{8 \times 10^{-16} + 9 \times 10^{-25} \times 9 \times 10^8}{4 \times 1.37 \times 10^{-23} \times 290 \times 30 \times 10^3}$$

$$= 4.35$$

and the minimum noise figure is

$$NF = 6.4 \text{ dB}$$

Low-Noise Design

The concept of optimizing the source resistance to minimize the noise figure must be used with caution. A better criterion of a network's noise performance is the output signal-to-noise ratio. Minimizing a network's noise figure does not necessarily minimize the output signal-to-noise ratio, since changing the source resistance also changes the input signal-to-noise ratio. The signal-to-noise ratio is readily determined by replacing the input to the noiseless network by its Thévenin equivalent. If the noise sources are independent, the equivalent input circuit is as shown in Fig. 3.17. There are three noise sources due to e_n, i_n, and the thermal noise generated in the source resistor R_s. Since the voltage sources are in series with the signal e_s, they will have the same input-to-output transfer function, and the signal-to-noise ratio in any unit bandwidth is given by

$$\frac{S}{N} = \frac{e_s^2}{e_n^2 + i_n^2 R_s^2 + 4kTR_s} \tag{3.55}$$

If the applied signal is not a function of source resistance, then it is evident that the smaller the source resistance the larger will be the signal-to-noise ratio. It would be detrimental to circuit performance to increase the source resistance in order to optimize the noise figure. Although this would result in the minimum noise figure, it would not maximize the output signal-to-noise ratio.

Example 3.12 The previous example found that a source resistance of 9.42 kΩ resulted in a minimum noise factor of 4.35. For this value of source resistance and the values of e_n and i_n specified in Example 3.10 the

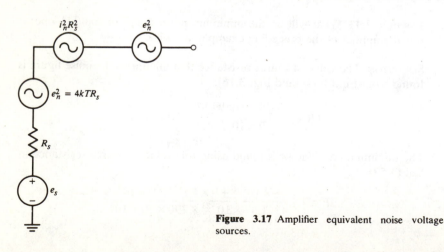

Figure 3.17 Amplifier equivalent noise voltage sources.

total noise is found to be

$$N = e_n^2 + i_n^2 R_s^2 + 4kTR_s = 17.48 \times 10^{-16} \, \text{V}^2/\text{Hz}$$

If the source resistance is zero, the total noise is

$$N = e_n^2 = 8 \times 10^{-16} \, \text{V}^2/\text{Hz}$$

which is less than 50 percent of the noise obtained when the noise factor is minimized.

If the signal voltage is a function of source resistance, then it is often possible to simultaneously minimize the noise figure and maximize the output signal-to-noise ratio. A most important case is that in which the source can be transformer-coupled to the amplifier input. In this case

$$R_s' = R_s N^2 \tag{3.56}$$

where N is the transformer turns ratio and R_s' is the reflected source impedance. The equivalent signal voltage will be

$$e_s' = e_s N \tag{3.57}$$

so the output signal-to-noise ratio will be (assuming the transformer does not add any noise)

$$\left(\frac{S}{N}\right)_o = \frac{N^2 e_s^2}{e_n^2 + i_n^2 N^4 R_s^2 + 4kTN^2 R_s} \tag{3.58}$$

The rate of change of the output signal-to-noise ratio as a function of N is given by

$$\frac{d(S/N)_o}{dN} = \frac{2N e_s^2 (e_n^2 + i_n N^4 R_s^2 + 4kTN^2 R_s - 2N^4 i_n^2 R_s^2 - 4kTN^2 R_s)}{e_n^2 + i_n^2 N^4 R_s^2 + 4kTN^2 R_s}$$

which is equal to zero for

$$N^2 R_s = \frac{e_n}{i_n} \tag{3.59}$$

This is also the value of source resistance that minimizes the noise figure. Therefore, if noiseless transformer coupling can be used to match the source to the amplifier input, then the turns ratio that minimizes the noise figure will also maximize the output signal-to-noise ratio.

A Low-Noise Amplifier

Figure 3.18 illustrates a low-noise amplifier design using the concepts discussed in the previous section. This amplifier is a common-gate amplifier selected for its relatively low input impedance so that it can be matched to the source resistance R_s. The resistor R_b serves to bias the circuit for linear operation, and

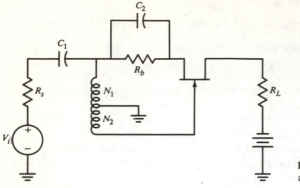

Figure 3.18 A low-noise FET amplifier.

capacitors C_1 and C_2 are short circuits in the frequency region of interest. The equivalent source impedance looking back from the gate to source terminals is found from the equivalent circuit, shown in Fig. 3.19, to be

$$Z'_s = R_s \left(\frac{N_1 + N_2}{N_1}\right)^2 \tag{3.60}$$

For the minimum noise factor and maximum signal-to-noise ratio the turns ratio is selected so that

$$R_s \left(\frac{N_1 + N_2}{N_1}\right)^2 = \frac{e_n}{i_n} = R_n$$

where e_n and i_n are the noise voltage and current of the transistor.

The midfrequency small-signal equivalent circuit, with r_d neglected, is shown in Fig. 3.20. If the reactance of the transformer is sufficiently large, little current will flow through the windings, and the input impedance will be

$$Z_i = \frac{V_g}{g_m V_{gs}} = \frac{V_g}{g_m[1 + (N_2/N_1)]V_g}$$

$$= \left[g_m \left(1 + \frac{N_2}{N_1}\right)\right]^{-1}$$

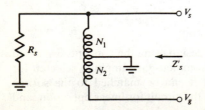

Figure 3.19 Input circuit of amplifier illustrated in Fig. 3.18.

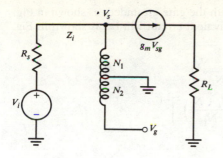

Figure 3.20 A small-signal equivalent circuit of amplifier shown in Fig. 3.18.

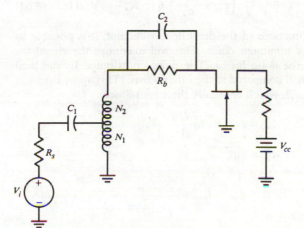

(a)

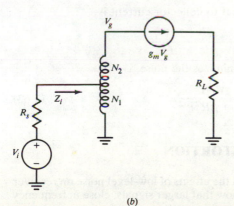

(b)

Figure 3.21 (a) A common-gate amplifier; (b) simplified equivalent circuit.

The circuit could also be realized with the gate grounded, as shown in Fig. 3.21a. In this case the small-signal equivalent circuit will be as shown in Fig. 3.21b. The source impedance is

$$Z'_s = R_s \left(\frac{N_1 + N_2}{N_1} \right)^2$$

and

$$Z_i = \left[g_m \left(\frac{N_1 + N_2}{N_1} \right)^2 \right]^{-1}$$

Optimization of BJT Bias Current

The total input noise of the common-emitter amplifier illustrated in Fig. 3.15 is

$$N_i = 4kTR_s + 4kTR'_b + \left(\frac{r_\pi + r'_b}{\beta} \right)^2 \left(i_{n_2}^2 + \frac{4kT}{R_L} \right) + i_{n_1}^2 R_s^2 \qquad \text{V}^2/\text{Hz} \qquad (3.61)$$

Since r_π, i_{n_1}, and i_{n_2} are functions of the dc collector current, it is possible to adjust the bias current for minimum noise. This will maximize the signal-to-noise ratio and minimize the noise factor. The noise contributed by the load resistor will be assumed small compared to the other terms. (This approximation is valid as long as $R_L \gg r_\pi/\beta$, which is usually the case.) Since

$$i_{n_1}^2 = 2qI_B = 2q \frac{I_c}{\beta}$$

$$i_{n_2}^2 = 2qI_c$$

and

$$\frac{r_\pi}{\beta} = \frac{1}{g_m} = \frac{V_T}{I_c}$$

the input noise can be written as

$$N_i = 4kTR_s + 4kTR'_b + \frac{2qI_c}{\beta} R_s^2 + \frac{2qV_T^2}{I_c} + \frac{4qV_T r_{b'}}{\beta} + \frac{2qI_c r_{b'}^2}{\beta^2} \qquad \text{V}^2/\text{Hz} \qquad (3.62)$$

The derivative of the noise with respect to collector current is

$$\frac{dN_i}{dI_c} = \frac{2qR_s^2}{\beta} + \frac{2qr_{b'}^2}{\beta^2} - \frac{2qV_T^2}{I_c^2}$$

The value of direct current, which minimizes the noise, is

$$I_c = \frac{V_T \beta}{(r_{b'}^2 + \beta R_s^2)^{1/2}} \qquad (3.63)$$

3.5 INTERMODULATION DISTORTION

The previous sections have considered the effects of low-level noise on receiver performance. In this section we will show that larger signals, close in frequency to the desired signal, can also affect receiver performance.

All communications receivers contain some degree of nonlinearity which can cause a change in the frequencies of the input signals and/or a change in the network gain. For these reasons the network nonlinearities need to be clearly delineated and considered during the design phase. The network nonlinearities can be described by the expansion

$$y(x) = k_1 f(x) + k_2 [f(x)]^2 + k_3 [f(x)]^3 + \text{higher-order terms} \qquad (3.64)$$

It is assumed that the nonlinearity is frequency-independent and can be adequately described by the first three terms; the higher-order terms will be ignored. Let $f(x)$ consist of two sinusoidal signals:

$$f(x) = A_1 \cos \omega_1 t + A_2 \cos \omega_2 t$$

If ω_1 and ω_2 are sufficiently close together, k_i can be considered the same for both signals. Also, for simplicity we will assume that all of the k_i are real. If Eq. (3.64) describes the network's response to an input $f(x)$, the response will be

$$
\begin{aligned}
y ={}& k_1(A_1 \cos \omega_1 t + A_2 \cos \omega_2 t) + k_2(A_1 \cos \omega_1 t + A_2 \cos \omega_2 t)^2 \\
& + k_3(A_1 \cos \omega_1 t + A_2 \cos \omega_2 t)^3 \\
={}& k_1(A_1 \cos \omega_1 t + A_2 \cos \omega_2 t) \\
& + k_2\left[A_1^2 \frac{1 + \cos 2\omega_1 t}{2} + A_2^2 \frac{1 + \cos 2\omega_2 t}{2} \right. \\
& \left. + A_1 A_2 \frac{\cos (\omega_1 + \omega_2)t + \cos (\omega_1 - \omega_2)t}{2} \right] \\
& + k_3 \left\{ \left[A_1^3 \left(\frac{\cos \omega_1 t}{2} + \frac{\cos \omega_1 t}{4} + \frac{\cos 3\omega_1 t}{4} \right) + A_2^3 \left(\frac{3 \cos \omega_2 t}{4} + \frac{\cos 3\omega_2 t}{4} \right) \right] \right. \\
& + A_1^2 A_2 [\tfrac{3}{2} \cos \omega_2 t + \tfrac{3}{4} \cos (2\omega_1 - \omega_2)t + \tfrac{3}{4} \cos (2\omega_1 - \omega_2)t] \\
& \left. + A_2^2 A_1 [\tfrac{3}{2} \cos \omega_1 t + \tfrac{3}{4} \cos (2\omega_2 + \omega_1)t + \tfrac{3}{4} \cos (2\omega_2 - \omega_1)t] \right\} \qquad (3.65)
\end{aligned}
$$

Gain Compression

One effect of the nonlinearity that can be deduced from Eq. (3.65) is that the amplitude of the $\cos \omega_1 t$ signal has become

$$A_1' = k_1 A_1 + k_3(\tfrac{3}{4} A_1^3 + \tfrac{3}{2} A_1 A_2^2) \qquad (3.66)$$

k_3 will normally be negative, and a large signal $A_2 \cos \omega_1 t$ can effectively mask a smaller signal $A_1 \cos \omega_1 t$, since it results in a reduced gain because of the third-order coefficient k_3. To avoid the "gain compression," the third-order coefficient k_3 must be reduced. Also, multiple signals will result in a further reduction of the gain. If only one signal is present, the ratio of gain with distortion to the idealized (linear) gain is

$$A_1' = \frac{k_1 + k_3(\tfrac{3}{4} A_1^2)}{k_1} \qquad (3.67)$$

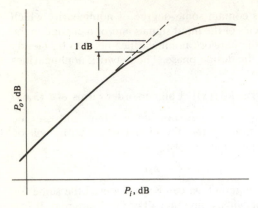

Figure 3.22 Idealized amplifier power-transfer characteristic illustrating the 1-dB compression point.

and is referred to as the *single-tone gain-compression factor*. Figure 3.22 illustrates how the k_3 term causes the gain to deviate from the idealized curve. The point at which the power gain is down 1 dB from the ideal is referred to as the *1-dB compression point*. Receivers must be operated below their gain-compression point if the nonlinear gain region is to be avoided.

Second Harmonic Distortion

Second harmonics will occur at the receiver output because of the k_2 term. If a single signal is present at the receiver input, the amplitude of the second harmonic will be

$$\frac{k_2 A_1^2}{2} \tag{3.68}$$

Intermodulation Distortion Ratio

Another important effect of receiver nonlinearity is the intermodulation distortion caused by the cubic term in Eq. (3.65). Equation (3.65) shows that the cubic term creates the intermodulation frequencies $2\omega_1 \pm \omega_2$ and $2\omega_2 \pm \omega_1$. If ω_1 and ω_2 are of approximately the same frequency, the higher frequencies $2\omega_1 + \omega_2$ and $2\omega_2 + \omega_1$ will normally be outside the passband and can be eliminated with filtering, but the two frequencies $2\omega_1 - \omega_2$ and $2\omega_2 + \omega_1$ can lie in the system passband and appear at the output as signal distortion. The *intermodulation distortion ratio* (*IMR*) is defined as the ratio of the amplitude of one of the intermodulation terms to the amplitude of the desired output signal. For the two-tone input signals, Eq. (3.66) yields

$$\text{IMR} = \frac{\frac{3}{4} k_3 A_1^2 A_2}{k_1 A_1} = \frac{3}{4} \frac{k_3}{k_1} A_1 A_2 \tag{3.69}$$

Intercept point The intermodulation distortion (IMD) power is defined as

$$P_d = \frac{(\frac{3}{4}k_3 A_1^2 A_2)^2}{2} \tag{3.70}$$

If the two input amplitudes are the same, the intermodulation distortion power varies as the cube of the input power; that is, for every 1-dB change in input power there is a 3-dB change in the power of the intermodulation terms. In this case,

$$P_d = (k_d P_i)^3 \tag{3.71}$$

where $P_i = A_1^2/2$, the power in one signal component, and k_d is the scale factor. The ratio (P_{IMR}) of the IMD power to the desired output power for the case where the two input-signal amplitudes are the same is defined as

$$P_{IMR} = \frac{P_d}{P_o} \tag{3.72}$$

Since the distortion power is proportional to the cube of the input power and the output power is directly proportional to the input power,

$$P_{IMR} = (K_i P_i)^2 \tag{3.73}$$

A normalized plot of the desired output and intermodulation powers is shown in Fig. 3.23. On a logarithmic scale, the IMD power increases three times as fast as the desired output power. The value of input power for which the IMD power is equal to the output power contributed by the linear term $(k_1 A_1)^2/2$ is referred to as the *intercept point* P_I, a term which is finding increasing usage, especially in describing the distortion characteristics of frequency mixers. In order to express P_I in terms of P_{IMR} and P_i, note that when the output distortion power and the desired output power are equal (the intercept point), the IMR ratio is, by

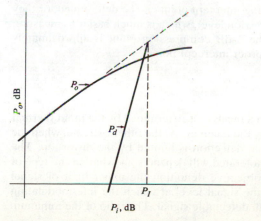

Figure 3.23 Power-transfer characteristic, including the third-order intermodulation distortion P_d and the two-tone third-order intercept point P_I.

definition, unity, and thus Eq. (3.73) becomes

$$1 = (K_i P_i)^2$$

Since $\qquad\qquad\qquad P_i = P_I$

at this signal level,

$$K_i = (P_I)^{-1}$$

and Eq. (3.73) can be written as

$$P_{\text{IMR}} = \left(\frac{P_i}{P_I}\right)^2 \tag{3.74}$$

where P_i is the input power $A_1^2/2$.

Example 3.13 If the system intercept point is +20 dBm, what is the IMR for an input signal power of 0 dBm?

SOLUTION To solve this problem Eq. (3.74) is used. Thus $P_{\text{IMR}} = 0 - 2 \times 20 = -40$ dBm.

A receiver's intercept point is a measure of the distortion created in the receiver, and it is also a measure of its ability to reject large amplitude signals that lie in close frequency proximity to a weak signal targeted for reception. The receiver intercept point is primarily determined by the intercept point of the input mixer. Double-balanced diode ring mixers with intercept points of +15 to +27 dBm are readily available and relatively inexpensive, but they are used in only the more expensive receivers because they require higher oscillator drive levels (+7 to +23 dBm) than do other types of mixers (such as integrated-circuit mixers which use field-effect transistors). Higher local-oscillator levels will usually require additional shielding of the system components. Mixer specifications normally list either the two-tone third-order distortion at some level, from which the corresponding intercept point can be determined, or they specify the 1-dB RF input compression level, which is much easier to measure. As a practical rule of thumb, the 1-dB compression point is approximately 15 dB below the two-tone third-order intercept point.

Dynamic Range

The minimum detectable signal in a receiver is determined by the input thermal noise and the noise contributed by the receiver. At the other extreme, when the input signal is too large the signal detection is limited by the distortion. The amount of distortion that can be tolerated will depend somewhat on the type of signals, but for purposes of an objective definition the upper limit of signal detectability will be considered the signal level at which the intermodulation distortion is equal to the minimum detectable signal. The ratio of the minimum detectable signal to the signal power that causes the distortion power (in one frequency component) to be equal to the noise floor N_f is referred to as the receiver's *dynamic range*. Since the ideal power out is

$$P_o = k_1^2 P_i \tag{3.75}$$

the intermodulation distortion ratio can be written as

$$P_{\text{IMR}} = \frac{P_d}{P_o} = \frac{P_d}{k_1^2 P_i} = \left(\frac{P_i}{P_I}\right)^2$$

Define $P_{di} = P_d / k_1^2$ (the distortion referred to the input); then

$$P_{\text{IMR}} = \frac{P_{di}}{P_i} = \left(\frac{P_i}{P_I}\right)^2 \tag{3.76}$$

When P_{di} is equal to the noise floor N_f,

$$\frac{N_f}{P_i} = \frac{P_i^2}{P_I^2}$$

or

$$P_i = (P_I^2 N_f)^{1/3} \tag{3.77}$$

Therefore the dynamic range is

$$\text{DR} = \frac{(P_I^2 N_f)^{1/3}}{N_f} = \left(\frac{P_I}{N_f}\right)^{2/3} \tag{3.78}$$

It must be kept in mind that the intercept point and noise floor are measured at the same point in the system. Also, the noise floor depends upon the specified output signal-to-noise ratio, and thus so does the dynamic range.

Example 3.14 The receiver of Example 3.6 has an intercept point of 20 dBm. What will be the dynamic range for an output signal-to-noise ratio of 10 dB?

SOLUTION From Example 3.7 it is known that the required available input signal power $S_i = -123$ dBm for a 10-dB output signal-to-noise ratio of 10 dB. The receiver's dynamic range is thus [using Eq. (3.78)].

$$\text{DR} = 0.67(20 + 123) = 95.3 \text{ dB}$$

If a linear preamplifier with a voltage gain A_v is added before a network that has an intercept point P_I, as illustrated in Fig. 3.24, then the overall intercept

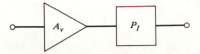

Figure 3.24 Circuit with a power intercept point P_I preceded by a preamplifier with a voltage gain A_v.

point is P_I/A_v^2. The output power due to the linear term is

$$P_o = \frac{(k_1 A_1)^2}{2} A_v^2 \tag{3.79}$$

and the intermodulation distortion output will be

$$P_d = \frac{\frac{3}{4}(k_3 A_v^3 A_1^3)^2}{2} \tag{3.80}$$

and the IMD distortion ratio is

$$P_{\text{IMR}} = \frac{\frac{3}{4}(k_3 A_1^3)^2}{(k_1 A_1)^2} A_v^4$$

$$= (K_i P_i)^2 A_v^4$$

when $P_{\text{IMR}} = 1$, $P_i = P_I$, so

$$P_I = (K_i A_v^2)^{-1} \tag{3.81}$$

The addition of a linear preamplifier reduces the intercept point. Unless the preamplifier can reduce the noise floor by the same amount the intercept point is reduced, the dynamic range will be decreased by the addition of the pre-amplifier.

SINAD

Another figure of merit which is becoming widely used for commercial (particularly stereo) receivers is the SINAD ratio. SINAD is the ratio of signal plus noise plus distortion powers to noise and distortion powers.

$$\text{SINAD} = \frac{S + N + D}{N + D} \tag{3.82}$$

For many applications the distortion, even at low levels, is an important factor in describing performance. Also, the SINAD ratio is easy to measure using the method illustrated in Fig. 3.25. The measurement procedure consists of applying an RF signal modulated by an audio signal (usually 1 kHz) and measuring signal plus noise plus distortion. The audio signal is then filtered out, and the noise plus distortion are determined. A SINAD measurement is the same as a total harmonic distortion measurement. The SINAD ratio can also be

Figure 3.25 SINAD-measuring network.

used to define receiver sensitivity. One possibility is to define receiver sensitivity as the amount of RF signal needed to get a specified SINAD ratio.

PROBLEMS

3.1 Determine an expression for the total noise voltage squared across the output terminals of the circuit shown in Fig. P3.1.

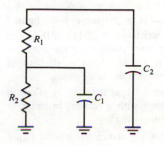

Figure P3.1 Frequency-dependent network including two noise sources.

3.2 Calculate the 3-dB and noise bandwidths of the circuit shown in Fig. P3.2.

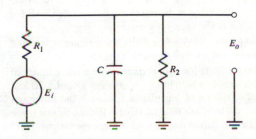

Figure P3.2 Frequency-dependent network with two noisy resistors.

3.3 Show that the amount of excess noise $E_n^2 = (k/f) \, df$ generated in each decade of frequency is constant, independent of frequency.

3.4 Derive the equation for the noise factor of n cascaded networks, each with a noise factor F_i and power gain G_i.

3.5 Determine the equivalent input noise sources of a common-base amplifier in terms of the thermal and shot-noise sources.

3.6 A receiver has a 3-kHz bandwidth, a 50-Ω input impedance, and a 5-dB noise figure. It is connected to an antenna by means of a 50-Ω coaxial cable that has an equivalent gain (loss) of -3 dB and an NF of 2 dB. What is the overall noise figure?

3.7 A receiver with an 8-dB noise figure, a 50-Ω input impedance, and a 3-kHz bandwidth is connected to an antenna that has a noise temperature of 2000 K. What is the minimum detectable signal for a 10-dB output signal-to-noise ratio? If a preamplifier with a gain of 10 dB, an NF of 5 dB, and bandwidth of 4 kHz (which overlaps the receiver's frequency response) is added between the antenna and receiver, what is the minimum detectable signal for a 10-dB output signal-to-noise ratio?

3.8 A receiver is to be designed to have an overall noise figure of 4 dB. The input mixer has a noise figure of 8 dB and the preamplifier (which is to be located at the input) has a noise figure of 3 dB. What must be the minimum preamplifier gain?

3.9 An amplifier with a 10-dB noise figure and a 4-dB power gain is cascaded with a second amplifier which has a 10-dB noise figure and a 10-dB power gain. What are the overall noise figure and power gain?

3.10 Calculate the value of source resistance which will minimize the noise figure of a 741 operational amplifier at a frequency of 500 Hz. What is the minimum noise figure?

3.11 A 741 operational amplifier is used with a source resistance of 20 kΩ. If the input signal level is 1 mV, what is the output signal-to-noise ratio? Assume that the amplifier bandwidth is 1 Hz and that the center frequency is 1 kHz. What is the output signal-to-noise ratio with the source resistance that minimizes the noise figure?

3.12 Consider an amplifier in which the source resistance is larger than that required to minimize the noise figure. Show that shunting the source with a resistor in order to minimize the noise figure will reduce the output signal-to-noise ratio.

3.13 A receiver with a 3-kHz bandwidth and 50-Ω input impedance has a noise figure (NF) of 8 dB. What is the minimum detectable signal for an output signal-to-noise ratio of 10 dB? If the two-tone intercept point is +20 dBm, what is the receiver's dynamic range? What will be the dynamic range if a linear noiseless preamplifier with a voltage gain of 10 is added at the input?

3.14 A receiver has a 10-dB noise figure, a 50-Ω input impedance, a −5-dBm two-tone intercept point (P_t), and 3.5-kHz bandwidth. What is the minimum detectable signal for a 0-dB output signal-to-noise ratio? What is the receiver's dynamic range?

3.15 A linear preamplifier with a voltage gain of 5 and a 4-dB noise figure is inserted before the receiver of Prob. 3.13. What is the overall dynamic range?

3.16 A receiver has a 3-kHz bandwidth, a 70-Ω input impedance, and a noise figure of 6 dB. It is connected to an antenna with a cable which has an equivalent loss of 6 dB and an NF of 3 dB. (The cable is matched to the input impedance.) What is the minimum detectable input signal for an output signal-to-noise ratio of 10 dB? If the antenna noise temperature is 3000 K, what is the minimum detectable signal for the same output S/N?

3.17 Use a transistor with a $\beta = 100$ to design a common-emitter amplifier to couple a 100-Ω source to a 1-kΩ load resistance. The base spreading resistance can be neglected. What is the optimum value of collector current for lowest noise figure?

REFERENCES

3.1 Netzer, Y.: The Design of Low-Noise Amplifiers, *Proc. IEEE*, **69**:728–741 (1981).
3.2 IRE Standards on Methods of Measuring Noise in Linear Twoports, *Proc. IRE*, **48**:60–68 (1960).

ADDITIONAL READING

Das, M. B.: FET Noise Sources and Their Effects on Amplifier Performance at Low Frequencies, *IEEE Trans. Electron Devices*, ED-19, 1972.

Fisk, J. R.: Receiver Noise Figure, Sensitivity and Dynamic Range—What the Numbers Mean, *Ham Radio*, 8–25 (Oct. 1975).

Fris, H. T.: Noise Figures of Radio Receivers, *Proc. IRE*, **32**:410–422 (1944).

Goldman, S.: *Frequency Analysis, Modulation and Noise*, McGraw-Hill, New York, 1948.

Motchenbacher, C. D., and F. C. Fitchen: *Low-Noise Electronic Design*, Wiley, New York, 1973.

North, D. D.: The Absolute Sensitivity of Radio Receivers, *RCA Review*, 1942.

Perlow, S. W.: Third Order Distortion in Amplifiers and Mixers, *RCA Review*, **37**:234–265 (1976).

Representation of Noise in Linear Twoports, *Proc. IRE*, **48**:69–74 (1960).

Sherwin, J.: Noise Specs Confusing, National Semiconductor Application Note #104 (May 1974).

FOUR

FREQUENCY-SELECTIVE NETWORKS AND TRANSFORMERS

4.1 INTRODUCTION

Communication networks must frequently select a band of frequencies and attenuate other undesired frequencies. Modern filter theory now provides methods for designing such filters to meet virtually any specification, but the most commonly used frequency-selective circuits are still the rather simple series and parallel tuned resonant circuits. Even these simple circuits lead to complex equations when nonideal elements, such as the nonzero resistance of inductors are considered. While standard procedures for circuit analysis can be used, the resulting equations are often too complex to provide insight into the design process. However, years of research have yielded many approximations that greatly facilitate the design and analysis of simple resonant circuits.

This chapter first describes the analysis of these circuits and provides approximations that allow one to design and analyze such circuits with a minimum of mathematics. We then show how transformers are incorporated into resonant circuits. Methods are provided here for analyzing frequency-selective circuits containing simple transformers, one of the most useful components of communication circuits. And finally we demonstrate how frequency-selective methods can be used for impedance matching and how a combination of inductors and capacitors can be used to alter the input impedance of the network over a limited frequency range.

4.2 SERIES RESONANT CIRCUITS

One of the simplest frequency-selective circuits is the series resonant circuit illustrated in Fig. 4.1. At low frequencies the current is blocked by the

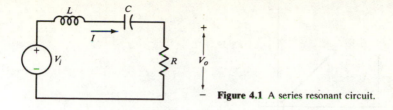

Figure 4.1 A series resonant circuit.

capacitor, and the inductor blocks the current at high frequencies. At some intermediate frequency the impedance of the inductor is equal in magnitude and opposite in sign to the impedance of the capacitor. At this frequency, referred to as the *resonant frequency*, maximum current flows and is in phase with the applied voltage. This circuit will now be analyzed in order to quantitatively describe its frequency performance. The circuit current is

$$I(s) = \frac{V_i(s)}{sL + (sC)^{-1} + R}$$

and the output voltage $V_o(s)$ is

$$V_o(s) = \frac{RV_i(s)}{sL + (sC)^{-1} + R} = \frac{(R/L)sV_i(s)}{s^2 + Rs/L + (LC)^{-1}} \tag{4.1}$$

The voltage gain $V_o(s)/V_i(s)$ has a zero at the origin, and two poles s_1 and s_2 located at

$$s_{1,2} = -\frac{R}{2L} \pm \left[\frac{(R/L)^2 - 4/LC}{4} \right]^{1/2} \tag{4.2}$$

Both poles may be real or they can occur as a complex conjugate pair. If

$$R > 2 \left(\frac{L}{C} \right)^{1/2}$$

the poles are both real, and the circuit is said to be overdamped. If the two poles are equal, that is,

$$s_1 = s_2 = -\frac{R}{2L}$$

the circuit is critically damped. In most frequency-selective networks, the circuit is undamped, that is,

$$R < 2 \left(\frac{L}{C} \right)^{1/2}$$

and the poles form a complex conjugate pair. The transfer function is frequently written as

$$A(s) = \frac{RCs}{s^2/\omega_o^2 + (2\zeta/\omega_o)s + 1} \tag{4.3}$$

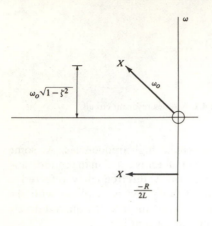

Figure 4.2 Pole-zero diagram of the series resonant circuit.

where $\omega_o = [(LC)^{1/2}]^{-1}$ is called the *undamped natural frequency* and

$$\zeta = \frac{R}{2}\left(\frac{C}{L}\right)^{1/2}$$

is referred to as the *damping ratio*. If $\zeta < 1$, the circuit is underdamped. The pole-zero plot of this transfer function for the underdamped case is given in Fig. 4.2. The real part of each pole is equal to $-R/2L = \zeta\omega_o$, and the magnitude of the imaginary part of each pole is

$$\frac{1}{2}\left[\frac{4}{LC} - \left(\frac{R}{L}\right)^2\right]^{1/2} = \omega_o\left(1 - \frac{R^2 C}{4L}\right)^{1/2} = \omega_o(1 - \zeta^2)^{1/2}$$

Note that the distance of the pole from the origin is equal to

$$\left\{\left(\frac{R}{2L}\right)^2 + \left[\frac{1}{2}\left(\frac{4}{LC}\right)^{1/2} - \left(\frac{R}{L}\right)^2\right]^2\right\}^{1/2} = [(LC)^{1/2}]^{-1} = \omega_o \qquad (4.4)$$

and the angle θ is determined by

$$\cos\theta = \frac{R(LC)^{1/2}}{2L} = \zeta$$

ω_o and ζ are the most convenient parameters for describing the transient response of the circuit. For a unit step input voltage $V_i(t) = U(t)$, it is readily shown (by taking the inverse Laplace transform) that if ζ is less than 1,

$$V_o(t) = \frac{2\zeta}{(1 - \zeta^2)^{1/2}} e^{-\zeta\omega_o t} \sin[\omega_o(1 - \zeta^2)^{1/2} t] U(t) \qquad (4.6)$$

where $U(t)$ is the unit step function

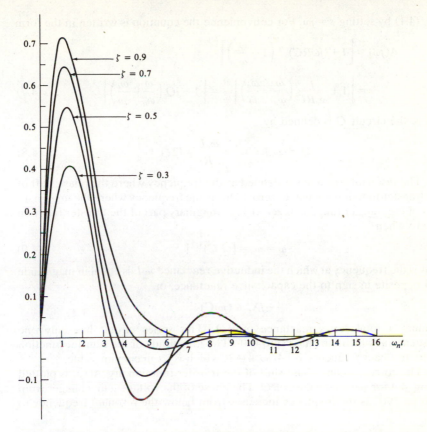

Figure 4.3 Transient response of underdamped series resonant circuits to a step input, for selected values of damping.

$$U(t) = \begin{cases} 1 & t \geq 0 \\ 0 & t < 0 \end{cases}$$

The step response (plotted in Fig. 4.3 for selected values of ζ) is an exponentially damped sinusoid. The ringing frequency is

$$\omega_o(1 - \zeta^2)^{1/2} \qquad \text{rad/s}$$

which is equal to the imaginary part of each pole, and the envelope's damping factor $\zeta\omega_o = -R/2L$ is equal to the magnitude of the real part of each pole. The smaller the value of circuit resistance (for a fixed inductance), the smaller the damping factor and the longer it takes for the transient to die out.

The steady-state (frequency) analysis of the circuit is readily evaluated from

Eq. (4.1) by setting $s = j\omega$. For convenience the equation is written in the form

$$A(j\omega) = \left[1 + (j\omega RC)^{-1}\left(1 - \frac{\omega^2}{\omega_o^2}\right)\right]^{-1}$$

$$= \left[1 + \frac{j}{\omega_o RC}\left(\frac{\omega}{\omega_o} - \frac{\omega_o}{\omega}\right)\right]^{-1} = \left[1 + jQ\left(\frac{\omega}{\omega_o} - \frac{\omega_o}{\omega}\right)\right]^{-1} \tag{4.7}$$

where the circuit Q is defined by

$$Q = (\omega_o RC)^{-1} = \frac{\omega_o L}{R} = (2\zeta)^{-1} \tag{4.8}$$

The resonant frequency is defined as the frequency where the phase shift of the transfer function is equal to zero. (This is the frequency where the imaginary part of the transfer function is zero.) The imaginary part of the transfer function is zero when

$$\omega = \omega_o = [(LC)^{1/2}]^{-1} \tag{4.9}$$

This is the frequency at which the inductive reactance $j\omega L$ is equal in magnitude and opposite in sign to the capacitance reactance, or

$$-jX_c = (j\omega C)^{-1}$$

yielding a total series reactance of $0\,\Omega$. This series circuit has only one frequency at which the phase shift is 0°. The magnitude of the transfer function at the resonant frequency is $|A(j\omega_o)| = 1$, which is its maximum value.

The corresponding phase shift of the transfer function arg $A(j\omega)$ is plotted in Fig. 4.4 for selected values of Q. The phase of the voltage gain changes from +90 to −90° as the frequency increases from below the resonant frequency to

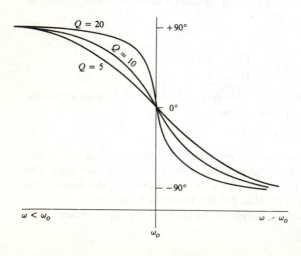

Figure 4.4 Phase response of underdamped series resonant circuits.

above the resonant frequency. The higher the circuit Q the more abrupt is the transition in phase. This phenomenon is particularly important in oscillator design and is discussed further in Chap. 7.

The half-power frequencies ω_1, ω_2 at which $|A(j\omega)|$ is reduced to 0.707 times its maximum value (the -3-dB frequencies) can be found by solving

$$|A(j\omega)| = (2^{1/2})^{-1} = \left| \left[1 + jQ\left(\frac{\omega}{\omega_o} - \frac{\omega_o}{\omega}\right) \right]^{-1} \right|$$

or

$$\frac{1}{2} = \left[1 + Q^2\left(\frac{\omega}{\omega_o} - \frac{\omega_o}{\omega}\right)^2 \right]^{-1}$$

which is equivalent to

$$Q\left(\frac{\omega}{\omega_o} - \frac{\omega_o}{\omega}\right) = 1 \tag{4.10}$$

The solution of this equation can be facilitated by noting that $|A(j\omega)|$ displays geometric symmetry. That is,

$$|A(j\omega)| = \left| A\left(j\frac{\omega_o^2}{\omega}\right) \right|$$

One of the half-power frequencies ω_1 is determined by solving

$$\omega_1 - \frac{\omega_o^2}{\omega_1} = \frac{\omega_o}{Q}$$

From the symmetry argument it follows that the other half-power frequency is found from

$$\omega_2 = \frac{\omega_o^2}{\omega_1} \quad \text{and} \quad \omega_1 - \omega_2 = \frac{\omega_o}{Q}$$

The difference between the two half-power frequencies $\omega_1 - \omega_2$ is by definition the circuit bandwidth B. Therefore, the bandwidth is

$$B = \frac{\omega_o}{Q} \tag{4.11}$$

It will be shown in subsequent sections that this relationship between bandwidth, center frequency, and Q also holds true for parallel resonant circuits with the bandwidth being inversely related to the circuit Q. The bandwidth can be specified independently of the center frequency ω_o since

$$B = \frac{\omega_o}{Q} = \omega_o^2 RC = \frac{R}{L} \tag{4.12}$$

and R is not a function of ω_o. The steady-state behavior of this circuit is completely described in terms of its center frequency ω_o and its Q.

Example 4.1 Design a filter to couple a voltage source, with negligible

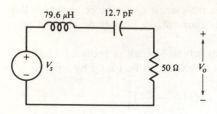

Figure 4.5 Series resonant circuit described in Example 4.1.

source impedance, to a 50-Ω load resistance. The specifications are that the filter center frequency be 5 MHz and the bandwidth be 100 kHz.

SOLUTION A series resonant LC circuit can be used to meet the specifications. Since the source impedance is negligible, the total series resistance is 50 Ω. The inductance is determined using Eq. (4.12)

$$L = \frac{R}{B} = \frac{50}{2\pi(10^5)} = 79.6\ \mu\text{H}$$

The capacitance is determined from Eq. (4.8)

$$C = [(79.6 \times 10^{-6})(2\pi \times 5 \times 10^7)^2]^{-1} = 12.7\ \text{pF}$$

The complete filter, including the load resistor, is shown in Fig. 4.5.

The frequency response of the magnitude of Eq. (4.7) is plotted in Fig. 4.6 for selected values of Q. For the overdamped circuit the gain rolls off at -6 dB per octave at both high and low frequencies, but for the high-Q case the attenuation rate is much greater near the resonant frequency. The frequency response becomes more selective (sharper) as the Q is increased. The pole-zero diagram shown in Fig. 4.7 provides a means for estimating the roll-off

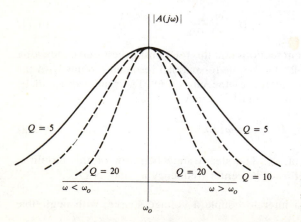

Figure 4.6 Magnitude of the series resonant circuit gain as a function of frequency.

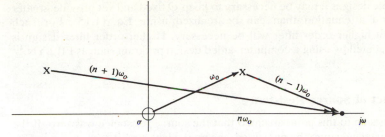

Figure 4.7 Pole-zero diagram used to evaluate the magnitude of the frequency response at the nth harmonic of the resonant frequency.

characteristics of the high-Q filter. In the high-Q case the circuit poles are close to the $j\omega$ axis, and their distance from the origin is approximately ω_o. The circuit gain at any frequency is the product of the distances from the zeros to the desired frequency, divided by the product of the distances from the poles to the desired frequency. Thus at any frequency $n\omega_o$ where n is an integer, the zero distance is $n\omega_o$ and the pole distances are $(n-1)\omega_o$ and $(n+1)\omega_o$. Therefore, the magnitude of the gain is

$$|A(jn\omega_o)| = \frac{n\omega_o R/L}{(n-1)\omega_o(n+1)\omega_o} = \frac{n\omega_o R/L}{(n^2-1)\omega_o^2} \tag{4.13}$$

Since

$$\frac{R}{\omega_o L} = Q^{-1} \tag{4.14}$$

the attenuation at any harmonic of the circuit's resonant frequency is given by

$$\left|\frac{A(jn\omega_o)}{A(j\omega_o)}\right| = \frac{n}{Q(n^2-1)} \tag{4.15}$$

As the Q of the circuit increases, the attenuation at the harmonic frequencies increases. This equation can be most useful in circuit design, as illustrated by the following example.

Example 4.2 A series-tuned circuit is to be used to filter out the harmonics of a waveform. What must be the minimum circuit Q for the amplitude of the fifth harmonic to be 40 dB below the amplitude of the fundamental frequency?

SOLUTION Forty decibels corresponds to a voltage ratio of $100:1$. Therefore, since $|(A(j\omega_o)| = 1$,

$$|A(j\omega_o 5)| = 0.01 = \frac{5}{Q(25-1)}$$

or

$$Q_{\min} = 20.83$$

In some designs it may be necessary to keep Q fixed and yet provide greater out-of-band attenuation than can be realized using Eq. (4.15). For such problems a higher-order filter will be necessary. Higher-order filter design is readily executed by using a computer-aided design program such as FILSYN.[4.1]

The Effect of Source Resistance

The analysis up to this point assumed that the source resistance was zero. If the source resistance cannot be neglected, the equivalent circuit is as shown in Fig. 4.8; the transfer function for this circuit is

$$A(s) = \frac{V_o}{V_i} = \frac{R_L}{R_L + R_s + sL + (sC)^{-1}}$$

$$= \frac{R_L}{R_L + R_s} \frac{(R_L + R_s)sC}{s^2 LC + (R_L + R_s)sC + 1} \qquad (4.16)$$

This equation is identical to Eq. (4.1) except that R is replaced by $R_L + R_s$, and the transfer condition is multiplied by the frequency-independent attenuation factor

$$K = \frac{R_L}{R_L + R_s} \qquad (4.17)$$

The effect of the nonzero source resistance is to reduce the amplitude of the transfer function at all frequencies and to reduce the Q of the circuit from $\omega_o L/R_L$ to $\omega_o L/(R_L + R_s)$, which is equivalent to widening the bandwidth by the same factor by which the gain is reduced. The analysis is obviously the same if the inductance L is modeled as an ideal inductor L in series with a resistance R_s. Since

$$Q = \frac{\omega_o L}{R_s + R_L}$$

any additional series resistance reduces the circuit Q (and increases the bandwidth) without changing the resonant frequency.

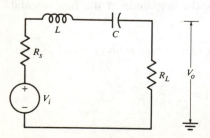

Figure 4.8 A series resonant circuit including a source resistance R_s.

Voltage Application

Another property of the series resonant circuit is that the voltage across the reactive components can be much larger than the applied voltage. The voltage across the capacitor in the circuit shown in Fig. 4.1 is

$$V_c(s) = \frac{V_i(s)}{s^2 LC + RsC + 1}$$

At the resonant frequency, the magnitude of the capacitor voltage is

$$|V_c(j\omega_o)| = \frac{V_i}{\omega_o RC} = QV_i \qquad (4.18)$$

At the resonant frequency, the voltage across the capacitor is Q times the input voltage. Since the magnitude of the reactance of the inductor is the same as that of the capacitor at the resonant frequency, the voltage across the inductor will be the same.

4.3 PARALLEL RESONANT CIRCUITS

In parallel resonance (actually antiresonance), two equal and opposite susceptances are added in parallel so that the admittance, instead of the impedance, is a minimum at the resonant frequency. For the parallel resonant circuit shown in Fig. 4.9, the transfer function (transfer impedance) is

$$A(s) = \frac{V_o(s)}{I(s)} = [sC + (sL)^{-1} + (R_p)^{-1}]$$

$$= \frac{sL/R_p}{s^2 LC + sL/R_p + 1} R_p$$

This equation is identical to Eq. (4.1), except for the scale factor, provided $L/R_p = RC$, where R and C denote the resistance and capacitance of the series resonant circuit. Thus the results for the transient and steady-state responses of the series resonant circuit can be used for the parallel resonant circuit, provided

$$\frac{2\zeta}{\omega_o} = \frac{L}{R_p}$$

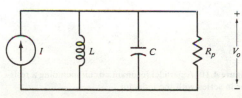

Figure 4.9 A parallel resonant circuit.

Table 4.1 Relation of series and parallel resonant circuits

	Series	Parallel
ω_o	$[(LC)^{1/2}]^{-1}$	$[(LC)^{1/2}]^{-1}$
Q	$\dfrac{\omega_o L}{R}$	$\dfrac{R_p}{\omega_o L}$
ζ	$\dfrac{R}{2}\left(\dfrac{C}{L}\right)^{1/2}$	$(2R_p)^{-1}\left(\dfrac{L}{C}\right)^{1/2}$

the only difference is that the magnitude of Eq. (4.18) is equal to R_p at the resonant frequency, whereas it is equal to unity for the series resonant circuit driven by a voltage source. The resonant frequency

$$\omega_o = [(LC)^{1/2}]^{-1} \tag{4.19}$$

is the same for both circuits. The Q of the parallel resonant circuit is defined as

$$Q_p^{-1} = 2\zeta_p = R_p^{-1}\left(\frac{L}{C}\right)^{1/2} = R_p^{-1}\frac{L}{(LC)^{1/2}} = \frac{\omega_o L}{R_p}$$

or

$$Q_p = \frac{R_p}{\omega_o L} \tag{4.20}$$

Equations (4.19) and (4.20) permit one to determine the transient and steady-state responses for the parallel resonant circuit using the results derived for the series resonant circuit. Table 4.1 summarizes the relations for the parallel and series resonant circuits.

In the simple two-pole resonant circuits described here, Q and ω_o completely describe the network. For the series resonant circuit, the higher the series resistance the lower the Q. Just the opposite is true for the parallel resonant circuit. Any resistance added in parallel with the tuned circuit reduces the resistance across the circuit and hence reduces the Q.

Nonideal inductors always possess finite resistance in series with the inductance, and so a more accurate model of a parallel LC circuit is as shown in Fig. 4.10. It will now be shown, with certain assumptions, that the analysis of this

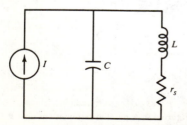

Figure 4.10 A parallel resonant circuit including a resistor in series with the inductor.

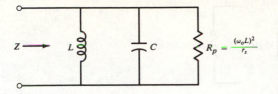

Figure 4.11 Approximate equivalent circuit of the circuit illustrated in Fig. 4.10.

circuit can be reduced to that presented in the previous section for the parallel circuit with a lossless inductor. The transfer function for the circuit of Fig. 4.10 is

$$\frac{V_o(s)}{I(s)} = A(s) = \frac{sL + r_s}{s^2 LC + r_s Cs + 1}$$

This equation is of the same form as Eq. (4.16) except that the zero is now located at $-r_s/L$ instead of at the origin. If this zero distance is small compared to ω, that is, if $\omega \gg r_s/L$, then to a first approximation the zero can be assumed to be at the origin, and Eq. (4.20) can be accurately approximated by

$$A(s) \approx \frac{sL}{s^2 LC + sCr_s + 1}$$

which is identical to Eq. (4.20), provided

$$\frac{L}{R_p} = r_s C$$

or

$$R_p = \frac{L}{Cr_s} = \frac{L}{r_s}\omega_o^2 L = \frac{(\omega_o L)^2}{r_s} \qquad (4.22)$$

If $\omega L \gg r_s$, the parallel resonant circuit (Fig. 4.10) with a resistor in r_s series with the inductor can be replaced by the parallel resonant circuit shown in Fig. 4.11. This approximation is almost always valid in the frequency region of interest since one would not select a low-Q inductor for use in a high-Q circuit.

The approximation is not valid at low frequencies where $\omega L \ll r_s$. $\omega_o L/r_s$ is referred to as the *inductor Q*. Every inductor has a finite Q, so the overall Q of the tuned circuit must also be finite. Note that the Q of the circuit in Fig. 4.11 is

$$Q = \frac{R_p}{\omega_o L} = \frac{\omega_o L}{r_s} \qquad (4.23)$$

which is the same as the Q of the coil. This is referred to as the *unloaded Q*, or Q_u, of the circuit. Any additional load resistance added across the circuit will further reduce the Q, so the circuit (or loaded) Q will always be less than the inductor Q unless active components are used.

Example 4.3 The tuned circuit shown in Fig. 4.12 employs an inductor with a Q_u of 100. If a 100-kΩ load resistor is added across the circuit, what is the loaded Q_L?

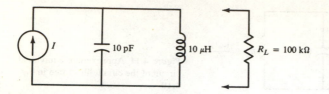

Figure 4.12 Parallel resonant circuit discussed in Example 4.3.

SOLUTION The resonant frequency of the circuit (ignoring the finite resistance) is

$$\omega_o = [(LC)^{1/2}]^{-1} = 10^8 \text{ rad/s}$$

The Q of the coil is 100; therefore

$$r_s = \frac{\omega_o L}{Q} = 10$$

Also, since $\omega_o L \gg r_s$ for frequencies near resonance, the series resistor can be replaced by a parallel resistor R_p, where

$$R_p = \frac{(\omega_o L)^2}{r_s} = 10^5 \ \Omega$$

The total parallel resistance, including R_L, is equal to two 100-kΩ resistors in parallel, or 50 kΩ, and the loaded Q is

$$Q_L = \frac{50 \text{ k}\Omega}{\omega_o L} = 50$$

The parallel load resistor reduces the Q of the circuit by a factor of 2 below that of the unloaded Q of the inductor.

Branch Currents

At the resonant frequency of the parallel RLC circuit, the output voltage is $V_o = I_i R_p$. The magnitude of the current through the capacitor is then

$$I_c = V_o \omega_o C = I_i \omega_o C R_p$$

$$= I_i \frac{R_p}{\omega_o L} = Q I_i \tag{4.24}$$

This equation shows that at frequencies near resonance the branch currents through the inductor and capacitor can be much larger than the applied current.

4.4 PARALLEL RESONANT CIRCUITS INCLUDING TRANSFORMERS

Transformers are extensively used in resonant circuits to provide phase inversion, dc isolation, and impedance-level shifting. Since the transformers already contain an inductor, it is possible to form a parallel resonant circuit with a transformer by simply adding a capacitor. The parallel resonant circuit Q is directly proportional to the parallel load resistance and inversely proportional to the magnitude of the inductive reactance. High-frequency circuits will usually have a low input impedance. For example, the equivalent load resistance of an antenna is small (on the order of 50 kΩ). These small impedance values make it difficult to realize high-Q circuits unless some method is used to transform the load to a larger value. There are several methods available to realize this transformation; the magnetically coupled transformer is a frequently used technique. For the transformer circuit shown in Fig. 4.13 the equilibrium equations are

$$V_1(t) = L_1 \frac{di_1}{dt} + M \frac{di_2}{dt} \tag{4.25}$$

and

$$V_2(t) = M \frac{di_1}{dt} + L_2 \frac{di_2}{dt} \tag{4.26}$$

or

$$V_1(s) = sL_1 I_1(s) + sM I_2(s) \tag{4.25a}$$

and

$$V_2(s) = sM I_1(s) + sL_2 I_2(s) \tag{4.26a}$$

where M is the mutual inductance between the input and secondary of the transformer. The standard dot convention is used, which places a dot on one terminal of a coil and another on the terminal of a second coil such that if current is sent into the two dotted terminals, the magnetic fluxes linking the coils will reinforce each other. With this convention M is always positive.

There are many equivalent circuits for this transformer, but the one which has the greatest utility in communication circuits is shown in Fig. 4.14. The ideal

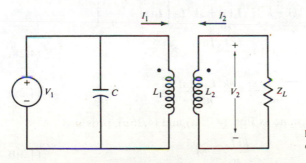

Figure 4.13 A magnetically coupled tuned circuit.

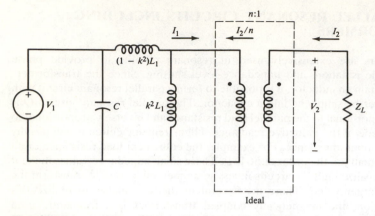

Figure 4.14 An equivalent circuit to the one shown in Fig. 4.13. Here the transformer has been replaced by an ideal transformer plus two inductors.

transformer contained in this circuit is a mathematical abstraction that facilitates the analysis by simplifying the transformer circuit models. For the ideal transformer,

$$V_1 = nV_2$$

and

$$I_1 = -\frac{I_2}{n}$$

independent of frequency. Also,

$$\frac{V_1}{I_1}(s) = Z_1 = -n^2 \frac{V_2}{I_2}(s) = n^2 Z_L \tag{4.27}$$

The circuit model for the transformer of Fig. 4.14 is equivalent to that of Fig. 4.13, provided the terminal voltage and current relations are the same as given by Eqs. (4.25a) and (4.26a). The equilibrium equations for the circuit of Fig. 4.14 are

$$V_1(s) = s(1 - k^2)L_1 I_1(s) + sk^2 L_1 \left[I_1(s) + \frac{I_2(s)}{n} \right]$$

$$= sL_1 I_1(s) + \frac{sk^2 L_1 I_2(s)}{n} \tag{4.28}$$

and

$$V_2(s) = \frac{sk^2 L_1 I_1(s)}{n} + \frac{sk^2 L_1 I_2(s)}{n^2} \tag{4.29}$$

These equations are the same as Eqs. (4.25a) and (4.26a), provided

$$\frac{k^2 L_1}{n} = M \tag{4.30}$$

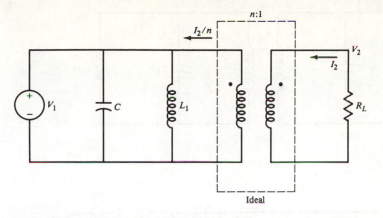

Figure 4.15 A simplified version of the circuit shown in Fig. 4.14, valid when the coefficient of coupling $k \approx 1$.

and

$$\frac{k^2 L_1}{n^2} = L_2 \qquad (4.31)$$

That is, the equivalent turns ratio is

$$n = k \left(\frac{L_1}{L_2}\right)^{1/2} \qquad (4.32)$$

and k, referred to as the *coefficient of coupling*, is

$$k = \frac{M}{(L_1 L_2)^{1/2}} \qquad (4.33)$$

One reason that this two-inductor model of the real transformer is so useful is that in narrowband circuits, transformers are used which have a coefficient of coupling k near unity. If $k \approx 1$, the model can be simplified to that shown in Fig. 4.15; the transformer is represented by an inductor in parallel with an ideal transformer (which simply reflects the secondary impedance, multiplied by the turns ratio squared, to the primary). The model shown in Fig. 4.14 provides an easily analyzable circuit when the transformer coefficient of coupling is close to unity.

Example 4.4 A tightly coupled transformer with a primary inductance L_1 of 25 μH and a secondary inductance L_2 of 400 μH is used in the circuit shown in Fig. 4.16a. What is the overall frequency response of the circuit?

SOLUTION The tightly coupled transformer circuit can be replaced by the

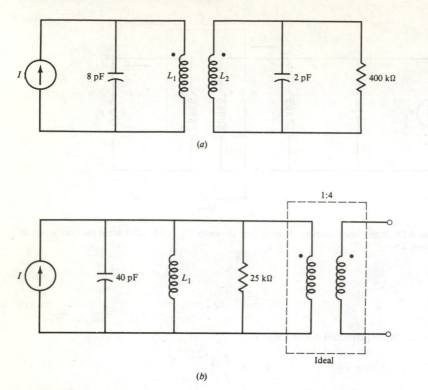

(a)

(b)

Figure 4.16 Equivalent circuits discussed in Example 4.4.

equivalent circuit shown in Fig. 4.15 (since $k \approx 1$). Therefore,

$$n = k \left(\frac{L_1}{L_2}\right)^{1/2} \approx \left(\frac{L_1}{L_2}\right)^{1/2} = \frac{1}{4}$$

and the total primary capacitance is

$$C_T = 8 + \frac{2}{n^2} = 8 + 2(4)^2 = 40 \text{ pF}$$

and the load resistance reflected to the primary is

$$n^2 R_L = \frac{400}{16} \times 10^3 = 25 \text{ k}\Omega$$

Thus, with the load reflected to the primary, the equivalent circuit is a parallel-tuned circuit as shown in Fig. 4.16b. The resonant frequency is

$$\omega_o = \{[(40 \times 10^{-12})(25 \times 10^{-6})]^{1/2}\}^{-1} = 31.6 \times 10^6 \text{ rad/s}$$

and
$$Q = \frac{R}{\omega_o L} = \frac{25 \times 10^3}{(31.6 \times 10^6)(25 \times 10^{-6})} = 31.68$$

Since the secondary in this circuit is coupled to the primary with an ideal transformer, the output voltage is

$$V_o(t) = 4 V_i(t)$$

Transformers have a voltage gain equal to the turns ratio $1/n$ so, as in the preceding example, the transformer can provide voltage amplification. Transformers with a turns ratio greater than 1 are frequently used in amplifier design for circuits with a small load resistance. The transformer then increases the impedance seen at the amplifier input. If the amplifier gain without a transformer is $A_v = g_m R_L$, then the gain with the transformer is $A'_v = n g_m R_L$.

Transformers with Tuned Secondaries

Transformers with a turns ratio less than 1 are also used with voltage amplifiers. The reduced voltage gain results in a smaller Miller capacitance (see Chap. 5) and thus a wider bandwidth. Consider the amplifier with a tuned secondary circuit as illustrated in Fig. 4.17a. If the transformer is replaced by the equivalent circuit model, the equivalent circuit seen from the transistor collector is as shown in Fig. 4.17b.

The equilibrium equations are

$$n V_o(s) = sk^2 L_1 \left[I_1(s) + \frac{I_2(s)}{n} \right] \tag{4.34}$$

and
$$V_o(s) = -I_2(s) \frac{R_L}{R_L sC + 1} \tag{4.35}$$

The transfer impedance $Z_{12}(s)$ can be found by eliminating I_2:

$$Z_{12}(s) = \frac{V_o}{I_1} = \frac{sk^2 L_1/n}{s^2 k^2 L_1 C/n^2 + sk^2 L_1/n^2 R_L + 1} \tag{4.36}$$

Since from Eq. (4.31)

$$\frac{k^2 L_1}{n^2} = L_2$$

the transfer impedance is

$$Z_{12}(s) = \frac{V_o}{I_1} = \frac{nsL_2}{s^2 L_2 C + sL_2/R_L + 1} \tag{4.37}$$

which is also the equation of a parallel resonant circuit with

$$\omega_o = [(L_2 C)^{1/2}]^{-1} \tag{4.38}$$

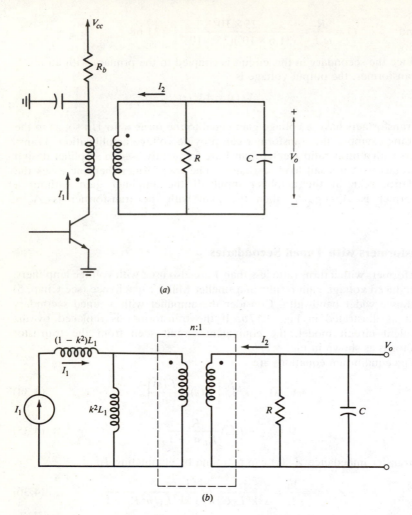

Figure 4.17 (a) A transistor amplifier including a transformer-coupled circuit with a tuned secondary; (b) simplified equivalent circuit.

Equation (4.38) states that when the capacitor is placed in the transformer secondary, the resonant frequency is determined by this capacitance in parallel with the inductance of the transformer secondary. The input impedance seen by the current source I_1 is determined from

$$V_1(s) = sI_1(1 - k^2)L_1 + nV_o(s) \qquad (4.39)$$

with V_o replaced using Eq. (4.37)

$$V_1(s) = sI_1(1 - k^2)L_1 + nZ_{12}(s)I_1(s) \qquad (4.40)$$

The circuit input impedance (at the collector of the transistor) is

$$Z_i = \frac{V_i(s)}{I_1(s)} = sL_1(1 - k^2) + nZ_{12}(s)$$

$$Z_i(s) = \frac{V_i(s)}{I_1(s)} = sL_1(1 - k^2) + \frac{n^2 sL_2}{s^2 L_2 C + sL_2/R_L + 1} \qquad (4.41)$$

At the resonant frequency ω_o, the input impedance is

$$Z_i(j\omega_o) = j\omega_o L_1(1 - k^2) + n^2 R_L \qquad (4.42)$$

The transformer is tightly coupled if $(k \approx 1)$

$$Z_i(j\omega_o) = n^2 R_L \qquad (4.43)$$

At the resonant frequency the tightly coupled transformer reflects back to the collector of the transistor, the load resistance R_L amplified by n^2. If the turns ratio n is 1, the collector voltage is less than the output voltage. This has the net effect of reducing the base-to-collector voltage gain and hence reducing the Miller capacitance.

If the capacitance is added to the primary, then

$$C_1 = (L_1 \omega_o^2)^{-1}$$

and if it is added to the secondary, then

$$C_2 = (L_2 \omega_o^2)^{-1}$$

So the ratio of possible capacitors is

$$\frac{C_1}{C_2} = \frac{L_2}{L_1} = (n^2)^{-1} \qquad (4.44)$$

in a tightly coupled transformer. This equation shows that if $n > 1$, the transformer circuit can be tuned by a smaller capacitor across the input, but if $n < 1$ a tuned secondary will result in a smaller capacitor C_2.

Double-Tuned Circuits

In addition to tuning either the primary or secondary of a transformer-coupled circuit, it is also possible to include tuned circuits in both the primary and secondary (doubled-tuned circuits). A transformer-coupled circuit with parallel resonant circuits in the primary and secondary is shown in Fig. 4.18. The transadmittance of this network is

$$\frac{V_o}{I_i} = \frac{-k\omega_1\omega_2 s}{(1 - k^2)(C_1 C_2)^{1/2}(s^4 + a_3 s^3 + a_2 s^2 + a_1 s + a_0)} \qquad (4.45)$$

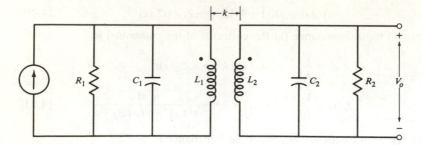

Figure 4.18 A double-tuned magnetically coupled circuit.

Where k is the coefficient of coupling, the primary resonant frequency is

$$\omega_1 = [(L_1 C_1)^{1/2}]^{-1} \tag{4.46}$$

and the resonant frequency of the secondary circuit is

$$\omega_2 = [(L_2 C_2)^{1/2}]^{-1} \tag{4.47}$$

The values of the coefficients of Eq. (4.45) are

$$a_3 = \frac{\omega_1}{Q_1} + \frac{\omega_2}{Q_2} \qquad a_2 = \frac{\omega_1 \omega_2}{Q_1 Q_2} + \frac{\omega_1^2 + \omega_2^2}{1 - k^2}$$

$$a_1 = \frac{\omega_1^2 \omega_2}{Q_2(1 - k^2)} + \frac{\omega_2^2 \omega_1}{Q_1(1 - k^2)} \qquad a_0 = \frac{\omega_1^2 \omega_2^2}{1 - k^2} \tag{4.48}$$

where

$$Q_1 = \frac{R_1}{\omega_1 L_1} \qquad \text{and} \qquad Q_2 = \frac{R_2}{\omega_2 L_2}$$

   This transformer-coupled double-tuned network is a bandpass network with four poles and a single zero at the origin. One method of double-tuned network design is to obtain numerical values for the coefficients that correspond to an approximation of a desired frequency response (such as the Butterworth or Chebyshev approximation) and then solve for the circuit values with Eq. (4.48).

   For narrowband circuits the analysis can be simplified. If $Q_1 = Q_2$ and both circuits are tuned to the same frequency, $\omega_1 = \omega_2 = \omega_o$, then for the loosely coupled case ($k^2 \ll 1$) the approximate pole positions are found to be

$$s_1, s_1^*, s_2, s_2^* = \frac{-\omega_o}{2Q} \pm j\omega_o \left(1 \pm \frac{k}{2}\right) \tag{4.49}$$

For this circuit with equal input and output Q's, it is easy to find the desired circuit parameters for a maximally flat (Butterworth) filter (or any other filter requiring two pairs of complex-conjugate poles) where the real parts of the poles $-\omega_o/2Q$ are the same. For the Butterworth filter, the coefficient of coupling $k = 1/Q$. For this value of k, the circuit is said to be "critically coupled." The bandwidth is

$$B = \frac{2^{1/2}\omega_o}{Q} \tag{4.50}$$

and the gain-bandwidth product is

$$GB = [(2C_1C_2)^{1/2}]^{-1} \tag{4.51}$$

For a single-tuned circuit, the gain-bandwidth product is

$$GB = (C_1 + C_2)^{-1} \tag{4.52}$$

(C_1 and C_2 are the primary and secondary capacitances, respectively.) Therefore the double-tuned (equal Q) method has a better gain-bandwidth product by at least a factor of $2^{1/2}$.

Double-tuned circuits can also be designed for wideband operation.[4.2]

Autotransformers

The Q of a parallel-tuned circuit is directly proportional to the load resistance shunting the tuned circuit. In many applications this resistance is too small to realize the desired Q, and some method is needed to increase the load resistance shunting the tuned circuit. One method is to use a separate transformer, but a simple method which often suffices is to place the load resistance across only a portion of the inductor, as shown in Fig. 4.19. For analysis, the autotransformer can also be replaced by the equivalent circuit shown in Fig. 4.20. The inductor serves as an autotransformer. For the two circuits to be equivalent the equilibrium equations must be the same. For the circuit of Fig. 4.19 the equations are

$$V_1 = s(L_1 + L_2 + 2M)I_1 + s(L_2 + M)I_2 \tag{4.53}$$

$$V_2 = s(L_2 + M)I_1 + sL_2 I_2 \tag{4.54}$$

where M is the mutual inductance between the two sections of the coil. For the

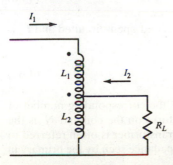

Figure 4.19 A step-up autotransformer.

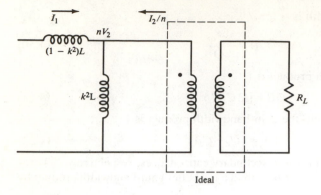

Figure 4.20 An equivalent circuit for the step-up autotransformer.

circuit shown in Fig. 4.20 the equilibrium equations are

$$V_1 = sLI_1 + \frac{sk^2L}{n} I_2 \tag{4.55}$$

$$V_2 = \frac{sk^2L}{n} I_1 + \frac{sk^2L}{n^2} I_2 \tag{4.56}$$

These two sets of equations will be equivalent if

$$L = L_1 + L_2 + 2M \tag{4.57}$$

$$n = \frac{k^2L}{L_2 + M} \tag{4.58}$$

and

$$n^2 = \frac{k^2L}{L_2} \tag{4.59}$$

In a closely coupled transformer ($k \approx 1$), the equivalent turns ratio simplifies to

$$n \approx \left(\frac{L}{L_2}\right)^{1/2} \tag{4.60}$$

where L is the total inductance measured with the load open-circuited and L_2 is measured with the input open-circuited:

$$L_2 = \frac{V_2}{sI_2}\bigg|_{I_1=0} \tag{4.61}$$

Since L and L_2 are proportional to the square of the corresponding number of turns $n \approx N/N_2$ where N is the total number of turns on the coil and N_2 is the number of turns on the lower section. This autotransformer is often referred to as an *impedance step-up transformer* since the impedance seen by the primary in a closely coupled transformer is n^2 times the load impedance and n is greater than 1. The derivation of the equivalent circuit for an impedance step-down transformer is left as an exercise (Prob. 4.8).

Capacitive Transformers

Impedance level shifting can also be accomplished in narrowband circuits without incurring the cost of expensive and often bulky transformers. One simple method of increasing the impedance level is to use the capacitive transformer illustrated in Fig. 4.21a. To analyze this circuit, consider first the impedance Z in parallel with the inductor L (Fig. 4.21b). This impedance will be equal to the parallel combination of a resistor and capacitor if certain conditions, which will now be derived, hold.

The impedance of the circuit shown in Fig. 4.21b is

$$Z = (j\omega C_2)^{-1} + \frac{R}{Rj\omega C_1 + 1} = \frac{Rj\omega(C_1 + C_2) + 1}{j\omega C_2(Rj\omega C_1 + 1)} \tag{4.62}$$

and the admittance is

$$Y(\omega) = [Z(j\omega)]^{-1} = \frac{j\omega C_2(j\omega R C_1 + 1)}{1 + j\omega R(C_1 + C_2)} = \frac{j\omega C_2(j\omega R C_1 + 1)[1 - j\omega R(C_1 + C_2)]}{1 + \omega^2 R^2(C_1 + C_2)^2}$$

$$\tag{4.63}$$

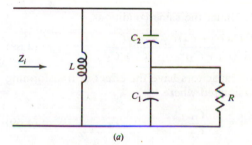

(a)

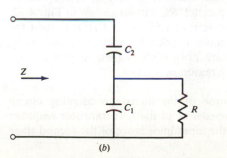

(b)

Figure 4.21 (a) A tuned circuit. (b) Impedance in parallel with the inductor containing a capacitive autotransformer.

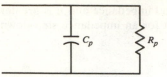

Figure 4.22 A narrowband equivalent circuit for the capacitive autotransformer.

This circuit can be replaced (at some frequency) by the simpler parallel circuit shown in Fig. 4.22, provided the admittances are equal at that frequency. This requires that both the real and imaginary parts are equal. That is,

$$G_p = \frac{1}{R_p} = \frac{-\omega^2 RC_1C_2 + \omega^2 RC_2(C_1 + C_2)}{1 + \omega^2 R^2(C_1 + C_2)^2} = \frac{\omega^2 RC_2^2}{1 + \omega^2 R^2(C_1 + C_2)^2} \quad (4.64)$$

and

$$\omega C_p = \frac{\omega C_2 + \omega^3 R^2 C_1(C_1 + C_2)C_2}{1 + \omega^2 R^2(C_1 + C_2)^2} \quad (4.65)$$

The parallel resistance

$$R_p = \frac{1 + \omega^2 R^2(C_1 + C_2)^2}{\omega^2 RC_2^2} \quad (4.66)$$

is approximately

$$R_p \approx R\left(\frac{C_1 + C_2}{C_2}\right)^2 \quad (4.67)$$

provided that $\omega^2 R^2(C_1 + C_2)^2 \gg 1$. Under the same conditions,

$$C_p = \frac{C_2 + \omega^2 R^2 C_1(C_1 + C_2)C_2}{1 + \omega^2 R^2(C_1 + C_2)^2} \approx \frac{C_1 C_2}{C_1 + C_2} \quad (4.68)$$

These equations show that the two capacitors have the effect of transforming the resistance up by the turn ratio squared where

$$n = 1 + \frac{C_1}{C_2} \quad (4.69)$$

provided $\omega^2 R^2(C_1 + C_2) \gg 1$. In this case the circuit of Fig. 4.21b can be replaced (for analysis) by the simpler parallel RC circuit shown in Fig. 4.22.

If the approximation cannot be made that $\omega^2 R^2(C_1 + C_2)^2 \gg 1$, then the circuit is best analyzed by using computer-aided techniques. Fortunately, narrowband capacitance transformers are normally designed so that the approximations are valid in the frequency region of interest.

Example 4.5 Determine the response of the interstage coupling circuit shown in Fig. 4.23a. The output impedance of the first transistor amplifier can be assumed to be infinite, and the input impedance of the second stage is $R\,\Omega$.

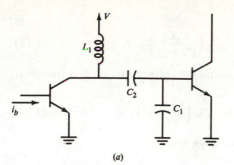

(a)

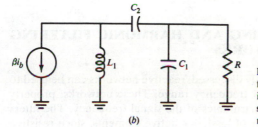

(b)

Figure 4.23 (a) A two-transistor amplifier with a capacitive transformer in the interstage coupling network; (b) a small-signal equivalent circuit for the interstage network.

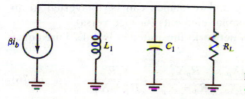

Figure 4.24 The interstage network of Fig. 4.23a modeled as a parallel resonant circuit.

SOLUTION The equivalent circuit is shown in Fig. 4.23b. If $\omega^2 R^2 (C_1 + C_2)^2 \gg 1$, the circuit is equivalent to that shown in Fig. 4.24, where

$$C = \frac{C_1 C_2}{C_1 + C_2} \qquad (4.70)$$

and

$$R_L = \left(1 + \frac{C_1}{C_2}\right)^2 R \qquad (4.71)$$

The response of this equivalent parallel-tuned circuit is now well known. The center frequency is

$$\omega_o = [(LC)^{1/2}]^{-1} \qquad (4.72)$$

and the circuit

$$Q = \frac{R_L}{\omega_o L} = \frac{(C_1 + C_2)^2}{C_2^2 L^{1/2}} \left(\frac{C_1 C_2}{C_1 + C_2} \right)^{1/2} R \tag{4.73}$$

At the resonant frequency the voltage at the collector of the first transistor is

$$V_c = -(\beta i_b) R_L = -\beta i_b \left(\frac{C_1 + C_2}{C_2} \right)^2 R \tag{4.74}$$

and the input voltage to the second stage is

$$V_i = -(\beta i_b) R_L \frac{C_2}{C_2 + C_1} = -(\beta i_b) R \frac{C_1 + C_2}{C_2} \tag{4.75}$$

4.5 IMPEDANCE MATCHING AND HARMONIC FILTERING USING REACTIVE NETWORKS

Besides the transformers previously discussed, reactive networks can be used to match impedances over a narrow frequency range. These networks, properly designed, also serve to filter out harmonics of the signal frequency. The filters described here are composed only of lossless reactive elements, since resistive components would result in power being dissipated in the coupling network while no power is dissipated in a network consisting solely of inductors and capacitors. The simplest design method utilizes the fact that at any frequency any series combination of resistance and reactance can be converted to an equivalent parallel combination of similar elements (or vice versa).

The input impedances of the two networks illustrated in Fig. 4.25 are equal provided

$$Z_i = R_s \pm jX_s = \frac{\pm R_p jX_p}{R_p \pm jX_p} = \frac{R_p X_p^2}{R_p^2 + X_p^2} \pm j \frac{X_p R_p^2}{R_p^2 + X_p^2} \tag{4.76}$$

Since both the real and imaginary parts must be equal, Eq. (4.76) is equivalent to

$$R_s = \frac{R_p X_p^2}{R_p^2 + X_p^2} \tag{4.77}$$

and

$$X_s = \frac{X_p R_p^2}{R_p^2 + X_p^2} \tag{4.78}$$

It is seen that the series and parallel reactances will be of the same type (since they are of the same sign). If one is capacitive, the other will be also. The equations for converting from series to parallel impedances can be derived in a like manner (equating admittances). They are

$$R_p = \frac{R_s^2 + X_s^2}{R_s} \tag{4.79}$$

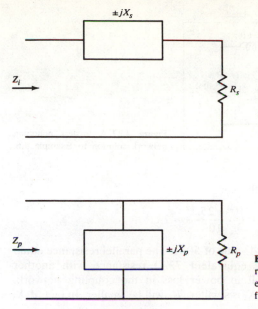

Figure 4.25 Any series combination of resistance and reactance is equivalent to an equivalent parallel combination at a given frequency.

and

$$X_p = \frac{R_s^2 + X_s^2}{X_s} \qquad (4.80)$$

The application of these relationships is best illustrated by an example.

Example 4.6 The input impedance of a transistor amplifier is equal to $10\,\Omega$ in series with $0.2\,\mu\text{H}$. Design a matching network so that the input impedance is $50\,\Omega$ at $20\,\text{MHz}$.

SOLUTION At $20\,\text{MHz}$ the inductive reactance of a $0.2\text{-}\mu\text{H}$ inductor is $25.1\,\Omega$. The problem then is to convert the series impedance $10 + j25.1$ to a resistance of $50\,\Omega$, using a lossless matching network as illustrated in Fig. 4.26. Equation (4.79) shows that the equivalent parallel resistance is larger

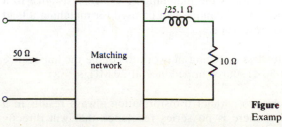

Figure 4.26 Network discussed in Example 4.6.

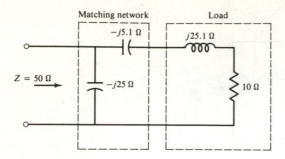

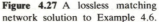

Figure 4.27 A lossless matching network solution to Example 4.6.

than the series resistance. In this case,

$$R_p = \frac{10^2 + 25.1^2}{10} = 73$$

which is larger than the desired value of 50 Ω. The parallel resistance could be reduced by shunting the equivalent 73-Ω resistance with another resistor, but this would result in power loss in the coupling network. Equation (4.79) shows that if X_s is smaller, R_p will be smaller. In fact, if X_s satisfies the equation

$$50 = \frac{10^2 + X_s^2}{10}$$

or

$$jX_s = j400^{1/2} = j20$$

then the correct impedance level can be reached. The magnitude of X_s can be reduced by adding a capacitive reactance of $-j5.1$ (0.00156 μH) in series with the load. If this is done, the equivalent parallel resistance is 50 Ω, and the equivalent parallel reactance is

$$jX_p = j\frac{10^2 + 20^2}{20} = j25$$

The reactance can be cancelled by adding a parallel capacitor of $-j25$ (318 pF) in parallel. The complete circuit is shown in Fig. 4.27.

The preceding example illustrates the matching of a load resistance to a larger source resistance. The same technique can be used for matching a load resistance to a smaller source resistance.

Example 4.7 Design a lossless matching network to couple the impedance shown in Fig. 4.28 to a 50-Ω source impedance at 20 MHz.

SOLUTION Since the series-to-parallel transformation always results in a larger parallel resistance, there is no series reactance that will directly

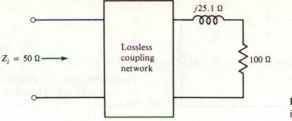

Figure 4.28 Network discussed in Example 4.7.

transform the 100-Ω resistor to a parallel equivalent of 50 Ω. There are, nevertheless, many possible matching networks that can be used. One network would consist of adding a capacitive reactance of $-j25.1\,\Omega$ in series and then adding a reactance in parallel (as shown in Fig. 4.29) that will transfer the 100 Ω to a series resistance of 50 Ω. Using this approach, X_p is selected so that

$$R_s = \frac{R_p X_p^2}{R_p^2 + X_p^2} = 50 = \frac{100 X_p^2}{100^2 + X_p^2}$$

Therefore,

$$50 X_p^2 = 50(100)^2$$

or

$$X_p = 100$$

The magnitude of X_p must be 100 Ω, but it can be either capacitive or inductive. An inductor would be selected in cases where it is desired to filter the low-frequency components from the load, and a capacitor would be selected if it is desired to filter the high-frequency components. Once X_p is selected, then a series reactance of the opposite sign must be added to cancel the equivalent series reactance (in this case $X_s = -j50$). A completed circuit is shown in Fig. 4.30. If the input impedance is to be real at 20 MHz, the corresponding component values are $C_1 = 159$ pF, $C_2 = 317$ pF, $L_1 = 0.8\,\mu$H, and $L_2 = 0.2\,\mu$H. If one preferred to filter the high-frequency components instead, the $-j50\,\Omega$ capacitor would be

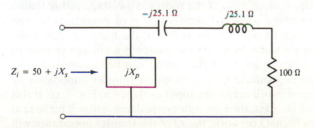

Figure 4.29 Intermediate solution to Example 4.7.

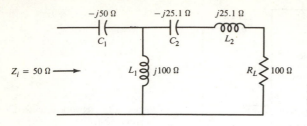

Figure 4.30 A lossless network for matching a 100-Ω load to a 50-Ω source.

replaced by a $+j50$-Ω inductor, and the shunt inductance would be replaced by a $-j100$-Ω capacitance.

In the design of the coupling networks in the preceding two examples, no consideration was given to the frequency response of the circuit. Circuit Q is often specified to provide a quantitative measure of the attenuation. Circuit Q is an ambiguous term, depending upon the transfer function under consideration. The transfer function of the circuit input impedance is generally different from that of the transimpedance (the latter is important when the circuit is driven by a current source), which is, in general, also different from the voltage-gain transfer function. In the previous example it was shown that the equivalent input impedance could be represented by a 50-Ω resistor in parallel with capacitive and inductive impedances of 100 Ω. The circuit Q is often interpreted as $Q = \frac{50}{100} = 0.5$. This is the Q of the circuit input impedance, and it is only valid at one frequency. The transfer impedance of this circuit (shown in Fig. 4.30) can be written as

$$\frac{V_o(s)}{I_i(s)} = Z_{12}(s) = \frac{R_L L_1 C_2 s^2}{s^2(L_1 + L_2)C_2 + R_L C_2 s + 1} \tag{4.81}$$

Note that the transfer impedance describes a high-pass transfer function. The magnitude of the transimpedance as a function of frequency is plotted in Fig. 4.31. The Q of a high-pass or low-pass network has no meaning in terms of bandpass.

The transfer impedance represents a high-pass transfer function, and the Q of such a filter cannot be interpreted as the center frequency divided by the bandwidth as in the bandpass filter. Also, if the source is a voltage rather than a current source, it is the voltage transfer function that is of interest. Since the voltage transfer function will be different from that of the transimpedance the Q's will not generally be the same. Whether the source is a voltage or current source must be specified before the Q of the circuit can be determined. The network of Fig. 4.30 can be made to have a bandpass transfer impedance by adding a parallel resonant network across the input as shown in Fig. 4.31. If this network is antiresonant at 20 MHz, then the input impedance will still be 50 Ω at that frequency. If this is a high-Q network, the Q of the transfer impedance will approximate the Q of the parallel network. If precise circuit impedance and

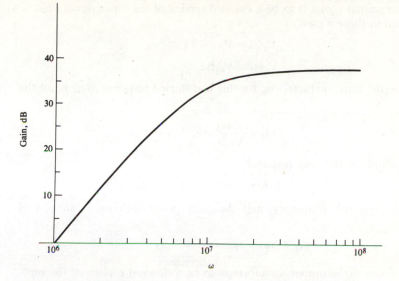

Figure 4.31 Frequency response of the magnitude of the transimpedance of the circuit shown in Fig. 4.30.

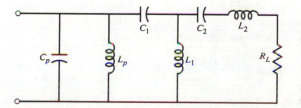

Figure 4.32 The high-pass transimpedance shown in Fig. 4.30 has been converted to a bandpass transfer function by the addition of the parallel resonant circuit at the input.

frequency characteristics are required, it is better to use network synthesis techniques and computer-aided design programs.

4.6 FILTER DELAY AND SIGNAL DISTORTION

The design of filter frequency-response characteristics has been largely concerned with the magnitude of the frequency response. Filter phase characteristics are also of importance in modern communication systems. The effect of phase shift on the signal is seen by determining the response of a linear filter $F(s)$ to an input signal $V_i(s)$. The filter output

$$V_o(s) = V_i(s)F(s)$$

If the filter output signal is to be a delayed replica of the input signal (that is, undistorted in shape), then

$$V_o(t) = V_i(t - T)$$

and
$$V_o(s) = V_i(s)e^{-sT}$$

Therefore, the filter characteristic for this undistorted response must be of the form

$$F(s) = \frac{V_o(s)}{V_i(s)} = e^{-sT} \qquad (4.82)$$

The magnitude of the filter response

$$|F(j\omega)| = |e^{-j\omega T}| = 1$$

must not vary with frequency, and the filter phase shift as a function of frequency

$$\arg F(j\omega) = \arg e^{-j\omega T} = -\omega T$$

must be linear if the output waveform is to be a delayed replica of the input signal. If the phase shift is not linear, then all frequency components of the input signal are not delayed to the same degree.

The filter delay has become an accepted method of describing the effect of phase nonlinearities on the input waveform. The common definitions for describing filter delay are group delay and phase delay. *Group delay* describes the delay of a group of frequencies; *phase delay* describes the delay of a single sinusoid. Group delay (also called *envelope delay*) is defined as

$$D(\omega) = -\frac{d \arg F(\omega)}{d\omega} \qquad (4.83)$$

and phase delay is defined as

$$\phi_D(\omega) = -\frac{\arg F(\omega)}{\omega} \qquad (4.84)$$

where $F(\omega)$ is the filter phase shift in radians. The group delay is the negative of the slope of the filter phase shift. The filter group delay and the phase delay will be equal at all frequencies if the filter phase shift is linear. That is, if $\arg F(\omega) = -\omega T$, then

$$D(\omega) = -\frac{d\omega T}{d\omega} = -T \qquad \text{and} \qquad \phi_D(\omega) = \frac{-\omega T}{\omega} = -T$$

For other filters the two delays are not the same.

Example 4.8 Compare the group and phase delays of a first-order low-pass filter.

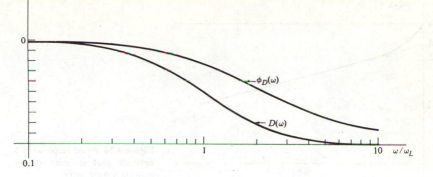

Figure 4.33 Phase delay $\phi_D(\omega)$ and group delay $D(\omega)$ of a first-order low-pass filter.

SOLUTION For a first-order low-pass filter the filter response is

$$F(j\omega) = \left(\frac{j\omega}{\omega_L} + 1\right)^{-1}$$

Where ω_L is the -3-dB frequency. The phase shift is

$$\arg F(j\omega) = -\tan^{-1} \frac{\omega}{\omega_L}$$

so the phase delay is

$$\phi_D(\omega) = \frac{\tan^{-1}(\omega/\omega_L)}{\omega} \tag{4.85}$$

and the group delay is

$$D(\omega) = \frac{\omega_L}{\omega^2 + \omega_L^2} \tag{4.86}$$

The two delays are plotted as a function of frequency in Fig. 4.33.

Phase-equalizer circuits, which have phase shifts but whose gain magnitudes do not change as a function of frequency, can be added to the filter in order to linearize the phase response over the filter passband.[4.3] Equalization can result in a phase-frequency diagram such as the one illustrated in Fig. 4.34. The phase is linear across the passband ($\omega_2 - \omega_1$) and can be approximated by

$$\arg F(\omega) = -T\omega - \theta_o$$

where $-\theta_o$ is the phase at $\omega = 0$. The nonzero value of the phase at $\omega = 0$ is known as *phase-intercept distortion*. This type of distortion can affect modulated signals whose carrier contains information. The problem will be described by showing the effect of phase-intercept distortion on an amplitude-modulated signal.

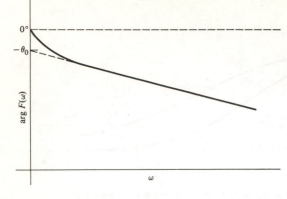

Figure 4.34 Phase response of a network that includes phase-equilization circuitry.

A carrier modulated by a sine wave is described by the equation

$$S(t) = (1 + m \cos \omega_m t) \cos \omega_c t$$

$$= \cos \omega_c t + \frac{m}{2} \cos (\omega_c - \omega_m)t + \frac{m}{2} \cos (\omega_c + \omega_m)t$$

where ω_m = modulating frequency
m = modulating index
ω_c = carrier frequency

$(1 + m \cos \omega_m t)$ is known as the *signal envelope*. The modulated signal has a frequency component at the carrier frequency plus upper and lower sidebands at the sum and difference frequencies $(\omega_c \pm \omega_m)$. If the modulated signal is passed through a linear phase filter that has phase-intercept distortion, the output signal (assuming the filter gain is constant) is

$$S_o = \cos \left[\omega_c(t - T) - \theta_o \right] + \frac{m}{2} \cos \left[(\omega_c - \omega_m)(t - T) - \theta_o \right]$$

$$+ \frac{m}{2} \cos \left[(\omega_c + \omega_m)(t - T) - \theta_o \right]$$

$$= [1 + m \cos \omega_m(t - T)] \cos \omega_c \left[t - \left(T + \frac{\theta_o}{\omega_c} \right) \right] \qquad (4.88)$$

The envelope of the signal is delayed by the group delay T. If the group delay had not been constant the two sideband signals would not have been delayed equally, and the resulting envelope would also have been distorted. For the modulated signal the carrier component has been delayed by phase delay. If synchronous detection is used (see Chap. 12), the output amplitude can be reduced because of the phase-intercept distortion.

PROBLEMS

4.1 (*a*) Determine the Q, bandwidth B, and resonant frequency of the circuit illustrated in Fig. P4.1.

 (*b*) Repeat (*a*) with a 50-Ω resistor added in series with the voltage source.

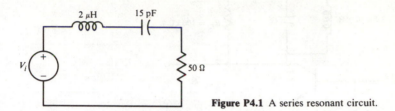

Figure P4.1 A series resonant circuit.

4.2 For the circuit of Prob. 4.1 find the value of capacitance that will make the voltage across the capacitor a maximum at the frequency ω_c. The maximum capacitor voltage at any frequency can be expressed in terms of the circuit Q. Is this also the value of capacitance required for circuit resonance? Explain.

4.3 Derive Eq. (4.6).

4.4 (*a*) Calculate the resonant frequency, Q, and bandwidth of the circuit illustrated in Fig. P4.4.

 (*b*) What is the attenuation of the third harmonic compared to the fundamental frequency amplitude? Of the fifth harmonic?

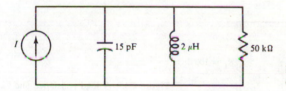

Figure P4.4 A parallel resonant circuit.

4.5 Repeat Prob. 4.4 for the case where a 5-Ω resistor is in series with the inductor.

4.6 A 0.1-μH inductor with a Q_u of 100 at 1 MHz is used in a series resonant circuit with a 50-Ω load resistor. What must be the capacitance for the circuit to resonate at 1 MHz? What is the circuit Q?

4.7 The output impedance of the transistor used in the amplifier shown in Fig. P4.7 is 50 kΩ. Does this affect the circuit bandwidth? What will the bandwidth be if the inductor has a Q_u of 100? The resonant frequency of the tuned circuit is 10 MHz. The capacitor $C = 160$ pF.

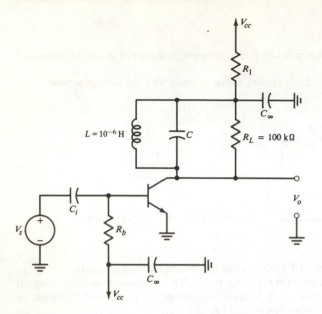

Figure P4.7 A transistor amplifier with a tuned circuit load.

4.8 Derive the equivalent turns ratio for a closely coupled step-down autotransformer.

4.9 If $C_2 = 100$ pF, for what value of C_1 will the input impedance of the circuit in Fig. P4.9 be real? At 20 MHz? Are there any values of C_2 for which C_1 cannot be adjusted to make the input impedance real at this frequency?

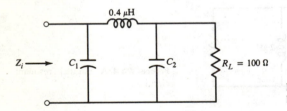

Figure P4.9 A load resistance and lossless matching network.

4.10 Design a lossless coupling network to match a 50-Ω load to a 20-Ω source resistance. What are the component values if the operating frequency is to be 20 MHz? Is the transfer impedance a low-pass bandpass or high-pass circuit? What about the input impedance?

4-11 Design a lossless coupling network that matches a $10 + 5j$-Ω load to a 50-Ω source impedance.

4.12 Design a lossless coupling network that matches a 10-Ω load in parallel with a $+5j$-Ω reactance to a 50-Ω source.

4.13 Design a lossless coupling network that matches a 50-Ω load impedance to a 15-Ω source impedance. Do the Q of the input impedance and the transfer impedance have practical significance? What is the bandwidth of the transimpedance?

4.14 Determine the value of C_2 for the circuit shown in Fig. P4.14 to be resonant at $\omega_o = 4 \times 10^6$ rad/s. Determine the circuit input impedance and also the output voltage at the resonant frequency if $I = 10^{-2} \sin 4 \times 10^6 t$. The transformer is tightly coupled. If the unloaded Q of the secondary Q_2 is 70, what is the circuit bandwidth? L_1 is 0.30 μH and M is 3 μH.

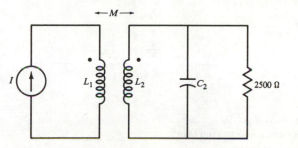
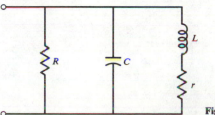

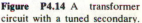

Figure P4.14 A transformer circuit with a tuned secondary.

4.15 Find (approximate) the half-power bandwidth of the circuit shown in Fig. P.4.15. C is 1000 pF, L is 10 μH, r is 1 Ω, and R is 10 kΩ.

Figure P4.15 A parallel resonant circuit.

4.16 In Prob. 4.15, if the circuit is excited by a 1-A peak sinusoidal current source with a frequency of 5 times the resonant frequency of the circuit, what will be the amplitude of the output.

4.17 Design a lossless coupling network that matches a load impedance of $10 + 5j$ Ω to a 50-Ω source impedance.

4.18 The transistor in the circuit shown in Fig. P4.18 has an input impedance of 1 kΩ and an output impedance of 80 kΩ. The unloaded Q of the inductor is 100. The transistor current gain β is 100.

    (a) What is the resonant frequency of the amplifier?
    (b) What is the amplifier bandwidth?
    (c) What is the ratio of the third harmonic output to the fundamental output?
    (d) What is the voltage gain at resonance?

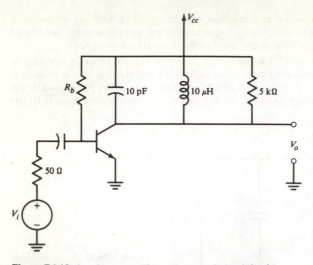

Figure P4.18 A voltage amplifier with a tuned circuit load.

4.19 Show that the gain-bandwidth product of the single-tuned inductively coupled circuit is given by Eq. (4.52).

4.20 For the circuit shown in Fig. P4.20, $L_1 = L_2 = 20\ \mu H$ and $M = 2\ \mu H$. What value of C is required for the input impedance to be resonant at 1 MHz? What will be the circuit Q?

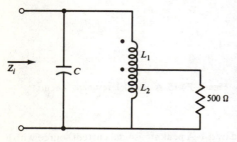

Figure P4.20 A capacitive transformer.

4.21 Derive expressions for the voltage gain as a function of frequency, center frequency, and bandwidth of the circuit shown in Fig. P4.21.

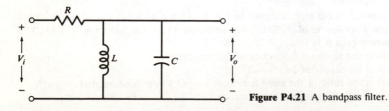

Figure P4.21 A bandpass filter.

4.22 A low-pass filter consists of two first-order low-pass filters connected in cascade. If the −3-dB bandwidth of each filter section is 100 kHz, what are the phase and group delays of the composite filter at 1000 kHz?

REFERENCES

4.1 Szentirmai, G.: FILSYN: A General Purpose Filter Synthesis Program, *Proc. IEEE*, **65**:1443–1458 (1977).

4.2 Valley, G., Jr., and H. Wallman (eds.): *Vacuum Tube Amplifiers*, McGraw-Hill, New York, 1948.

4.3 Blinchikoff, H. J., and A. I. Zverev: *Filtering in the Time and Frequency Domain*, Wiley, New York, 1976.

ADDITIONAL READING

Clarke, K. K., and D. T. Hess: *Communication Circuits: Analysis and Design*, Addison-Wesley, Reading, Mass., 1971.

Everitt, W. L.: *Communication Engineering*, 2d ed., McGraw-Hill, New York, 1937.

McIlwain, K., and J. G. Brainerd: *High-Frequency Alternating Currents*, 2d ed., Wiley, New York, 1939.

Szentirmai, G. (ed.): *Computer Aided Filter Design*, IEEE Press, New York, 1973.

Terman, F. E.: *Electronic and Radio Engineering*, 4th ed., McGraw-Hill, New York, 1955.

CHAPTER

FIVE

HIGH-FREQUENCY AMPLIFIERS AND AUTOMATIC GAIN CONTROL

5.1 INTRODUCTION

Various applications in high-frequency communication circuits require amplifiers. Amplifiers connected to the output of mixers, modulators, and demodulators serve as buffer amplifiers, where they provide a constant load impedance and gain. Other applications include IF amplifiers, repeaters for cable communications, power amplifiers, power-amplifier drivers, and video amplifiers. This chapter considers the design of amplifiers for applications in which the actual frequency response of the transistors must be considered. Transistors have an upper frequency limit beyond which they do not provide any gain; the frequency response of various transistor amplifier configurations is evaluated in this chapter, and then methods for extending the frequency range of the amplifiers are described. Amplifiers must often automatically adjust gain in response to variations in the input signal amplitude. This ability is called *automatic gain control or AGC*. Several AGC systems are analyzed in Sec. 5.4.

5.2 HIGH-FREQUENCY PERFORMANCE OF BIPOLAR AND FIELD-EFFECT TRANSISTOR AMPLIFIERS

The BJT High-Frequency Model

The bipolar transistor's frequency-dependent properties are accounted for by including the base-emitter capacitance C_π, the base-collector junction capacitance C_μ, and the collector-to-emitter terminal capacitance C_o in the small-signal equivalent circuit shown in Fig. 5.1. The capacitance C_π is actually

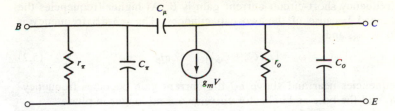

Figure 5.1 A small-signal, high-frequency equivalent circuit of the bipolar transistor.

composed of a diffusion capacitance plus a capacitance that represents the space charge layer present at the base-to-emitter junction. The junction capacitances vary inversely with the voltage across the junction. A detailed description of these capacitances can be found in the literature. For our purposes all parameters in the model can be assumed to be independent of frequency. Although this is not true for all transistors, there are a large number of devices for which the model is accurate at frequencies up to 100 MHz.

At higher frequencies parasitic components, such as lead inductance, become important. How the capacitances limit high-frequency performance can be seen by calculating the short-circuit current gain. To do this, the output is short-circuited as shown in Fig. 5.2. Since r_o and C_o are short-circuited, no current flows through them and

$$I_o = -(g_m - j\omega C_\mu) V$$

It is the case that in the frequency region of interest $I_o \approx -g_m V$. The voltage V is given by

$$V = \frac{I_b r_\pi}{j\omega r_\pi (C_\pi + C_\mu) + 1}$$

so the short-circuit current gain is

$$A_i = \frac{I_o}{I_b} = \frac{-g_m r_\pi}{j\omega r_\pi (C_\pi + C_\mu) + 1} = \frac{-\beta}{j\omega r_\pi (C_\pi + C_\mu) + 1} \tag{5.1}$$

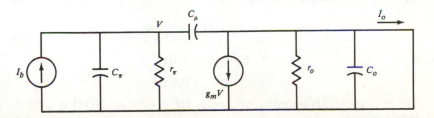

Figure 5.2 Equivalent circuit for calculating the high-frequency, short-circuit current gain.

The low-frequency short-circuit current gain is β. At higher frequencies the gain is reduced because of the two capacitances. The -3-dB frequency is referred to as ω_β, and

$$\omega_\beta = [r_\pi(C_\pi + C_\mu)]^{-1} = 2\pi f_\beta \qquad (5.2)$$

At the frequencies near and above ω_β the current gain becomes frequency-dependent, and the base and collector currents are no longer in phase.

An important figure of merit for the transistor is the frequency f_T at which the magnitude of short-circuit current gain becomes unity. That is,

$$|A_i| = 1 = \frac{\beta}{\{1 + [\omega_T r_\pi(C_\pi + C_\mu)]^2\}^{1/2}} \approx \frac{\beta}{\omega_T r_\pi(C_\pi + C_\mu)}$$

or

$$\omega_T = \frac{\beta}{r_\pi(C_\pi + C_\mu)} = \frac{g_m}{C_\pi + C_\mu} \qquad (5.3)$$

Note that both ω_β and ω_T depend upon the transistor's operating point. The transconductance g_m is directly proportional to the collector current: although C_π and C_μ are not as dependent on the operating point, they do depend on the junction voltages. Increasing the collector bias current will increase f_T. Although C_π is normally not specified on data sheets, C_μ is usually given as C_{cb} (collector-to-base capacitance) or C_{ob} (output capacitance, common-base configuration).

Example 5.1 The 2N3904 is specified to have an f_T of 3×10^8 Hz, a β of 100, and a $C_{ob} = 4$ pF. Determine the parameters to be used in the hybrid-π model. The dc collector current is 10 mA.

SOLUTION Since the collector current is 10 mA,

$$r_\pi = \frac{0.026 \times 100}{0.01} = 260\ \Omega$$

and

$$g_m = 40 \times 10^{-2} = 0.4\ S$$

C_π is then determined from Eq. (5.3):

$$\omega_T = 2\pi(3 \times 10^8) = \frac{0.4}{C_\pi + C_\mu}$$

or

$$C_\pi + C_\mu = 210\ pF$$

So

$$C_\pi = 206\ pF$$

The resistance r_o is normally not given, but a value of 50 kΩ is a good approximation.

Current Gain-Bandwidth Product

The unity gain frequency f_T is also referred to as the *current gain-bandwidth product* of the device, since

$$\omega_T = g_m r_\pi \omega_\beta$$

and $g_m r_\pi$ is the magnitude of the low-frequency short-circuit current gain:

$$A_i = \frac{-g_m r_\pi}{j\omega r_\pi (C_\pi + C_\mu) + 1} \approx \frac{-g_m}{j\omega (C_\pi + C_\mu)} = \frac{-\omega_T}{j\omega} \tag{5.4}$$

So, at any frequency greater than ω_β the magnitude of the short-circuit current gain is readily determined.

> **Example 5.2** What would the short-circuit current gain be at 10 MHz for the 2N3904 transistor of the preceding example?
>
> SOLUTION Since the gain-bandwidth product is 3×10^8 Hz, at 10 MHz the gain will be
>
> $$A_i = \frac{3 \times 10^8}{10^7} = 30$$

The gain-bandwidth product of the device serves as a useful rule of thumb for estimating the gain at any frequency. At frequencies below ω_β the magnitude of the current gain is equal to β.

The FET High-Frequency Model

The performance of a field-effect transistor is limited at high frequencies by parasitic capacitances, as is that of the bipolar transistor. The high-frequency performance can be described in terms of the equivalent circuit shown in Fig. 5.3. The gate-to-source capacitance C_{gs} and the gate-to-drain capacitance C_{gd} are junction capacitances that are inversely related to the voltage across the junctions. The actual relationship depends upon the particular manufacturing process used, but since the source voltage will be less than the drain voltage, C_{gs}

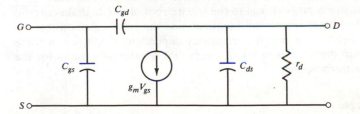

Figure 5.3 A small-signal, high-frequency equivalent circuit of a field-effect transistor.

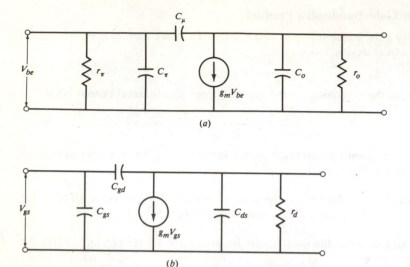

Figure 5.4 High-frequency equivalent circuits of (a) the BJT and (b) the FET.

will usually be larger than C_{gd}. The drain-to-source capacitance C_{ds} is the stray capacitance associated with the device package. The capacitances of junction and insulated-gate field-effect transistors differ in how they vary with the operating point, but the small-signal amplifier models are similar. We will now discuss the high-frequency behavior of these amplifiers.

The hybrid-π models of the junction and field-effect transistors are shown in Fig. 5.4. The only difference between the BJT and FET equivalent circuits is the resistor r_π included in the small-signal model for the bipolar transistor. The midfrequency input impedance of the common-emitter amplifier is $r_\pi\,\Omega$, whereas that of the common-source amplifier is very large and is considered to be infinite. The output circuits of the two devices are the same except that the bipolar transistor is essentially a current-controlled device, while the FET is a voltage-controlled device. Also, the transconductance of bipolar transistors is proportional to the dc collector current, while the transconductance of a field-effect transistor is proportional to the square root of the dc drain current. Nevertheless, the circuit models are identical, and the same equations can be used for both. Before studying the frequency-dependent properties of these amplifiers, we will discuss a network theorem that is extremely useful for the analysis of transistor amplifiers.

Miller's Theorem

Consider the network configuration indicated in Fig. 5.5. Nodes 1 and 2 are interconnected by the impedance Z. At any frequency the voltage gain between

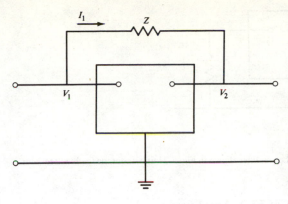

Figure 5.5 A voltage amplifier with impedance Z connected between the input and output.

these two nodes can be given as

$$\frac{V_2}{V_1} = K$$

K is generally frequency-dependent.

We will now show that the current I_1 drawn from node 1 through Z can be obtained by disconnecting Z from terminal 1 and by bridging an impedance $Z/(1 - K)$ from node 1 to ground. This follows since

$$I_1 = \frac{V_1 - V_2}{Z} = \frac{V_1 - KV_1}{Z} = \frac{V_1(1 - K)}{Z} = \frac{V_1}{Z_1} \tag{5.5}$$

where

$$Z_1 = \frac{Z}{1 - K}$$

Therefore, if Z is replaced at node 1 by an impedance Z_1 connected between node 1 and ground, the current leaving node 1 will be the same. The current leaving node 2 is

$$I_2 = \frac{V_2 - V_1}{Z} = \frac{V_2 - V_2/K}{Z} = \frac{V_2}{Z_2} \tag{5.6}$$

where

$$Z_2 = \frac{KZ}{K - 1}$$

If Z_2 is connected between node 2 and ground, the current leaving node 2 will be the same. Miller's theorem states that if the network illustrated in Fig. 5.5 has a voltage gain K between two nodes, then any impedance connected between those two nodes can be replaced by an impedance at each node connected to ground, as illustrated in Fig. 5.6. The values of the impedances are given by Eqs. (5.5) and (5.6).

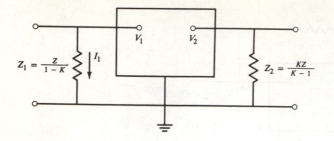

Figure 5.6 An equivalent circuit of the amplifier shown in Fig. 5.5 but without the feedback resistor.

High-Frequency Response of FET Amplifiers

The common-source amplifier. The utility of Miller's theorem can be appreciated by applying it to the analysis of the high-frequency performance of the FET amplifier illustrated in Fig. 5.7. The small-signal, high-frequency equivalent of the amplifier is given in Fig. 5.8. If $R_g \gg R$, $V_g \approx V_i$, and the midfrequency voltage gain is

$$A_v = \frac{V_2}{V_1} = -g_m R'_L$$

where

$$R'_L = \frac{R_L r_d}{R_L + r_d}$$

We will now use Miller's theorem to replace the gate-to-drain capacitance by two capacitors, one connected from gate to ground and one connected from drain to ground, as illustrated in Fig. 5.9. The impedance from gate to drain is

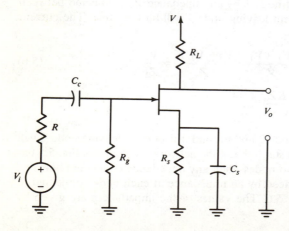

Figure 5.7 A common-source amplifier.

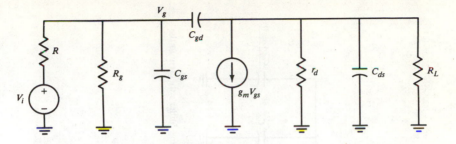

Figure 5.8 A small-signal, high-frequency equivalent circuit of the common-source amplifier.

$Z = (j\omega C_{gd})^{-1}$, so the impedance to be added from gate to ground is [Eq. (5.5)]

$$Z_1 = [j\omega C_{gd}(1 - A_v)]^{-1} = [j\omega C_{gd}(1 + g_m R'_L)]^{-1}$$

and the impedance to be added from drain to ground is [Eq. (5.6)]

$$Z_2 = \frac{A_v}{j\omega C_{gd}(A_v - 1)} = \left[j\omega C_{gd}\left(\frac{1 + g_m R'_L}{g_m R'_L}\right) \right]^{-1}$$

The gate-to-drain capacitance is replaced by a capacitor

$$C_1 = C_{gd}(1 + g_m R'_L) \tag{5.7}$$

connected from gate to ground, and a capacitor

$$C_2 = C_{gd}\frac{1 + g_m R'_L}{g_m R'_L} \tag{5.8}$$

connected from drain to ground. If the voltage gain is large, the capacitance shunting the input is large. In a high-gain inverting amplifier any capacitance connected from input to output is equivalent to the same capacitance shunting the input, but with its value amplified by the voltage gain. This is known as the *Miller effect*. The Miller effect limits the high-frequency performance of high-voltage-gain amplifiers.

If we again assume that $R \ll R_g$, the complete frequency-dependent transfer function for the circuit of Fig. 5.9 is

$$\frac{V_o}{V_i} = \frac{-g_m R'_L}{(j\omega RC_T + 1)(j\omega R'_L C_o + 1)} \tag{5.9}$$

where

$$C_T = C_{gs} + C_{gd}(1 + g_m R'_L)$$

and

$$C_o = \frac{C_{gd}(1 + g_m R'_L)}{g_m R'_L}$$

The amplifier has two high-frequency poles in the left-hand plane:

$$\omega_1 = -(RC_T)^{-1} \tag{5.10}$$

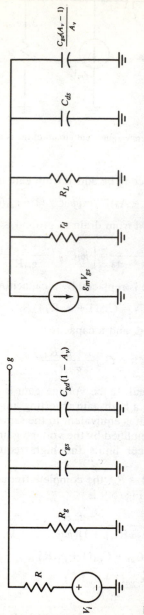

Figure 5.9 A small-signal, high-frequency equivalent circuit of the common-source amplifier obtained by applying Miller's theorem to the circuit shown in Fig. 5.8.

and
$$\omega_2 = -(R'_L C_o)^{-1} \qquad (5.11)$$

Unless R is very small, the magnitude of ω_1 will usually be much smaller than the magnitude of ω_2 because C_T is much bigger than C_o. Actually, the approximation used in applying Miller's theorem is such that the frequency response is not accurate beyond the lower of the two poles. At higher frequencies the voltage gain A_v is complex, but it was assumed to be real in deriving Eq. (5.9).

Example 5.3 A 2N5486 FET with a transconductance g_m of 2×10^{-3} S and an r_d of 13 kΩ is used in the amplifier shown in Fig. 5.10. $C_{gs} = 5$ pF and $C_{gd} = 1$ pF $= C_{ds}$. Determine the high-frequency response of the amplifier.

SOLUTION The small-signal, high-frequency equivalent circuit is illustrated in Fig. 5.11. Since $R \le R_g$, the midfrequency voltage gain is given as

$$A_v = -g_m R_L = \frac{(-2 \times 10^{-3})(6 \text{ k}\Omega \times 13 \text{ k}\Omega)}{6 \text{ k}\Omega + 13 \text{ k}\Omega} = -8.2$$

Therefore,
$$C_T = 5 + 1(1 + 8.2) = 14.2 \text{ pF}$$

and
$$C_o = 1 + \frac{1 + 8.2}{8.2} = 2.1 \text{ pF}$$

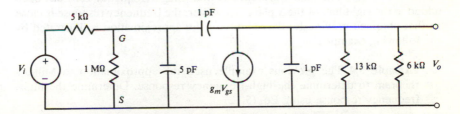

Figure 5.10 The common-source amplifier discussed in Example 5.3.

Figure 5.11 The high-frequency equivalent circuit of the amplifier shown in Fig. 5.10.

The high-frequency pole due to the input circuit is [Eq. (5.10)]

$$-\omega_1 \approx (RC_T)^{-1} = [(5 \times 10^3)(14.2 \times 10^{-12})]^{-1} = 1.41 \times 10^7 \text{ rad/s}$$

and the high-frequency pole due to the output circuit is [Eq. (5.11)]

$$-\omega_2 = (R_L C_o)^{-1} = [(4.1 \times 10^3)(2.1 \times 10^{-12})]^{-1} = 11.6 \times 10^7 \text{ rad/s}$$

The upper corner frequency is approximately 1.4×10^7 rad/s and is determined by the input circuit. The estimate for a second high-frequency pole obtained using the approximation is not accurate.

The application of Miller's theorem in the preceding example was somewhat of an approximation. The approximation involved using the midband gain, which is frequency-independent, as an approximation for the voltage gain at all frequencies. To be more precise, the frequency-dependent gain should be used. Since the gain will generally be complex, the capacitance must be replaced by complex impedances in the application of Miller's theorem. However, the approximation is sufficiently accurate if Miller's theorem is used with the midband gain. To illustrate the accuracy, the circuit of Fig. 5.8 can be analyzed directly. If we again assume that $R_g \geq R$, the equilibrium equations can be written as

$$\frac{V_g - V_i}{R} + V_g s C_{gs} + (V_g - V_o)s C_{gd} = 0$$

and

$$-V_g s C_{gd} + V_o s C_{gd} + \frac{V_o}{R'_L} + g_m V_g + V_o s C_{gs} = 0$$

where

$$R'_L = \frac{r_d R_L}{r_d + R_L}$$

If V_g is eliminated, the voltage gain is found to be

$$\frac{V_o}{V_i} = A_v = \frac{-(g_m - sC_{gd})R'_L}{s^2 RR'_L[C_{gd}C_{gs} + C_{ds}(C_{gd} + C_{gs})] + s[g_m C_{gd}RR'_L + (C_{gd} + C_{ds})R'_L + (C_{gd} + C_{gs})R] + 1}$$

$$(5.12)$$

This transfer function differs from that obtained using the approximation for Miller's theorem. The main difference is that a high-frequency zero has been added in the right half of the s plane. However, the frequency response is close to that obtained previously; the validity of the approximation is illustrated by the following example.

Example 5.4 The previous example used the approximation to Miller's theorem to determine the high-frequency response. Determine the high-frequency response using Eq. (5.12).

SOLUTION For this amplifier $g_m = 2 \times 10^{-3}$, $C_{gs} = 5 \times 10^{-12}$, $C_{gd} = 10^{-12}$,

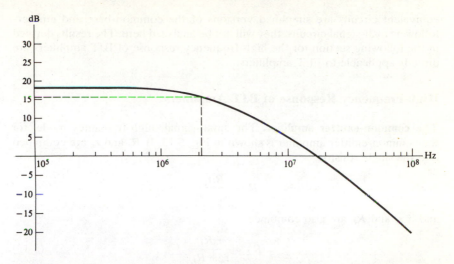

Figure 5.12 Frequency response of the voltage gain (magnitude) of the amplifier circuit shown in Fig. 5.11.

$R = 5 \times 10^3$, and $R'_L = 4.1 \times 10^3$. If these values are substituted into Eq. (5.12) the transfer function zero is found to be located at.

$$\omega_z = 2 \times 10^9 \text{ rad/s}$$

and the two poles are located at

$$\omega_{p_1} = -1.3 \times 10^7 \text{ rad/s}$$

and $$\omega_{p_2} = -33.7 \times 10^7 \text{ rad/s}$$

The actual frequency response obtained from a computer simulation of the amplifier is plotted in Fig. 5.12. The -3-dB corner frequency is located at 1.3×10^7 rad/s (2.1×10^6 Hz), whereas a corner frequency of 1.4×10^7 rad/s was obtained with the approximation. The zero is located at a much higher frequency and has a negligible effect on the transfer function within the passband of the amplifier, but it does affect the rate of attenuation and phase shift at high frequencies.

In general, the midfrequency gain can be used with Miller's theorem to obtain a sufficiently accurate high-frequency amplifier model. This approach greatly simplifies the analysis since it isolates the input and output stages of the amplifier. At frequencies above the -3-dB corner frequency the approximation error increases. At sufficiently high frequencies the amplifier rolls off at -6 dB per octave, not at -12 dB per octave as predicted by Eq. (5.9).

The common-gate and source-follower small-signal, high-frequency equivalent circuits are simplified versions of the common-base and emitter-follower small-signal circuits, they will not be analyzed here. The results derived in the following section for the high-frequency response of BJT amplifiers are directly applicable to FET amplifiers.

High-Frequency Response of BJT Amplifiers

The common-emitter amplifier. The small-signal, high-frequency model for the common-emitter amplifier is shown in Fig. 5.13. If R_s and r_π are combined into the single resistor

$$R = \frac{R_s r_\pi}{R_s + r_\pi}$$

and if r_o and R_L are also combined

$$R'_L = \frac{r_o R_L}{r_o + R_L} \tag{5.13}$$

the equivalent circuit is as shown in Fig. 5.14. This is identical to the high-frequency circuit for the common-source amplifier; hence the results derived for the frequency response of the common-source amplifier apply equally well to the common-emitter amplifier.

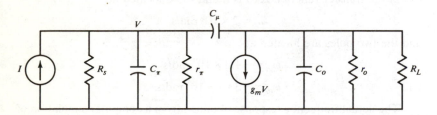

Figure 5.13 A small-signal, high-frequency equivalent circuit of a common-emitter amplifier.

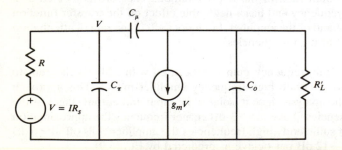

Figure 5.14 A simplified equivalent circuit of a common-emitter amplifier.

Example 5.5 The amplifier shown in Fig. 5.15 uses a 2N3904 BJT with an f_T of 3×10^8 Hz, a β of 100, and a C_{ob} of 4 pF. C_{ob} is the output capacitance of a common-base amplifier. It is usually included on a company's transistor specification sheet because it is easy to measure. The capacitance C_μ of the hybrid-π model is appreciably less than C_{ob},[5.1] but 4 pF has been assumed for C_μ in this example. This means that the actual high-frequency band-width will be greater than predicted. Calculate the upper 3-dB frequency of the amplifier. The dc collector current is 10 mA. Here C_o is assumed to have a negligible effect on the frequency response.

SOLUTION Example 5.1 showed that a 2N3904 transistor with a dc collec-tor current of 10 mA had $r_\pi = 260\ \Omega$, $g_m = 0.4$ S, and $C_\pi = 206$ pF.

The small-signal, high-frequency equivalent circuit is shown in Fig. 5.16. The parallel combination of the two base biasing resistors is shown as a 2.66-kΩ resistor. The parallel combination of this resistor and the 260-Ω emitter resistance is 237 Ω. It has also been assumed that the transistor output impedance is much greater than the 600-Ω load im-

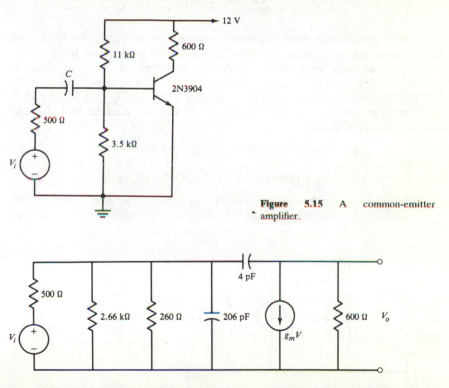

Figure 5.15 A common-emitter amplifier.

Figure 5.16 A high-frequency equivalent circuit of the amplifier shown in Fig. 5.15.

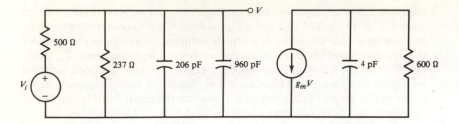

Figure 5.17 An approximate equivalent circuit of the amplifier shown in Fig. 5.15.

pedance. The midfrequency output voltage is

$$V_o = \frac{-g_m R_L \times 237}{237 + 500} V_i$$

$$= \frac{-0.4 \times 600 \times 237}{737} V_i = -77.2 V_i$$

The midfrequency base-to-emitter voltage gain is $-g_m R_L = 240$, so the approximate high-frequency, small-signal circuit is as shown in Fig. 5.17. The equivalent input Miller capacitance is 960 pF. If the input circuit is replaced by its Thévenin's equivalent, the circuit is as shown in Fig. 5.18. This circuit topology is the same as that given in Fig. 5.9 for the FET amplifier. Hence Eqs. (5.10) and (5.11) can be used to find the two high-frequency poles

$$-\omega_1 = [161(1166 \times 10^{-12})]^{-1} = 5.3 \times 10^6 \text{ rad/s}$$

and

$$-\omega_2 = [600(4 \times 10^{-12})]^{-1} = 33.2 \times 10^7 \text{ rad/s}$$

Since $|\omega_1| \ll |\omega_2|$, the high-frequency performance is determined by the input circuit. The upper -3-dB frequency of this amplifier is equal to $|\omega_1|$.

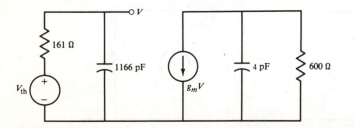

Figure 5.18 A simplified equivalent of the circuit shown in Fig. 5.17.

The high-frequency common-emitter circuit of Fig. 5.14 is identical in topology to the small-signal, high-frequency equivalent circuit given in Fig. 5.8 for the common-source amplifier. Therefore, Eq. (5.12) is also correct for the common-emitter circuit. The corresponding equation for the BJT is found by changing the notation to correspond to that of the BJT circuit:

$$\frac{V_o}{V_{th}} = A_v = \frac{-(g_m - sC_\mu)R'_L}{s^2 RR'_L[C_\pi C_\mu + C_o(C_\pi + C_\mu)] + s[R(C_\pi + C_\mu) + R'_L(C_\mu + C_o) + g_m C_\mu RR'_L] + 1}$$

(5.14)

R'_L is given by Eq. (5.13) and R is equal to the parallel combination of R_s, r_π, and any bias resistors connected between the base terminal and ground. V_{th} is the Thévenin equivalent voltage of the input circuit consisting of these resistors.

Example 5.6 Calculate the exact frequency response of the amplifier shown in Fig. 5.15 using Eq. (5.14).

SOLUTION If the values from the previous example are substituted into Eq. (5.14) the transfer function is

$$\frac{V_o}{V_{th}} = \frac{-(0.4 - s \times 4 \times 10^{-12})(966 \times 10^3)}{s^2(79.6 \times 10^{-18}) + s(190.2 \times 10^{-9}) + 1}$$

The transfer function has a zero

$$\omega_z = 4 \times 10^{11} \text{ rad/s}$$

and two left-half-plane poles

$$\omega_1 = -5.5 \times 10^6 \text{ rad/s}$$

and

$$\omega_2 = -2.41 \times 10^9 \text{ rad/s}$$

In this example the bandwidth approximation obtained using the Miller theorem is almost the same as that obtained from the more complex formula. This approximation does provide an easy method for estimating the frequency response. Precise estimates require a great deal more calculation or, preferably, some method of simulation.

Common-Base Amplifiers

Common-base amplifiers are used in many high-frequency circuits because they have a wider bandwidth than does the equivalent common-emitter amplifier. A typical common-base amplifier circuit is shown in Fig. 5.19. The small-signal, high-frequency model for this amplifier is shown in Fig. 5.20. The reasons for the wider bandwidth of the common-base amplifier are (1) the midfrequency

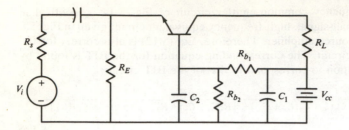

Figure 5.19 A common-base amplifier.

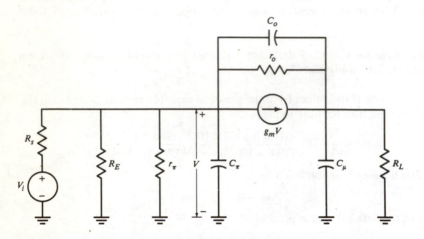

Figure 5.20 A small-signal, high-frequency equivalent circuit of the common-base amplifier.

input resistance, given by Eq. (2.17)

$$Z_i = \frac{r_\pi}{\beta} = (g_m)^{-1}$$

is $1/\beta$ times that of the common-emitter amplifier and (2) the equivalent Miller capacitance connected across the input is

$$C_m = C_\mu(1 - A_v)$$

as for the common-emitter amplifier, but in the common-base amplifier A_v is positive. If the voltage gain is greater than one, the Miller capacitance will be negative and reduce the total capacitance from emitter to base. The total input capacitance can, in fact, become negative, which means that the input impedance is inductive. Because of the small input impedance the transistor collector-to-emitter resistance r_o can be ignored with no sacrifice in accuracy.

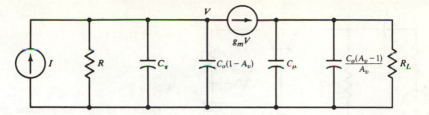

Figure 5.21 An equivalent circuit of the common-base amplifier, obtained using Miller's theorem.

The equivalent circuit can then be drawn as shown in Fig. 5.21. In this figure the voltage source has been replaced by an equivalent current source $I = V_i/R_s$, and the resistor R is the parallel combination of the three resistors

$$R = R_s \| R_E \| r_\pi$$

The emitter voltage V is then

$$V = (I - g_m V)Z$$

or

$$V = \frac{IZ}{1 + g_m Z}$$

where

$$Z = \frac{R}{1 + j\omega R C_T}$$

and

$$C_T = C_\pi + C_\mu(1 - A_v)$$

Therefore

$$\frac{V}{I} = \frac{R/(1 + g_m R)}{1 + j\omega C_T R/(1 + g_m R)} \tag{5.15}$$

The corner frequency of the input circuit is then

$$\omega_1 = \frac{1 + g_m R}{|C_T| R} \tag{5.16}$$

and the corner frequency of the output circuit is

$$\omega_2 = [R_L(C_\mu + C'_m)]^{-1} \tag{5.17}$$

where

$$C'_m = C_\mu \frac{A_v - 1}{A_v}$$

For the common-base amplifier the output-circuit corner frequency can be lower than the corner frequency of the input circuit. The particular characteristics of each amplifier must be evaluated.

Example 5.7 Calculate the gain and frequency response of the common-base amplifier illustrated in Fig. 5.22. The same transistor (and dc collector

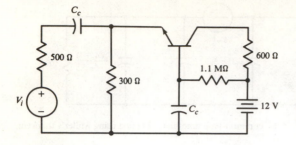

Figure 5.22 The common-base amplifier discussed in Example 5.7.

current) is used as in Example 5.4 (where the frequency response of the common-emitter amplifier was calculated). The collector-to-emitter capacitance is assumed to be $C_o = 1$ pF.

SOLUTION The high-frequency equivalent circuit is given in Fig. 5.23. The midband emitter-to-collector voltage gain is

$$\frac{V_o}{V} = g_m R_L = 240$$

so the Miller capacitance is

$$C_m = 1 - 240 = -239 \text{ pF}$$

and the collector-to-ground Miller capacitance is

$$C'_m = \frac{239}{240} \approx 1 \text{ pF}$$

Since $R_i = (g_m)^{-1} = 2.5 \, \Omega$, the midfrequency voltage gain is

$$A_v = 240 \left(\frac{2.5}{500}\right) = 1.2$$

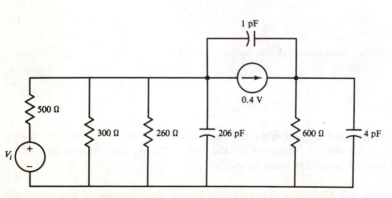

Figure 5.23 A high-frequency equivalent circuit of the amplifier shown in Fig. 5.22.

The voltage gain is much less than that of the common-emitter amplifier because of the low input resistance.

The high-frequency pole due to the output circuit is [Eq. (5.17)]

$$-\omega_2 = (600 \times 5 \text{ pF})^{-1} = 3.33 \times 10^8 \text{ rad/s}$$

The high-frequency pole due to the input circuit can be determined using Eq. (5.16). The equivalent resistance is

$$R = 500 \| 300 \| 260 = 109 \ \Omega$$

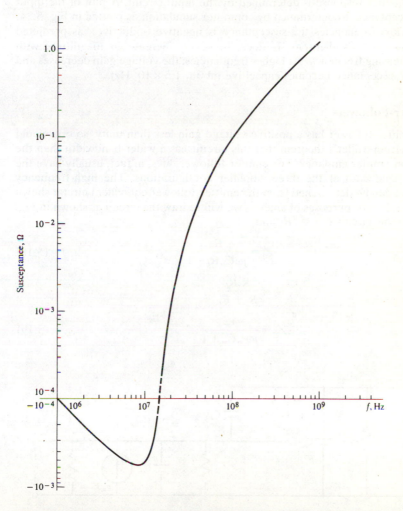

Figure 5.24 Input susceptance of the circuit shown in Fig. 5.23.

and the total capacitance

$$C_T = 206 - 239 = -33 \text{ pF}$$

is negative. The sign of the capacitance affects the phase shift but not the magnitude of the frequency response nor the corner frequency. The high-frequency pole of the input circuit [Eq. (5.16)]

$$-\omega_1 = (RC_T)^{-1} = 2.78 \times 10^8 \text{ rad/s}$$

is slightly smaller than the pole due to the output circuit, so the high-frequency response is determined by the input circuit. A plot of the input susceptance, as determined by computer simulation, is plotted in Fig. 5.24. At low frequencies the susceptance is negative (inductive), as predicted using Miller's theorem. It does, however, increase in magnitude with increasing frequency. At higher frequencies, the voltage gain decreases and the susceptance becomes capacitive (about 1.5×10^7 Hz).

Emitter-Followers

The emitter-follower has a positive voltage gain less than unity, so one would predict from Miller's theorem that this circuit has a wider bandwidth than the common-emitter amplifier. The emitter-follower does, in fact, usually have the widest bandwidth of the three amplifier configurations. The high-frequency response can be determined from the emitter-follower equivalent circuit shown in Fig. 5.25. For purposes of analysis we will redraw the circuit as shown in Fig. 5.26. In this circuit $I = V_i / R_s$ and

$$Z_i = \frac{R_i}{j\omega C_\mu R_i + 1} \tag{5.18}$$

where

$$R_i = \frac{R_s R_b}{R_s + R_b}$$

Also,

$$Z_\pi = \frac{r_\pi}{j\omega r_\pi C_\pi + 1} \tag{5.19}$$

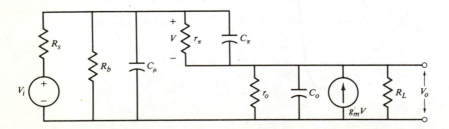

Figure 5.25 A small-signal, high-frequency equivalent circuit of the emitter-follower.

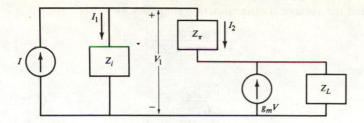

Figure 5.26 An emitter-follower equivalent circuit.

and
$$Z_L = \frac{R_o}{j\omega R_o C_o + 1} \qquad (5.20)$$

where
$$R_o = \frac{r_o R_L}{r_o + R_L}$$

The input current divides into the two currents I_1 and I_2. The amplifier response is derived from the three equations:

$$I = I_1 + I_2$$

$$V_1 = I_1 Z_i = I_2 Z_\pi + V_o$$

and
$$V_o = (1 + g_m Z_\pi) I_2 Z_L$$

If I_1 and I_2 are eliminated from these three equations the transimpedance is found to be

$$\frac{V_o}{I} = Z_T = Z_i \frac{Z_L(1 + g_m Z_\pi)}{Z_i + Z_\pi + (1 + g_m Z_\pi)Z_L} \qquad (5.21)$$

This complicated equation can be simplified for most practical applications. It will usually be the case that $Z_L \approx R_L$. If $Z_i \gg Z_\pi$, which will be the situation if the amplifier is driven by a current source, the transimpedance is then

$$Z_T \approx R_L(1 + g_m Z_\pi) = R_L \frac{1 + g_m r_\pi + j\omega r_\pi C_\pi}{j\omega r_\pi C_\pi + 1} \qquad (5.22)$$

The amplifier will have a pole at

$$-\omega_p = (r_\pi C_\pi)^{-1} \qquad (5.23)$$

which is approximately equal in magnitude to the current gain-bandwidth product of the device. The amplifier also has a zero at

$$\omega_z = -(1 + g_m r_\pi)/r_\pi C_\pi \qquad (5.24)$$

If the amplifier is driven by a voltage source with a source resistance $R_s \ll R_b$, then $Z_i \approx R_s$, and the voltage transfer function simplifies to

$$\frac{V_o}{V_i} = \frac{R_L(1 + g_m Z_\pi)}{R_s + Z_\pi + R_L(1 + g_m Z_\pi)} \qquad (5.25)$$

which has a zero at

$$-\omega_Z = \frac{1 + g_m r_\pi}{r_\pi C_\pi} \qquad (5.26)$$

and a pole at the lower frequency

$$-\omega_p = \frac{R_s + r_\pi + (1 + g_m r_\pi)R_L}{r_\pi C_\pi(R_s + R_L)} \qquad (5.27)$$

These equations contain several approximations, but they do provide good estimates for the design and analysis of high-frequency emitter-follower amplifiers. The following example illustrates their application.

Example 5.8 Estimate the frequency response of the amplifier illustrated in Fig. 5.27. The same values for load and source resistance are used as in the preceding example with the common-base amplifier. The transistor is biased so that g_m is again 0.4 S.

SOLUTION For this amplifier $R_b \gg R_s$ and $R_L \ll r_o$, so the amplifier bandwidth can be estimated using Eq. (5.27):

$$-\omega_p = \frac{500 + 260 + 105(600)}{260(206 \times 10^{-12})(500 + 600)} = 1.112 \times 10^9 \text{ rad/s}$$

The zero frequency is found from Eq. (5.26) to be

$$-\omega_Z = 2.01 \times 10^9 \text{ rad/s}$$

This model predicts that the frequency response does not decrease above

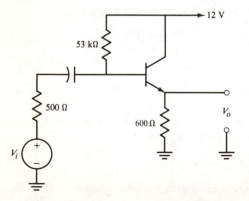

Figure 5.27 The emitter-follower discussed in Example 5.8.

ω_Z, but as with the Miller approximation, these equations are also an approximation and the high-frequency gain will decrease at $-6\,dB$ per octave.

Cascade of Identical Stages

If a large voltage gain is required, it is often convenient to cascade several identical stages. If each stage has a voltage gain A and a single pole at the upper corner frequency ω_p, the voltage transfer function of each stage is

$$A_v = \frac{A}{j\omega/\omega_p + 1}$$

If n of these stages are cascaded, the overall transfer function will be

$$A_T = \frac{A^n}{(j\omega/\omega_p + 1)^n} \tag{5.28}$$

The overall low-frequency gain is the nth power of the voltage gain of a single stage. The overall bandwidth ω_1 is that frequency at which the gain magnitude is 0.707 times its low-frequency value. That is,

$$\frac{A^n}{|j\omega/\omega_p + 1)|^n} = 0.707\,A^n = \frac{A^n}{2^{1/2}}$$

or

$$\left[\left(\frac{\omega_1}{\omega_p}\right)^2 + 1\right]^{n/2} = 2^{1/2}$$

The bandwidth is

$$\omega_1 = \omega_p(2^{1/n} - 1)^{1/2} \tag{5.29}$$

The bandwidth ω_1 obtained by cascading n identical stages each of bandwidth ω_p is reduced by the factor $(2^{1/n} - 1)^{1/2}$ of the single-stage bandwidth. This factor is referred to as the *bandwidth-reduction factor*.

> **Example 5.9** Two identical stages, each with a voltage gain of 20 dB and a bandwidth of 10 MHz, are cascaded. What are the overall gain and bandwidth?
>
> SOLUTION Since each stage has a gain of 20 dB the overall low-frequency gain is 40 dB. The overall bandwidth is
>
> $$f_1 = 10^7(2^{1/2} - 1)^{1/2} = 6.43 \text{ MHz}$$

5.3 BROADBANDING TECHNIQUES

The preceding analysis emphasized that the most troublesome factor in the design of high-frequency, high-gain amplifiers is the Miller capacitance con-

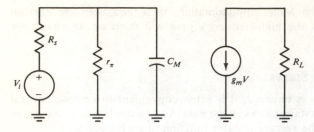

Figure 5.28 A simplified equivalent circuit of a common-emitter amplifier.

nected between the input and output. As the gain is increased the Miller capacitance increases; it is this capacitance that invariably limits the upper frequency response. Applications can require narrowband amplifiers, which include some type of tuned circuit, or broadband amplifiers, such as video amplifiers, in which the gain must be kept relatively constant over several decades of frequency. The following section describes methods for improving amplifier frequency response by the modification of the basic amplifier circuit.

Consider the simplified small-signal equivalent circuit of the common-emitter voltage amplifier shown in Fig. 5.28. Here the output capacitance has been neglected and the equivalent Miller capacitance is

$$C_M = C_\pi + C_\mu(1 + g_m R_L)$$

For this amplifier

$$\frac{V_o}{V_i}(j\omega) = \frac{-g_m r_\pi R_L}{r_\pi + R_s}\left(\frac{j\omega C_M r_\pi R_s}{r_\pi + R_s} + 1\right)^{-1}$$

The midband voltage gain is

$$A_{mid} = \frac{-g_m r_\pi R_L}{r_\pi + R_s} \tag{5.30}$$

and the bandwidth is

$$B = \frac{r_\pi + R_s}{C_M r_\pi R_s} \qquad \text{rad/s} \tag{5.31}$$

This expression shows that the bandwidth can be increased by decreasing R_s. There is a limit below which the bandwidth can be reduced by reducing R_s, since the analysis ignored the base-spreading resistance r_b', which is in series with R_s (for bipolar transistors). For low values, R_s should be replaced by $R_s + r_b'$ in Eq. (5.31). The analysis does imply that if the source resistance is very large, the bandwidth can be increased by using a source or emitter-follower for the first stage. There is also a limit beyond which the approximate circuit is no longer valid and the bandwidth is determined by the output circuit capacitance.

However, the expression does show that the source impedance should be as small as possible in order to maximize the voltage gain and bandwidth. The bandwidth can often be extended by modifying the amplifier input circuit. Input compensation is a method for extending the circuit bandwidth for applications in which the source resistance cannot be reduced.

Input Compensation

The bandwidth of a single-stage amplifier can be extended by adding additional frequency-sensitive components to the circuit. The method is often referred to as *broadbanding*. Figure 5.29 illustrates a method of broadbanding using input compensation. For this circuit the voltage gain is

$$A_v = \frac{-g_m R_L Z_\pi}{Z_\pi + Z_s}$$

where

$$Z_\pi = \frac{r_\pi}{j\omega r_\pi C_M + 1}$$

and

$$Z_s = \frac{R_s}{j\omega R_s C_s + 1}$$

If C_s is selected so that

$$R_s C_s = r_\pi C_M \tag{5.32}$$

then the voltage gain

$$A_v = \frac{-g_m R_L r_\pi}{r_\pi + R_s}$$

is independent of frequency. Input compensation can be used to cancel the effect of the Miller capacitance on the input with no reduction in voltage gain. In this case the bandwidth will be determined by the output circuit capacitance; it is

$$B = [R_L(C_o + C'_M)]^{-1} \tag{5.33}$$

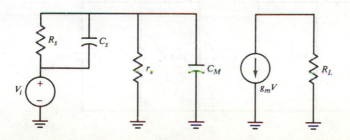

Figure 5.29 A broadband version of Fig. 5.28 obtained by adding C_s.

where C'_M is the equivalent Miller capacitance reflected across the output and C_o is the output capacitance (collector-to-emitter capacitance plus any external shunt capacitance).

Example 5.10 Consider the voltage amplifier shown in Fig. 5.30 which uses a 2N3904 transistor. The biasing is such that the transistor parameters are the same as in Example 5.5. Calculate the voltage gain and bandwidth Then determine the value of C_s required to cancel the effects of the Miller capacitance on the input side of the circuit and determine the resulting gain and bandwidth.

SOLUTION The results of Example 5.5 showed that the equivalent circuit for this particular amplifier can be simplified to that shown in Fig. 5.31. The midfrequency voltage gain is -77.2, and the calculated bandwidth of the

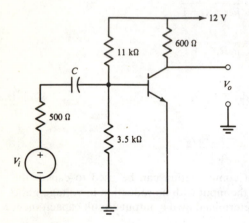

Figure 5.30 The common-emitter amplifier of Example 5.10.

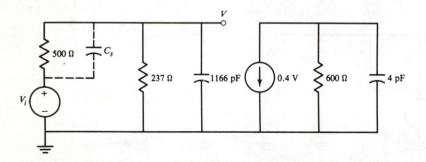

Figure 5.31 The equivalent circuit of Fig. 5.30 with the input compensation capacitor C_s added.

input circuit $\omega_1 = 5.3 \times 10^6$ rad/s. If C_s is selected so that

$$500\,C_s = 237 \times 1166 \text{ pF} \quad \text{or} \quad C_s = 553 \text{ pF}$$

then the bandwidth will be determined by the pole of the output circuit

$$B = (600 \times 4 \times 10^{-12})^{-1} = 4.17 \times 10^8 \text{ rad/s}$$

Input compensation has significantly widened the bandwidth with no reduction in midfrequency gain. A knowledge of r_b' is required for a more accurate prediction of the increase in bandwidth.

The preceding analysis has ignored the base-spreading analysis r_b' of bipolar transistors; r_b' cannot be shunted with a capacitor, since one terminal of the resistor is internal to the device and so limits the minimum value of input impedance. This base-spreading resistance limits the improvement that can be realized with input compensation of BJT amplifiers.

Feedback

Input compensation does increase the system bandwidth, but it requires that the transistor parameters be known. If this is not possible, or is undesirable, as in designs to be mass-produced, feedback is a good alternative. Negative feedback widens the bandwidth at the cost of a proportional reduction in loop gain. Feedback also has the advantage that it makes the system response much less sensitive to parameters over which the designer has little or no control. The block diagram representation of a negative-feedback amplifier is shown in Fig. 5.32. G is the transfer function of the system without the feedback factor H, and the transfer function with feedback is

$$\frac{V_o}{V_i} = \frac{G}{1 + GH} \tag{5.34}$$

The forward loop gain G and/or the feedback ratio H can be frequency-dependent. GH is known as the open-loop gain and $-GH$ is referred to as the *loop transmission*. If the magnitude of the open-loop gain $|GH| \gg 1$, then the closed-loop gain is

$$\frac{V_o}{V_i} \approx H^{-1} \tag{5.35}$$

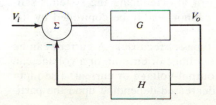

Figure 5.32 Block diagram representation of a negative-feedback amplifier.

The loop response can be designed to depend essentially on the feedback factor H, which is under the designer's control, and be independent of the forward gain G, which depends on the transistor parameters. The feedback formulation is usually difficult to apply directly to transistor amplifiers because of the circuit complexity, but it is an accurate approximation that if the amplifier bandwidth without feedback is B, then the bandwidth B_f of the system with negative feedback is

$$B_f \approx B|1 + GH| \approx B|GH| \qquad (5.36)$$

Consider the amplier with gain A_o and bandwidth ω_L which can be represented by the transfer function

$$G(j\omega) = \frac{A_o}{1 + j\omega/\omega_L}$$

The gain-bandwidth product of this amplifier is $GB = A_o\omega_L$. If a frequency-independent negative-feedback factor H is included, the closed-loop transfer function becomes

$$\frac{V_o}{V_i} = \frac{G(j\omega)}{1 + G(j\omega)H} = \frac{A_o}{1 + A_oH}\left[1 + \frac{j\omega}{\omega_L(1 + A_oH)}\right]^{-1}$$

The bandwidth has been increased from ω_L to $B_f = \omega_L(1 + A_oH)$, but the gain has been reduced. The gain-bandwidth product is still

$$G = \frac{A_o}{1 + A_oH}\,\omega_L(1 + A_oH) = A_o\omega_L$$

Example 5.11 An amplifier has an open-loop gain of 40 dB and a bandwidth of 10^7 rad/s. If 20 dB of feedback is added to the amplifier, what are the closed-loop gain and bandwidth?

SOLUTION The closed-loop gain has been reduced by 20 dB (a factor of 10), so the closed-loop gain is $40 - 20 = 20$ dB, and the closed-loop bandwidth has increased proportionally:

$$B_f = 10 \times 10^7 = 10^8 \text{ rad/s}$$

It is generally true that feedback will not increase the gain-bandwidth product. The increased bandwidth is obtained with a proportional reduction in gain. Feedback does have the advantage of also reducing the system's sensitivity to the amplifier components that cannot readily be controlled by the designer.

In transistor amplifiers there are four types of feedback. A current can be fed back that is proportional to the output voltage or current, or a voltage can be fed back which is proportional to the output voltage or current. The input and output impedances are increased or decreased depending upon the particular type of feedback employed. The following discussion concerns the two most frequently used methods of single-stage feedback.

Current-to-Voltage Feedback

The amplifier shown in Fig. 5.33 illustrates a frequently used method of single-stage negative feedback. Here a voltage proportional to the load current is fed back to the input via the emitter resistor R_E. It is readily shown that this feedback increases the amplifier input impedance and decreases its output impedance. The simplified small-signal, high-frequency equivalent circuit is shown in Fig. 5.34. Although this circuit can be analyzed directly, the resulting frequency-dependent transfer function is too complicated to be of much general utility. A good approximation to the circuit's frequency-dependent behavior can be obtained by calculating the gain and bandwidth without feedback and the midfrequency gain with feedback, then determining the feedback ratio H from this. The upper-frequency bandwidth can be estimated using Eq. (5.34), provided the upper corner frequency is known for the case in which there is no feedback.

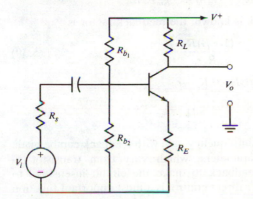

Figure 5.33 A common-emitter amplifier with current-to-voltage feedback.

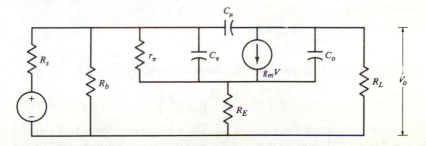

Figure 5.34 A simplified high-frequency equivalent circuit of the amplifier shown in Fig. 5.33.

The midfrequency gain without feedback is [Eq. (2.7)]

$$A_{\text{mid}} = \frac{V_o}{V'_i} = \frac{-g_m R_L r_\pi}{r_\pi + R}$$

where

$$R = R_s \| R_b$$

$$V'_i = \frac{V_i R_b}{R_b + R_s}$$

and

$$R_b = \frac{R_{b_1} R_{b_2}}{R_{b_1} + R_{b_2}}$$

It is assumed that $R_L \ll r_o$. The voltage gain with the feedback resistor R_E added is

$$\frac{V_o}{V'_i} = A_{\text{mid}} = \frac{-g_m R_L r_\pi / (r_\pi + R)}{1 + (1 + \beta) R_E / (R + r_\pi)} \tag{5.37}$$

The open-loop gain is

$$GH = \frac{(1 + \beta) R_E}{r_\pi + R} = \frac{g_m R_L r_\pi}{r_\pi + R} \frac{(1 + \beta) R_E}{g_m R_L r_\pi} \tag{5.38}$$

Since the gain G without feedback is known, the feedback factor is

$$H = \frac{-(1 + \beta) R_E}{g_m R_L r_\pi} \tag{5.39}$$

If

$$(1 + \beta) R_E \gg R_s + r_\pi$$

then

$$(A_{\text{mid}})_f \approx \frac{-g_m r_\pi R_L}{(1 + \beta) R_E} \approx H^{-1} \approx \frac{-R_L}{R_E}$$

This equation states that the midfrequency gain with feedback can be made independent of the transistor parameters, which vary from transistor to transistor. The use of negative feedback to make the circuit insensitive to parameters not under the designer's direct control is a most important function of negative feedback. Another important feature is that negative feedback increases the circuit bandwidth. For this circuit the upper -3-dB frequency without feedback is

$$\omega_L = \left\{ \frac{r_\pi R}{R + r_\pi} [C_\pi + C_\mu (1 + g_m R_L)] \right\}^{-1} \tag{5.40}$$

so the upper -3-dB frequency with feedback is estimated to be

$$\omega'_L = \omega_L \left[1 + \frac{(1 + \beta) R_E}{R + r_\pi} \right] \tag{5.41}$$

This is an approximation since GH is actually a frequency-dependent transfer function. The method and accuracy of the approximation are illustrated by the following example.

Example 5.12 Figure 5.35 illustrates the same amplifier as used in Example 5.5, only emitter feedback has been added. The small-signal equivalent circuit is as shown in Fig. 5.36. Here R_b is 2.655 kΩ and R is 421 kΩ.

The magnitude of the gain without feedback was found to be $A_v = 77.2$, and the corresponding bandwidth (obtained with computer simulation) is $B = 0.84 \times 10^6$ Hz. The gain-bandwidth product is

$$GB = 77.2(0.84 \times 10^6) = 65.15 \times 10^6 \text{ Hz}$$

The gain-bandwidth products for selected values of R_E were obtained by computer simulation and the results are plotted in Fig. 5.37. The transconductance g_m was kept constant and independent of R_E. As R_E approaches R_L, the gain V_o/V'_i approaches the asymptotic value of R_L/R_E. The gain-bandwidth product decreases as R_E increases. For $R_L = R_E$ the bandwidth is 36 MHz and the GB is 29.7×10^6 Hz. The bandwidth can be markedly improved with the addition of emitter feedback.

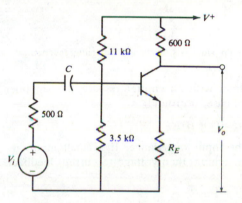

Figure 5.35 The common-emitter amplifier discussed in Example 5.12.

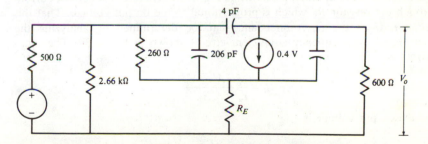

Figure 5.36 A small-signal equivalent circuit of the amplifier shown in Fig. 5.35.

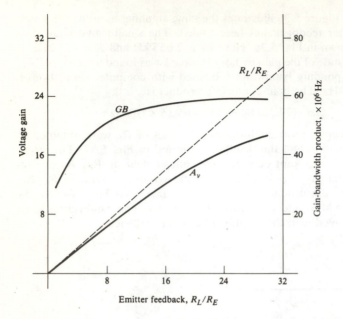

Figure 5.37 Voltage gain and gain-bandwidth product as a function of the emitter feedback.

For the common-emitter amplifier with an emitter resistor the amplifier midfrequency input resistance (from base to ground) is

$$R_i = r_\pi + (1 + \beta) R_E$$

Emitter feedback also increases the input impedance. If a small input impedance is needed for impedance matching, then voltage-to-current feedback can be used.

Voltage-to-Current Feedback

The common-emitter amplifier shown in Fig. 5.38 generates a feedback current through the resistor R_F which is proportional to the output voltage. That this circuit reduces the input impedance can be determined by analyzing the small-signal, midfrequency equivalent circuit shown in Fig. 5.39. The input current is

$$I_i = \frac{V}{r_\pi} + \frac{V - V_o}{R_F}$$

and the output voltage is

$$V_o = -g_m V R_L + \left(I_i - \frac{V}{r_\pi} \right) R_L$$

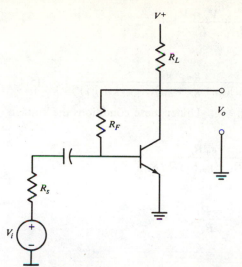

Figure 5.38 A common-emitter amplifier with voltage-to-current feedback.

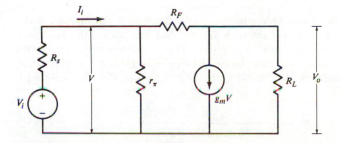

Figure 5.39 A small-signal equivalent circuit of the amplifier shown in Fig. 5.38.

If V_o is eliminated from these two equations, the input impedance is found to be

$$Z_i = \frac{V}{I_i} = r_\pi \| \frac{R_F + R_L}{1 + g_m R_L} \qquad (5.42)$$

This same expression holds true for FET amplifiers with $r_\pi \approx \infty$. The feedback resistor R_F reduces the input impedance; in addition, this form of feedback can reduce the circuit sensitivity to variations in the transistor parameters and can widen the amplifier bandwidth. The midfrequency voltage gain of the circuit shown in Fig. 5.39 can be determined from the two equations

$$\frac{V_i - V}{R_s} = \frac{V - V_o}{R_F} + \frac{V}{r_\pi}$$

and

$$g_m V + \frac{V_o}{R_L} + \frac{V_o - V}{R_F} = 0$$

The voltage gain is

$$A_v = \frac{V_o}{V_i} = R_s^{-1} \frac{g_m - g_F}{(g_L + g_F)(g_s + g_F + g_m)(-g_m g_F + g_F^2)} \quad (5.43)$$

Normally $g_m \gg g_F$, $g_m R_L \gg 1$, and $g_s \gg g_\pi$. Under these conditions the voltage gain expression simplifies to

$$A_v = \frac{-g_m R_L}{1 + g_F R_L + g_m R_L (R_s / R_F)}$$

which in turn simplifies to

$$A_v \approx \frac{-R_F}{R_s}$$

provided

$$g_m R_L \frac{R_s}{R_F} \gg 1$$

The gain can be made independent of the transistor parameters using feedback. The gain is a maximum with $R_F = \infty$ (no feedback); the feedback reduces the gain so the feedback is negative. The frequency response of this circuit can also be estimated by calculating the loop gain GH. If r_o is neglected, the gain without feedback is (assuming $R_s \ll r_\pi$) $A_v = -g_m R_L$. The gain with feedback is

$$A_v = \frac{-g_m R_L}{1 + g_F R_L + g_m R_L (R_s / R_F)}$$

So the open-loop gain

$$GH = g_F R_L + g_m R_L \frac{R_s}{R_F} \quad (5.44)$$

can be used for estimating the closed-loop bandwidth. The following example illustrates the accuracy of the approximations.

Example 5.13 The preceding example evaluated the effect of emitter feedback on the gain and bandwidth of the common-emitter amplifier. Here the effects of voltage-to-current feedback are considered by analyzing the same amplifier, except that now voltage-to-current feedback is used. The amplifier is shown in Fig. 5.40. The amplifier is again biased so that $g_m = 0.4$ S, and it is assumed to remain constant. The voltage gain and gain-bandwidth product as a function of R_F / R_s are plotted in Fig. 5.41. (In

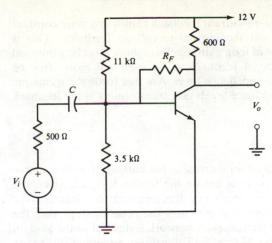

Figure 5.40 Feedback amplifier analyzed in Example 5.13.

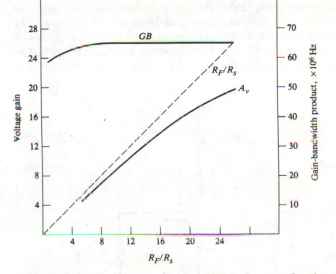

Figure 5.41 Voltage gain and gain-bandwidth product, as a function of shunt feedback of the amplifier shown in Fig. 5.40.

the ratio R_F/R_s, note the accuracy of the approximation even though r_π is not as large as R_s.) The gain-bandwidth product varies from a low of 59.5×10^6 Hz for a gain of unity to 65.1×10^6 Hz for a closed-loop gain of 17.9. The constancy of the gain-bandwidth product indicates that the effect of the negative feedback on bandwidth can be easily estimated using the approximations given in this section.

For this amplifier the voltage-to-current feedback results in a more constant gain-bandwidth product than did the current-to-voltage feedback. This is particularly true for small values of loop gain. Each amplifier must be evaluated on an individual basis; the type of feedback to be used will most often be determined by the desired input-impedance level. Another feedback technique for controlling the gain and impedance levels uses transformers as the feedback element.

Lossless Feedback Amplifiers[5.2]

The two methods already described for increasing the bandwidth with negative feedback used a resistor or resistors to create the feedback signal. Since any resistor adds additional noise to the circuit, resistive feedback also decreases the amplifier's noise performance (that is, it increases the noise figure). Also, the transistor output is delivered to the feedback network instead of to the load, in proportion to the feedback ratio. Figure 5.42 illustrates a method of lossless feedback that does not have these two limitations. The feedback is realized with a three-winding transformer connected to the common-base transistor amplifier so that it can simultaneously provide gain and impedance-matching. The transformer is wound with the polarities as shown, and the two turns ratios with respect to the primary are $m:1$ and $n:1$. The following analysis is simplified by making the close approximation that the common-base current gain is unity. If

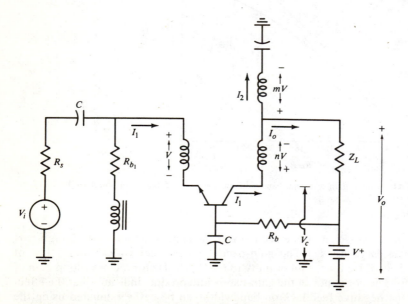

Figure 5.42 A common-base amplifier with lossless feedback.

the transformer is lossless, the input power is equal to the power out, or

$$I_1 V + I_1 n V + I_2 m V = 0 \quad \text{or} \quad I_2 = \frac{-(1+n)I_1}{m}$$

Therefore, the current to the load is

$$I_o = I_1 - I_2 = \frac{1+n+m}{m} I_1$$

The collector voltage is

$$V_c = (n+m) V = \frac{n+m}{m} V_o$$

Therefore

$$\frac{V_c}{I_1} = \frac{V_c}{I_o} \frac{1+n+m}{m}$$

$$= \frac{V_o}{I_o} \frac{(1+n+m)(n+m)}{m^2}$$

Since $V_o/I_o = Z_L$,

$$\frac{V_c}{I_1} = \frac{Z_L(1+n+m)(n+m)}{m^2}$$

If the transistor base-to-emitter voltage $V_{be} \ll V$, then the input impedance is

$$Z_i = \frac{V}{I_1} = \frac{1+n+m}{m^2} \cdot Z_L \tag{5.45}$$

The voltage gain is then

$$A_V = \frac{V_o}{V_i} = \frac{I_o Z_L}{I_1(Z_s + Z_i)} = \frac{1+n+m}{m} \frac{Z_L}{Z_s + Z_i} \tag{5.46}$$

Another advantage of this circuit is that the transformer turns ratios can be selected for impedance matching. For resistive loads

$$Z_i = \frac{1+m+n}{m^2} R_L$$

If

$$1+m+n = m^2 \tag{5.47}$$

then $Z_i = R_L$.

Also, the amplifier output impedance can be determined using the equivalent circuit shown in Fig. 5.43 (where it is again assumed that $V_{be} \ll V$). The output voltage $V_o = mV$, and the output current I_o equals $-I_1 + I_2$. If

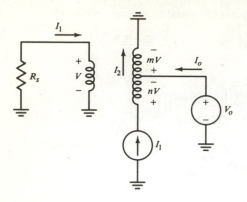

Figure 5.43 A small-signal equivalent circuit of the amplifier shown in Example 5.13.

the transformer is considered ideal, then

$$I_2 = \frac{-(1+n)I_1}{m}$$

and

$$I_o = \frac{-(m+1+n)}{m} I_1$$

so

$$Z_o = \frac{V_o}{I_o} = \frac{m^2}{m+1+n} \frac{V}{I_1} = \frac{m^2}{m+1+n} R_s$$

Therefore, if $R_s = R_L = R$ and $m^2 = m+1+n$, the transistor input and output impedances are also equal: $Z_i = Z_o = R$. The corresponding voltage gain is

$$A_V = \frac{m^2}{(m+n+1)2} = \frac{m}{2}$$

Since the common-base amplifier has a current gain close to unity, the power gain $A_p \approx A_V$.

For a common-base amplifier without feedback the voltage gain is approximately

$$A_V \approx \frac{R_L}{R_s}$$

If the load and source impedances are equal, the maximum voltage gain is unity. Since the voltage gain with feedback is $A_V = m/2$, the feedback can increase the voltage gain. This is positive feedback, which implies that the feedback will reduce the bandwidth by approximately the same factor that the gain is increased. This feedback technique employs the common-base configuration, which has a much wider bandwidth than the common-emitter configuration, then sacrifices some of the greater bandwidth for an increase in voltage gain. If

R_L is much larger than R_s, the feedback can reduce the voltage gain. The analysis always depends on the particular component values. The technique has the additional advantage of being able to match the amplifier input and output impedances. It is well suited as an element in a cascade of identical amplifiers.

Example 5.14 The same transistor with the same collector current ($g_m = 0.4$) as used in the two preceding examples is used to implement the common-base feedback amplifier shown in Fig. 5.44. Table 5.1 presents the results obtained from a computer simulation of the circuit.

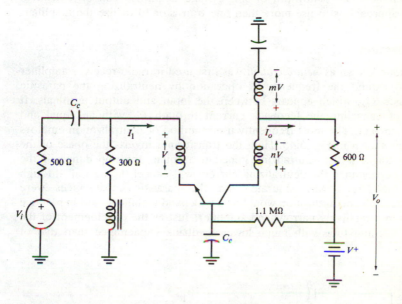

Figure 5.44 Lossless feedback amplifier discussed in Example 5.14.

Table 5.1 Gain-bandwidth product of a common-base amplifier with lossless feedback

Turns ratio	Voltage gain	Bandwidth, Hz	Gain-bandwidth product, Hz
0	0.976	79.4×10^6	77.5×10^6
4	1.984	11.22×10^6	22.2×10^6
6	2.975	4.67×10^6	13.9×10^6
10	4.95	1.6×10^6	7.92×10^6
20	9.615	4.22×10^5	4.06×10^6

The GB product decreases as the gain is increased. For voltage gains of 6 or more, the GB product is less than that obtained with the common-emitter amplifier using the same transistor and dc collector current. For this amplifier the input and output impedances are 500 Ω.

For the design of high-frequency amplifiers the most important step is to select a transistor with a sufficiently high gain-bandwidth product. The techniques of high-frequency design only become important when other transistor specifications (such as power level and noise figure) limit the selection to transistors with an insufficient gain-bandwidth product. The previous discussion has shown how the bandwidth of single-stage amplifiers can be extended. Another approach is to use more than one transistor to realize the amplifier.

Neutralization

A technique known as *neutralization* was first used in radio-receiver amplifier design to extend the frequency of operation by neutralizing the parasitic capacitance C_μ, which appears between the input and output terminals. It consists of canceling the feedback current through C_μ with an equal and opposite current. The most frequently used method of neutralization employs the circuit shown in Fig. 5.45. Here the transformer inverts the phase of the output voltage; if the neutralizing impedance equals the impedance of the parasitic capacitance the neutralizing current will cancel the current through C_μ. If the circuit was accurate, and the parasitic capacitances were known precisely, neutralization would be a wideband technique, but in practice it results in a relatively narrowband amplifier. Just as the development of the pentode vacuum tube with much lower feedback capacitance than that of

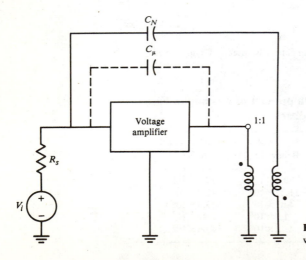

Figure 5.45 A voltage amplifier with neutralization.

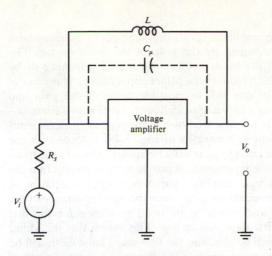

Figure 5.46 Another method of neutralization.

earlier triodes eliminated the need for neutralization in commercial radio receivers, the improvement in the high-frequency performance of transistors reduced the need for neutralization in transistor amplifiers. Another method of neutralization that can be used for narrowband amplifiers is given in Fig. 5.46, where the value of inductance is chosen so that it forms a parallel resonant circuit with the capacitance C_μ at the frequency of interest. Neutralization is particularly useful in tuned-input, tuned-output amplifiers where the feedback capacitance can cause oscillation (see Prob. 7.16). A frequently used neutralization circuit is shown in Fig. 5.47 where the phase inversion is obtained with a center-tapped output transformer.

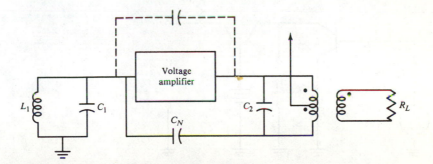

Figure 5.47 A neutralization circuit employing a center-tapped transformer.

Cascode Amplifiers

The analysis of the single-stage amplifiers shows that the upper frequency limit of the amplifier is usually determined by the pole of the input circuit. The bandwidth can be increased either by decreasing the source resistance or by decreasing the voltage gain (which reduces the Miller capacitance). If neither of these reductions is possible, it may still be possible to meet the gain and bandwidth specifications by using a two-stage amplifier consisting of a common-emitter (or common-drain) amplifier followed by a common-base (or common-gate) amplifier. This configuration, illustrated in Fig. 5.48, is known as the *cascode amplifier*. Although the circuit can readily be analyzed by a brute-force approach, we will use an intuitive approach as it provides more insight into the design process. The common-base amplifier has a low input resistance of approximately r_π/β Ω, and a unity current gain. Since the input resistance of the common-base stage, which is low, serves as the load resistance for the first stage, the voltage gain of the first stage will be low. This means that the Miller capacitance for the first stage will be low, and the first-stage bandwidth will be increased. The first-stage input impedance is r_π, so the voltage gain of the first stage will be

$$A_{v_1} \approx \frac{-r_{\pi_1}}{r_{\pi_1} + R_s} g_{m_1} \frac{r_{\pi_2}}{\beta}$$

Since the dc collector current is the same for both stages and the transistor current gains are assumed equal, $g_{m_1} = g_{m_2}$ and $r_{\pi_1} = r_{\pi_2}$; therefore

$$A_{v_1} = \frac{-r_\pi}{r_\pi + R_s}$$

The voltage gain of the first stage is less than unity. The ac base current of the

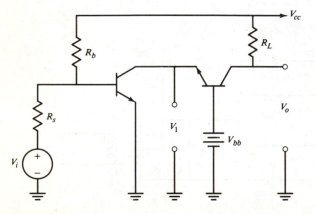

Figure 5.48 A BJT cascode amplifier.

first transistor is

$$I_b = \frac{V_i}{R_s + r_\pi}$$

Since the current gain of the first stage is β and that of the second stage is approximately 1, the output current is

$$I_o \approx -\beta I_b = \frac{-\beta V_i}{R_s + r_\pi}$$

and the output voltage is

$$V_o = I_o R_L = \frac{-\beta R_L V_i}{R_s + r_\pi}$$

So the voltage gain of the cascode circuit

$$A_v = \frac{V_o}{V_i} = \frac{-\beta R_L}{R_s + r_\pi} \tag{5.48}$$

is about the same as can be realized with a single-stage amplifier, but since the first stage no longer has a gain greater than unity, its bandwidth is increased even if a large source resistance is used. The second stage, which does have a voltage gain greater than 1, has an input impedance that is small (r_π/β). The bandwidth for the second stage can be determined from the small-signal model shown in Fig. 5.49. For this stage the midfrequency voltage gain is

$$A_{v_2} = \frac{V_o}{V} = g_m R_L$$

If the voltage gain is large, the small-signal equivalent circuit can be replaced by that of Fig. 5.50. Since the stage is noninverting, the equivalent Miller input capacitance is negative, reducing the input capacitance, and since the equivalent input impedance is r_π/β, the bandwidth of this stage will usually be determined by the capacitance across the output. The magnitude of the output pole is given by

$$\omega_p \approx [R_L(C_L + C_\mu)]^{-1} \tag{5.49}$$

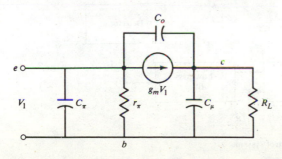

Figure 5.49 A small-signal model of the cascode amplifier output stage.

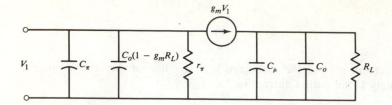

Figure 5.50 An equivalent small-signal model of the cascode amplifier output stage.

This two-stage cascode amplifier has a voltage gain equal to what can be obtained from a single-stage common-emitter amplifier, but it has an appreciably wider bandwidth. The cascode amplifier finds extensive application in high-frequency amplifiers and as the input stage of operational amplifiers.

5.4 AUTOMATIC GAIN CONTROL

Received signals are often subject to fading (slow variations in the amplitude of the received carrier). Fading required the first radio set operators to continually adjust the receiver gain in order to maintain a relatively constant output volume. This led to the design of the automatic volume-control circuit, which detects the amplitude of the received signal and automatically adjusts the gain in order to maintain a constant output signal level.[5.3] Automatic volume control, now generalized and known as *automatic gain control* (*AGC*), is one of the most useful circuits in modern communications receivers. In addition to maintaining constant output signal levels, it is used in a variety of applications, including signal encoding and decoding and oscillator amplitude stabilization.

The basic elements of an AGC system are illustrated in Fig. 5.51. The input

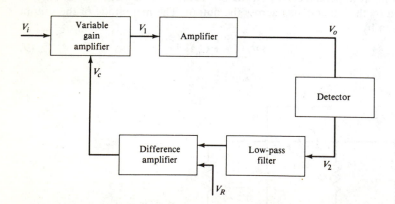

Figure 5.51 An automatic gain control system.

signal is amplified by a variable gain amplifier (VGA) whose gain depends on a control signal V_c. The amplified signal may be further amplified to produce the output signal V_o. Some parameter of the output signal, such as carrier amplitude, sideband power, or depth of modulation is detected and compared with a reference signal V_R. The difference between these two signals is then filtered and used to control the gain of the variable gain amplifier, the gain-controlling signal being either a voltage or current. If the input signal is amplitude-modulated, the AGC circuit must not respond to changes in amplitude modulation or the AGC loop will distort the modulated signal. This distortion is prevented by restricting the AGC circuit bandwidth so that it does not respond to the modulating frequencies; but AGC can still compensate for signal fading, which is relatively slow compared to the lowest modulating frequency.

The characteristics of an ideal AGC circuit are shown in Fig. 5.52. For low input signal levels the AGC is inoperative, and the output signal is linearly related to the input signal. When the input signal level is greater than V_1, the ideal AGC circuit maintains the output level constant until the input level reaches V_2. For larger signal levels the output is again a linear function of the input signal level. The elimination of the AGC action at very high gain levels is often done to prevent system instability problems.

The automatic gain control system described here is a negative-feedback system. The control signal represents an error signal between the amplitude of the output and the reference signal. As the input signal amplitude varies, the control signal varies in a manner to minimize the error signal. The requirements of an AGC loop differ depending upon the particular type of modulation of the input signal. The AGC requirements for AM receivers are usually more stringent than those for FM and pulse-modulation receivers.

The desired signal range at several points in the receiver is an essential design parameter. It is necessary to keep the signal levels sufficiently large so that receiver noise does not degrade the performance, but the signal levels must not exceed the linear range of the devices or distortion will occur. Signal-range requirements are the most critical in the input stage, where differences in signal strength are the greatest.

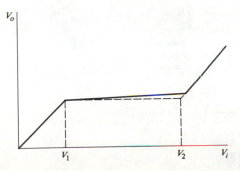

Figure 5.52 Idealized voltage-transfer characteristics of an AGC system.

Theory of Automatic Gain Control

An AGC system is an inherently nonlinear system, and there seldom are general solutions to the nonlinear equations that describe the system dynamics. There are, however, certain systems for which a closed-form solution can be found, and for most systems an approximate solution can be derived in terms of a small-signal model.

Figure 5.53 illustrates an AGC system that can be solved analytically. In this system the variable gain amplifier P obeys the control law

$$P = K_1 e^{+aV_c} \tag{5.50}$$

so
$$V_o = V_i K_1 e^{+aV_c}$$

where V_i and V_o represent the envelope amplitudes of the input and output, respectively, and the logarithmic amplifier gain is

$$V_2 = a \ln V_1 = \ln K_2 V_o$$

where K_2 is the gain of the envelope detector. The envelope detector output is always positive, so the output of the logarithmic amplifier is a real number (positive or negative). The control voltage is then

$$V_c = F(s)(V_r - V_2) = F(s)(V_r - \ln K_2 V_o)$$

where $F(s)$ is the filter transfer function.

Since the variable gain amplifier obeys an exponential law

$$\ln V_o = aV_c + \ln V_i K_1$$

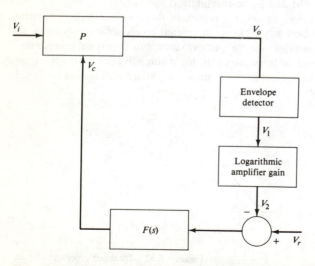

Figure 5.53 A linear (in decibels) AGC system.

the control voltage is

$$aV_c = \ln V_o - \ln V_i K_1$$

That is,

$$\ln V_o[1 + aF(s)] = \ln V_i + aF(s) V_r + \ln K_1 - aF(s) \ln K_2$$

The response to the input signal is

$$\ln V_o[1 + aF(s)] = \ln V_i + aF(s) V_r$$

Since $\ln V_o = 2.3 \log_{10} V_o$,

$$\ln V_o = \frac{2.3}{20} V_o = 0.115 V_o \qquad \text{dB}$$

Let e_o and e_i denote the output and input respectively, in decibels. Then

$$e_o = \frac{e_i}{1 + aF(s)} - \frac{8.7 \, aF(s) \, V_r}{1 + aF(s)} \qquad (5.51)$$

This particular AGC circuit is described by a linear differential equation, provided the input and output quantities are expressed in decibels. The system can then be represented by the linear negative-feedback system shown in Fig. 5.54. The loop dynamics are described by the filter $F(s)$ and the factor a of the variable gain amplifier. $F(s)$ will normally be a low-pass filter, since the loop bandwidth must be limited so that it does not respond to any amplitude modulation present on the input signal. Loop stability will depend on the order of the filter and the loop gain. The steady-state change in the output, in response to a change in the input, is

$$\frac{\Delta e_o}{\Delta e_i} = [1 + aF(0)]^{-1} \qquad (5.52)$$

where $F(0)$ is the dc gain of the filter. It is desired to keep the change Δe_o as small as possible in response to changes in the input amplitude. This is accomplished by making the dc loop gain as large as possible.

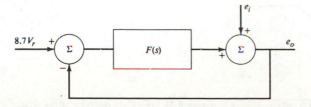

Figure 5.54 An equivalent block diagram representation of the AGC system shown in Fig. 5.53.

If $F(s)$ is a first-order filter described by

$$F(s) = \frac{K}{s/B + 1}$$

where K is the dc gain of the filter and B is the filter bandwidth, then the dc characteristic is

$$\Delta e_o = \frac{\Delta e_i}{1 + aK} \tag{5.53}$$

The total dc output of the system shown in Fig. 5.54 is

$$e_o = \frac{e_i}{1 + aK} + \frac{8.655\, V_r aK}{1 + aK} \tag{5.54}$$

Since the loop transmission aK is normally much greater than 1, the output e_o equals $8.655\, V_r$. The output amplitude in decibels is proportional to the reference voltage V_r. AGC loops containing a reference voltage are referred to as *delayed AGC*. This term does not imply that the gain control is delayed because of bandwidth limitations, but rather that the AGC loop contains a reference signal. Simple AGC loops, which do not contain a reference voltage, are common in inexpensive radio receivers.

The closed-loop transfer function for the loop with the first-order low-pass filter is

$$\Delta e_o = \frac{\Delta e_i}{1 + aK} \frac{s/B + 1}{s/B(1 + aK) + 1} \tag{5.55}$$

This system is inherently stable for all $aK > 0$, since the closed-loop pole will be in the left-half plane for all $aK > 0$. The magnitude of the frequency response of the closed-loop system is plotted in Fig. 5.55. AGC loops have a high-pass filter characteristic in response to changes in the amplitude of the input signal. That is, at high frequencies there is little AGC. For amplitude-modulated signals the corner frequency ω_L should be lower than the lowest modulating frequency ω_M:

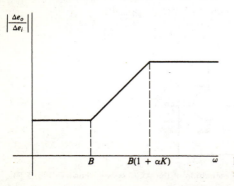

Figure 5.55 Frequency response of the AGC system shown in Fig. 5.53.

$$\omega_L = B(1 + aK) < \omega_M$$

This implies that the filter bandwidth should be much less than the lowest modulating frequency, since the negative feedback increases the closed-loop bandwidth.

It was stated that it is desirable to keep the dc loop gain as large as possible in order to maintain a constant output level. One method is to use an integrator as the filter $F(s)$. That is, $F(s) = C/s$. An ideal integrator has infinite gain at dc, so the steady-state output amplitude will not change in response to slowly varying changes in the input amplitude. The output for this particular filter is

$$e_o(s) = \frac{e_i(s)s}{s + aC} + \frac{8.6\,V_r a}{s + aC}$$

For constant inputs the steady-state output is again proportional to the reference voltage:

$$\lim_{t \to \infty} e_o(t) = \frac{8.6\,V_r}{C}$$

Example 5.15 Determine the time response of the AGC loop illustrated in Fig. 5.53 [with $F(s) = C/s$] to a step change in the input amplitude.

SOLUTION Since the system is described by linear differential equations, superposition can be used to calculate the change in the output. That is,

$$\Delta e_o(s) = \frac{s \Delta e_i(s)}{s + aC}$$

For a unit (decibel) step change in the input

$$\Delta e_i(s) = s^{-1}$$

and

$$\Delta e_o(s) = (s + aC)^{-1}$$

so,

$$\Delta e_o(t) = e^{-aCt}$$

The output (in decibels) exponentially decays toward zero in response to a step change in the input amplitude. The time constant of the decay is equal to the -3-dB frequency of the AGC loop ($\omega_n = aC$).

Another AGC Model

Many AGC loops do not contain the logarithmic amplifier which results in a linear model when used with an exponential-type variable-gain amplifier. For some AGC systems it is still possible to derive linear small-signal models. The small-signal limitation implies that the system analysis is valid for small changes from a particular operating point. Figure 5.56 illustrates a block diagram model for an AGC system in which the variable-gain amplifier and the

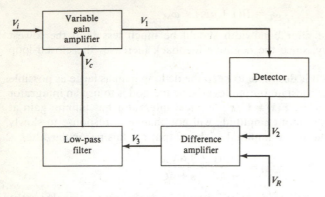

Figure 5.56 An AGC system with two nonlinear components.

detector are the only nonlinear components in the loop. To simplify the notation without loss of generality, the gains of the detector, the difference amplifier, and the amplifier following the variable-gain amplifier will be assumed to be unity. Then the system can be represented by the simplified block diagram shown in Fig. 5.57 where V_o and V_i now refer to envelope values and F is the frequency-dependent transfer function of the low-pass filter and amplifier combined. The output voltage V_o equals PV_i, where the gain P of the variable-gain amplifier is a function of the control voltage V_c. The control voltage is

$$V_c = (V_r - V_o)F \tag{5.56}$$

The differential change in output voltage is

$$\frac{dV_o}{dV_i} = \frac{d}{dV_i}(PV_i) = P + V_i\frac{dP}{dV_i} \tag{5.57}$$

Since
$$\frac{dP}{dV_i} = \frac{dP}{dV_c}\frac{dV_c}{dV_i} = \frac{dP}{dV_c}\frac{dV_c}{dV_o}\frac{dV_o}{dV_i} = \frac{dP}{dV_c}(-F)\frac{dV_o}{dV_i}$$

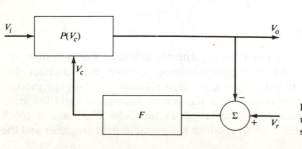

Figure 5.57 A simplified model of the AGC system shown in Fig. 5.56.

Equation (5.57) can be written as

$$\frac{dV_o}{dV_i}\left(1 + FV_i\frac{dP}{dV_c}\right) = P$$

or

$$\frac{dV_o/V_o}{dV_i/V_i} = \left(1 + FV_i\frac{dP}{dV_c}\right)^{-1} \tag{5.58}$$

This is the small-signal differential equation for the AGC loop illustrated in Fig. 5.56. It is valid for incremental changes about a particular control voltage. The loop transmission

$$L = -F(s)\,V_i\,\frac{dP}{dV_c} \tag{5.59}$$

is a function of the input signal, so the system is, in general, nonlinear. The transient behavior of the system shown in Fig. 5.55 in response to changes in the input amplitude is, in general, difficult to obtain because of the nonlinear nature of the system. Since the loop transmission depends on the input amplitude, so does the pole of the closed-loop system, and hence, the speed of the transient response. This feature of system behavior is illustrated by the following example.

Example 5.16 The AGC loop illustrated in Fig. 5.56 has a variable-gain amplifier with the linear characteristic $P(V_c) = V_c$ and includes an integrator as the low-pass filter, $F(s) = K/s$. Determine the transient response to small step changes in the input amplitude.

SOLUTION For this system the loop transmission is [Eq. (5.59)]

$$L = -\frac{K}{s}\,V_i$$

and for a small step change in the input signal

$$\frac{\Delta V_i}{V_i}(s) = \frac{\Delta_i}{s}$$

So the normalized change in output voltage is

$$\frac{\Delta V_o}{V_o}(s) = \frac{\Delta_i}{s + KV_i}$$

and

$$\frac{\Delta V_o}{V_o}(t) = \Delta_i e^{-KV_i t}$$

This last example illustrates how the loop dynamics can depend on the amplitude of the input signal. For AGC systems in which control of the transient response is critical, more complex loops are usually required.

If the variable-gain characteristic $P(V_c)$ is known, it is possible to numerically evaluate the dc characteristics of the loop by choosing a value of control voltage as the starting point.

Example 5.17 The AGC loop illustrated in Fig. 5.57 employs a variable-gain amplifier with a square-law characteristic

$$P(V_c) = V_c^2$$

Determine the dc voltage as a function of input voltage for a reference voltage of 1 V.

SOLUTION The simplest solution is to calculate the input and output voltages for a specified control voltage. For example, if $V_c = 0.5$ V, $V_o = V_r - V_c = 0.5$ V, and

$$V_i = \frac{V_o}{P(V_c)} = \frac{0.5}{0.5^2} = 2 \text{ V}$$

A plot of the static input-output characteristic is given in Fig. 5.58. The output voltage varies 20 dB as the input voltage increases from 0.1 V to an arbitrarily large voltage. Over this same range of input voltage the control voltage decreases from 1 to 0 V.

The AGC systems discussed here all provide a continuous monitoring of the output amplitude and a continuous adjustment of the variable-gain amplifier. There are many systems in which the output load is monitored intermittently and the gain adjusted during these intervals. For the remainder of the time the control loop is an open circuit, and the gain is held constant during this interval. Television receivers are an example of a gated AGC system. If the gated signal

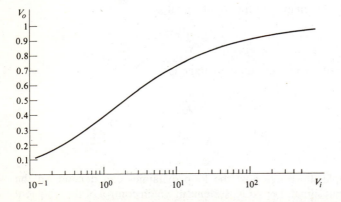

Figure 5.58 Voltage-transfer characteristic of the AGC system of Example 5.17.

that is used for the AGC action does not contain any modulation such as the TV synchronization pulses, then the AGC system bandwidth can be made quite wide to give very fast response, without suppressing the modulation between pulses. Pulse-type AGC systems have been analyzed using sampled-data techniques.[5.4] This approach has utility when a linear model can be obtained for the system.

AGC System Components

The designer of AGC systems has several variable-gain-amplifier (VGA) control laws from which to choose. The selection criteria include frequency response, available range of control voltage, and desired range of the variable-gain amplifier. A VGA whose gain is an exponential function of the control voltage can have a wider variation in gain than can, for example, a variable-gain amplifier with a linear control function. Multipliers have a linear control law by definition. Dual-gate MOSFETs and pin-diode attenuators are two of the many circuits that exhibit an exponential control law. Bipolar differential amplifiers, commonly used in integrated circuits, have a voltage gain that is proportional to the collector bias current, so the gain can be varied by adjusting the dc collector current. Figure 5.59 illustrates a simplified differential amplifier with the transistor Q_3 serving as a constant current source. The collector current of I_3 is

$$I_C = I_s e^{V_R / V_T}$$

so the gain of the amplifier (proportional to I_C) is an exponential function of the control voltage V_R. The circuit (with the bias network) shown in Fig. 5.60 can serve as a logarithmic amplifier for AGC loops that use one. Since the negative

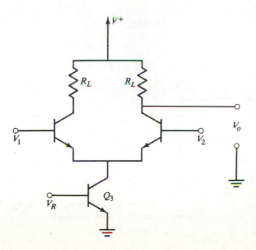

Figure 5.59 An exponential amplifier for AGC applications.

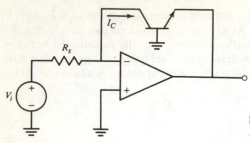

Figure 5.60 A logarithmic amplifier for AGC loops.

terminal of the operational amplifier is at ground potential,

$$I_1 = \frac{V_i}{R_1} = I_C = I_s e^{V_o/V_T}$$

and the output voltage is

$$V_o = V_T \ln \frac{V_i}{R_1 I_s}$$

which is a logarithmic function of the input voltage.

PROBLEMS

5.1 Add a capacitor to the amplifier shown in Fig. P5.1 which will make the upper 3-dB frequency 20 kHz. Indicate where the capacitor is to be connected. What is the smallest value of capacitance that can be used?

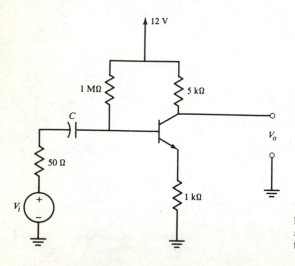

Figure P5.1 A common-emitter amplifier with voltage-to-current feedback.

5.2 Add a capacitor to the amplifier shown in Fig. P5.2 which will make the upper 3-dB frequency 20 kHz. Indicate where the capacitor is to be connected. What is the smallest value of capacitance that can be used?

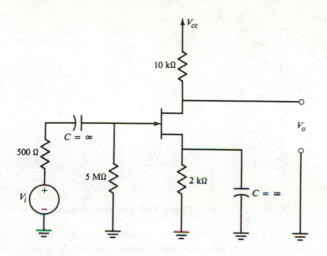

Figure P5.2 A common-source amplifier.

5.3 The transistor used in the amplifier shown in Fig. P5.3 has a β of 100, a C_π of 4 pF, and $C_\mu = 5$ pF. What is the current gain-bandwidth product of the transistor? What is the upper 3-dB frequency of the amplifier?

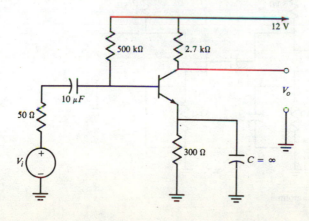

Figure P5.3 A common-emitter amplifier.

5.4 What is the upper 3-dB frequency of the amplifier shown in Fig. P5.4?

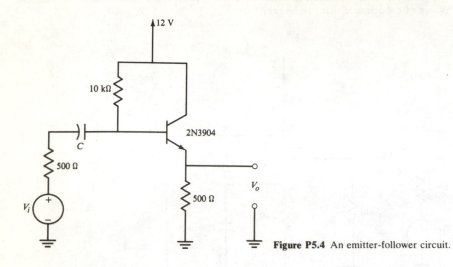

Figure P5.4 An emitter-follower circuit.

5.5 Design an amplifier (using discrete transistors) with a minimum voltage gain of 40 dB. The source impedance is 500 Ω, and the load impedance is 50 Ω. The upper 3-dB frequency is to be 30 kHz.

5.6 What are the midfrequency voltage gain and upper 3-dB corner frequency of the two-stage amplifier shown in Fig. P5.6? The transistors have identical characteristics: $g_m = 10^{-3}$ S, $r_d = 10^5\ \Omega$, $C_{ds} = 3$ pF, $C_{gd} = 1$ pF, and $C_{gs} = 3$ pF.

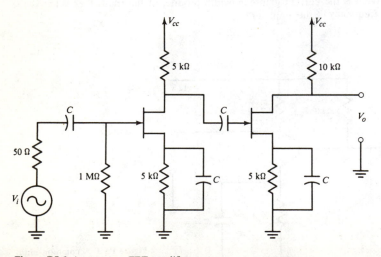

Figure P5.6 A two-stage FET amplifier.

5.7 Calculate the gain and frequency response of the common-gate amplifier illustrated in Fig. P5.7. See App. 2 for transistor specifications. The transistor is biased so that the transconductance $g_m = 12 \times 10^{-3}$ S.

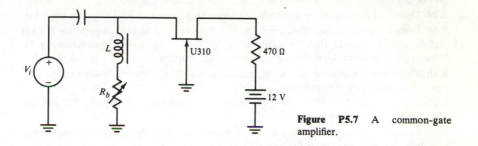

Figure P5.7 A common-gate amplifier.

5.8 Calculate the gain and high-frequency performance of the amplifier shown in Fig. P5.8. The U310 transistor is used. See App. 2 for transistor specifications. The transistor is biased for $g_m = 12 \times 10^{-3}$ S.

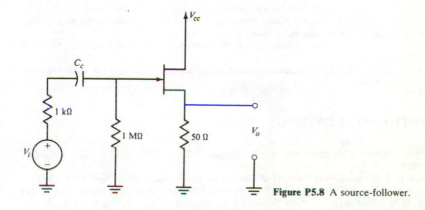

Figure P5.8 A source-follower.

5.9 It is desired to build an amplifier with a gain of 40 dB using amplifiers with a gain-bandwidth product of 10^7 Hz. Will a wider bandwidth be obtained using a single-stage amplifier or cascading two amplifiers, each with a gain of 20 dB?

5.10 Derive an approximate expression for the high-frequency performance of a Darlington amplifier.

5.11 Design a lossless feedback amplifier using a 2N3904 transistor that has a voltage gain of 6 and an input impedance of 50 Ω. The source resistance is 50 Ω and the load impedance is 300 Ω.

5.12 What will be the gain and bandwidth of the amplifier shown in Fig. P5.3 if the bypass capacitor is removed from the emitter resistor? (*Hint*: Estimate the bandwidth assuming the gain-bandwidth product remains constant.)

5.13 What will be the gain of the amplifier shown in Fig. P5.2 if the bypass capacitor is removed from the source resistance? Estimate the bandwidth assuming the gain-bandwidth product remains constant.

5.14 Derive an expression for the high-frequency response of an FET cascode amplifier.

5.15 Design a cascode amplifier using 2N3904 transistors. The load impedance is 1 kΩ and the source is 500 Ω. Bias the amplifier so that the voltage gain is approximately 25. What is the predicted upper-frequency limit of the amplifier?

5.16 Show that the control law K/V_c results in the AGC system shown in Fig. 5.56 having constant loop transmission.

5.17 Repeat Example 5.17 for the case in which the reference voltage in 2 V and the control voltage can vary from 0 to 2 V.

5.18 Plot the static input-output transfer characteristic of an AGC system in which the variable-gain amplifier obeys the control law

$$P(V_c) = e^{V_c} - 1$$

REFERENCES

5.1 Searle, C. L., et al.: *Elementary Circuit Properties of Transistors*, Wiley, New York, 1964, p. 103.

5.2 Norton, D. E.: High Dynamic Range Transistor Amplifiers using Lossless Feedback, *Proc. IEEE Int. Symp. on Circuits and Systems*, 438–440 (1975).

5.3 Wheeler, H. A.: Automatic Volume Control for Radio Receiving Sets, *Proc. IRE*, **16**:30–39 (1928).

5.4 Mercy, D. V.: A Review of Automatic Gain Control Theory, *Radio and Electronic Engineer*, 579–590 (1981).

ADDITIONAL READINGS

Gilbert, B.: A New Wide-Band Amplifier Technique, *IEEE J. Solid-State Circuits*, **3**:353–365 (1968).

Maclean, D. J. H.: *Broadband Feedback Amplifiers*, Research Studies Press, New York, 1982.

Oliver, B. M.: Automatic Volume Control as a Feedback Problem, *Proc. IRE*, **36**:466–473 (1948).

Ricker, D. W.: Non-linear Feedback System for the Normalization of Active Sonar Returns, *J. Acoust. Soc. Am.*, **59**:389–396 (1976).

CHAPTER SIX

HYBRID AND TRANSMISSION-LINE TRANSFORMERS

6.1 INTRODUCTION

In addition to the transformers described in Chap. 4, there are two other kinds of transformers that are frequently used in communication circuits. The three-winding and transmission-line transformers can both be configured for adding and/or subtracting multiple inputs. Another particularly useful transformer, called the *hybrid transformer*, consists of a four-port circuit that provides isolation between selected signals (ports) and at the same time provides for maximum transfer between the other two ports. In the following sections of this chapter the three-winding and transmission-line transformers will be analyzed, and we will demonstrate how they can both be used to create the hybrid transformer. The idealized behavior of these transformers, most useful for design and analysis, will be presented, along with the description of more detailed models that are best analyzed by computer methods.

6.2 THREE-WINDING TRANSFORMERS

It is often required that voltages in certain sections of an electric circuit induce voltages in selected branches and at the same time isolate other branches. The three-winding transformer is the most widely used means to accomplish this end. To illustrate the technique, the ideal three-winding transformer will first be evaluated.

Consider the ideal transformer of Fig. 6.1 with an $N:1$ turns ratio between each half of the primary and the secondary winding. Since the transformer is

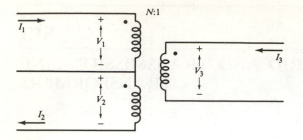

Figure 6.1 A center-tapped transformer.

ideal,

$$V_1 = V_2 = NV_3 \tag{6.1}$$

Also, no power is dissipated in an ideal transformer, so the power output is equal to the power supplied:

$$I_1 V_1 + I_2 V_2 = -I_3 V_3 \tag{6.2}$$

or
$$(I_1 + I_2)NV_3 = -I_3 V_3$$

Therefore, the current relation in this ideal transformer is

$$-I_3 = N(I_1 + I_2) \tag{6.3}$$

Consider now the ideal transformer used to couple the four voltage sources and source resistances as shown in Fig. 6.2. The three current-loop equations are

$$E_1 - E_4 = I_1(Z_1 + Z_4) - I_2 Z_4 + NV_3 \tag{6.4}$$

$$-E_2 + E_4 = -I_1 Z_4 + I_2(Z_2 + Z_4) + NV_3 \tag{6.5}$$

$$E_3 = I_3 Z_3 + V_3 \tag{6.6}$$

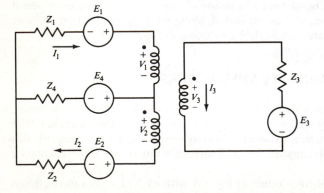

Figure 6.2 A circuit with four voltage sources and a center-tapped transformer.

If Eq. (6.6) is used to replace V_3, Eqs. (6.3) through (6.5) can be written (in matrix notation) as

$$\begin{bmatrix} E_1 - E_4 - NE_3 \\ -E_2 + E_4 - NE_3 \\ 0 \end{bmatrix} = A \begin{bmatrix} I_1 \\ I_2 \\ I_3 \end{bmatrix} \tag{6.7}$$

where the matrix

$$A = \begin{bmatrix} Z_1 + Z_4 & -Z_4 & -NZ_3 \\ -Z_4 & Z_2 + Z_4 & -NZ_3 \\ N & N & 1 \end{bmatrix} \tag{6.8}$$

It is possible to select the impedances so that there is isolation between ports. For example, to isolate port 4 from E_3, the current in response to E_3, through $Z_4(I_1 - I_2)$, should be zero. The current I_1 in response to E_3 is

$$I_1 = \frac{-NZ_4 E_3 - N(Z_2 + Z_4)E_3}{\det A} \tag{6.9}$$

where det A is the determinant of the matrix A. Also, the current I_2 in response to E_3 is

$$I_2 = -\frac{[N(2Z_4 + Z_1)]E_3}{\det A} \tag{6.10}$$

so the current through Z_4 in response to E_3 is

$$I_1 - I_2 = -\frac{N(Z_1 - Z_2)E_3}{\det A} \tag{6.11}$$

Port 4 will be isolated from port 3 if $Z_1 = Z_2$. That is, if $Z_1 = Z_2$, an applied voltage E_3 will not develop a voltage across Z_4. This property of the center-tapped transformer is one of the reasons this type of transformer is so often used. An example of its application can be found in the impedance-bridge circuit illustrated in Fig. 6.3. An unknown impedance can be inserted as Z_2, and then Z_1 is adjusted until the voltage across Z_4 is zero. This occurs only when Z_1 is equal to Z_2. Very fine resolution can be obtained using a null detector to

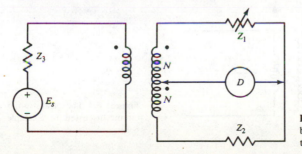

Figure 6.3 An impedance-bridge circuit that uses a center-tapped transformer.

measure and amplify the voltage across Z_4, or a current meter can be used in place of a voltmeter. If the voltage across Z_4 is zero, the current through Z_4 is also zero.

If, in addition, ports 1 and 2 are to be isolated from each other, I_2 in response to E_1 must be zero.

That is,

$$I_2 = \frac{E_1(N^2 Z_3 - Z_4)}{\det \mathbf{A}} \tag{6.12}$$

I_2 will be zero for all E_1 if

$$\frac{Z_4}{Z_3} = N^2 \tag{6.13}$$

Ports that are isolated from each other are said to be *conjugate*. We have seen that ports 3 and 4 will be conjugate if $Z_1 = Z_2$ and that ports 1 and 2 are conjugate if $Z_4/Z_3 = N^2$. If both sets of ports are conjugate the transformer is said to be *biconjugate*.

Example 6.1 If identical loads are placed on the transformer primary so that $I_1 = I_2$ and $V_1 = V_2$, what is the impedance R' seen looking into port 3 of the transformer circuit shown in Fig. 6.4?

SOLUTION Since $R_1 = R_2 = R$, ports 3 and 4 are conjugate; therefore, the current $(I_1 - I_2)$ is zero and terminals a and b can be either short- or open-circuited. The secondary resistance is

$$R' = \frac{V_3}{I_3} = \frac{V_1/N}{2NI_1} = \frac{R}{2N^2}$$

If only half of the primary is used ($R_1 = \infty$) so that $I_1 = 0$, then

$$R' = \frac{V_3}{I_3} = \frac{V_2/N}{NI_2} = \frac{R}{N^2}$$

provided that terminals a and b are short-circuited together.

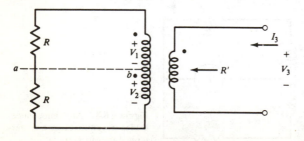

Figure 6.4 The transformer circuit discussed in Example 6.1.

A second property often required of center-tapped transformer circuits is that all four source impedances (Z_1, Z_2, Z_3, and Z_4) be matched for maximum power transfer. For the center-tapped transformer of Fig. 6.2, with the port impedances matched for isolation, Eq. (6.8) becomes

$$
\begin{bmatrix} E_1 - E_4 - NE_3 \\ -E_2 + E_4 - NE_3 \\ 0 \end{bmatrix} = \begin{bmatrix} Z_1 + Z_4 & -Z_4 & -Z_4/N \\ -Z_4 & Z_1 + Z_4 & -Z_4/N \\ N & N & 1 \end{bmatrix} \begin{bmatrix} I_1 \\ I_2 \\ I_3 \end{bmatrix} \tag{6.14}
$$

The input impedance at port 1 is defined to be

$$
Z_{i_1} = \frac{E_1}{I_1} - Z_1 \tag{6.15}
$$

since Z_1 is taken as the source impedance of E_1. If Eq. (6.14) is solved for I_1 in response to E_1, it is found that

$$
\frac{I_1}{E_1} = \frac{Z_1 + 2Z_4}{\det \mathbf{A}} \tag{6.16}
$$

where

$$
\det \mathbf{A} = (Z_1 + Z_4)^2 + 2Z_4(Z_1 + Z_4) + Z_4^2 = (Z_1 + 2Z_4)^2 \tag{6.17}
$$

Therefore,

$$
\frac{E_1}{I_1} = \frac{(Z_1 + 2Z_4)^2}{Z_1 + 2Z_4} = Z_1 + 2Z_4 \tag{6.18}
$$

and

$$
Z_{i_1} = 2Z_4 \tag{6.19}
$$

Thus for maximum power transfer from port 1,

$$
Z_4 = \frac{Z_1^*}{2} \tag{6.20}
$$

where Z_1^* is the complex conjugate of Z_1.

The input impedance of port 3 is likewise defined as

$$
Z_{i_3} = \frac{V_3}{I_3} = \frac{E_3}{I_3} - Z_3 \tag{6.21}
$$

and can be determined by solving Eq. (6.14) for I_3 in response to E_3. That is,

$$
I_3 = \frac{\begin{vmatrix} Z_1 + Z_4 & -Z_4 & -NE_3 \\ -Z_4 & Z_1 + Z_4 & -NE_3 \\ N & N & 0 \end{vmatrix}}{\det \mathbf{A}} \tag{6.22}
$$

or

$$
\frac{I_3}{E_3} = \frac{N^2 Z_4 + N^2 Z_4 + 2N^2(Z_1 + Z_4)}{(Z_1 + 2Z_4)^2} = \frac{2N^2(Z_1 + 2Z_4)}{(Z_1 + 2Z_4)^2} \tag{6.23}
$$

Therefore

$$Z_{i_3} = \frac{E_3}{I_3} - Z_3 = \frac{Z_1 + 2Z_4}{2N^2} - Z_3 \tag{6.24}$$

Since ports 1 and 2 are isolated, $Z_4 = N^2 Z_3$ and

$$Z_{i_3} = \frac{Z_1}{2N^2} \tag{6.25}$$

Thus for maximum power transfer from port 3,

$$Z_3 = \frac{Z_1^*}{2N^2} \tag{6.26}$$

which is the same requirement obtained for port 1.

It is easily shown that these relations will also result in maximum power transfer from ports 4 and 2. That is, if the transformer is biconjugate and one port is matched for maximum power transfer, they are all matched for maximum power transfer. To summarize these findings, if in the ideal transformer circuit of Fig. 6.2

$$Z_1 = Z_2 = Z$$

and

$$Z_3 = \frac{Z_4}{N^2} = \frac{Z^*}{2N^2}$$

the transformer will be biconjugate with all ports matched for maximum power transfer. A biconjugate transformer with all four input impedances matched for maximum power transfer is referred to as a *hybrid transformer*.

6.2 ASYMMETRICAL THREE-WINDING TRANSFORMERS

The ideal transformer we have been considering has equal turns ratios on both sides of the center tap. If the turns ratios are not equal, the impedances can still be matched for port isolation and maximum power transfer between ports. Consider the ideal transformer shown in Fig. 6.5. (Purely resistive impedances will be considered here without loss of generality.) The turns ratio from the secondary to the top section of the primary is kN and from the secondary to the bottom section of the primary is N. Since the transformer is ideal,

$$V_1 = kNV_3 \tag{6.27}$$

$$V_2 = NV_3 \tag{6.28}$$

and the power to the transformer is equal to the power output, so

$$V_1 I_1 + V_2 I_2 = -V_3 I_3 \tag{6.29}$$

or

$$kNI_1 + NI_2 = -I_3 \tag{6.30}$$

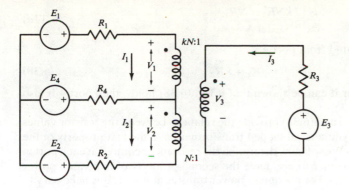

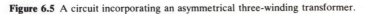

Figure 6.5 A circuit incorporating an asymmetrical three-winding transformer.

The loop equations are

$$-E_4 + E_1 - kNV_3 = I_1(R_1 + R_4) - I_2R_4 \tag{6.31}$$

$$E_4 - E_2 - NV_3 = -I_1R_4 + I_2(R_2 + R_4) \tag{6.32}$$

$$E_3 = I_3R_3 + V_3 \tag{6.33}$$

If Eq. (6.33) is used to eliminate V_3, Eqs. (6.31) through (6.33) become

$$
\begin{bmatrix} E_1 - E_4 - kNE_3 \\ -E_2 + E_4 - NE_3 \\ 0 \end{bmatrix}
=
\begin{bmatrix} R_1 + R_4 & -R_4 & -kNR_3 \\ -R_4 & R_2 + R_4 & -NR_3 \\ kN & N & 1 \end{bmatrix}
\begin{bmatrix} I_1 \\ I_2 \\ I_3 \end{bmatrix} \tag{6.34}
$$

or

$$E = AI \tag{6.34a}$$

In order for port 2 to be isolated from port 1, I_2 must be zero in response to E_1. Since

$$\frac{I_2}{E_1} = \frac{-kN^2R_3 + R_4}{\det \mathbf{A}} \tag{6.35}$$

port 2 will be isolated from port 1 (and port 1 from port 2), provided

$$R_4 = N^2R_3k \tag{6.36}$$

In order for port 3 to be isolated from port 4, I_3 must be zero in response to E_4. Since

$$
I_3 = \frac{\begin{vmatrix} R_1 + R_4 & -R_4 & -E_4 \\ -R_4 & R_2 + R_4 & E_4 \\ kN & N & 0 \end{vmatrix}}{\det \mathbf{A}} \tag{6.37}
$$

or
$$\frac{I_2}{E_4} = \frac{kNR_2 - NR_1}{\det \mathbf{A}}$$
(6.37a)

port 3 will be isolated from port 4 if

$$R_1 = kR_2$$
(6.38)

In the same manner it can be shown that if Eq. (6.38) holds, then port 4 is also isolated from port 3.

Equations (6.36) and (6.38) provide the relations between the resistor values in order that the asymmetrical tapped transformer have the first property of the hybrid transformer. We will now determine the resistance required at each port in order for the transformer to also have the second property of maximum power transfer between ports. For maximum power transfer at port 1, it is necessary to determine the port's input resistance

$$R_{i_1} = \frac{E_1}{I_1} - R_1$$
(6.39)

The response I_1 to the voltage E_1 can be found by solving Eq. (6.34) [with Eqs. (6.36) and (6.38) substituted for R_3 and R_2 respectively]

$$I_1 = \frac{\begin{vmatrix} E_1 & -R_4 & -R_4/N \\ 0 & R_1/k + R_4 & -R_4/Nk \\ 0 & N & 1 \end{vmatrix}}{\det \mathbf{A}}$$
(6.40)

Therefore,

$$\frac{E_1}{I_1} = \frac{\det \mathbf{A}}{R_1/k + R_4 + R_4/k}$$
(6.40a)

The value of the determinant $\mathbf{A}$ in the case in which the ports are isolated (biconjugate) is given by

$$\det \mathbf{A} = R_1 \left(\frac{R_1}{k} + R_4 + \frac{R_4}{k} + R_4 \right) + \frac{R_4 R_1}{k} + R_4^2$$

$$= \frac{[R_1 + R_4(1 + k)]^2}{k}$$
(6.41)

so
$$\frac{E_1}{I_1} = R_1 + R_4(1 + k)$$

Therefore, for maximum power transfer from port 1,

$$R_1 = R_4(1 + k)$$
(6.42)

Likewise, the resistance at port 3 can be calculated using Eqs. (6.34), (6.36), (6.38), and (6.41):

$$I_3 = \frac{\begin{vmatrix} R_1 + R_4 & -R_4 & -kNE_3 \\ -R_4 & R_1/k + R_4 & -NE_3 \\ kN & N & 0 \end{vmatrix}}{\det \mathbf{A}} \tag{6.43}$$

Thus
$$\frac{E_3}{I_2} = \frac{[R_1 + R_4(1+k)]^2}{k}\{N^2(1+k)[R_1 + R_4(1+k)]\}^{-1} \tag{6.44}$$

$$= \frac{R_1 + R_4(1+k)}{N^2 k(1+k)} \tag{6.44a}$$

Since ports 3 and 4 are conjugate [and using Eq. (6.42)],

$$R_{i_3} = \frac{E_3}{I_3} - R_3 = \frac{E_3}{I_3} \frac{R_4}{kN^2} = \frac{R_1}{N^2 k(1+k)} \tag{6.45}$$

That is,
$$R_3 = \frac{R_1}{N^2 k(1+k)} \tag{6.46}$$

for maximum power transfer from port 3. This is also the condition for maximum power transfer from port 1.

It should be observed that once the port isolation properties have been established, the equations can be solved in a much easier manner. For example, if it is desired to calculate the response to an applied voltage E_3, the network can be drawn as shown in Fig. 6.6 (since $I_1 - I_2 = 0$). The secondary voltage and current can be found from the two equations

$$-I_3 = kI_1 N + I_2 N = N(k+1)I_1 \tag{6.47}$$

Since
$$I_1 = \frac{-NkV_3}{Rk} \tag{6.48}$$

we have
$$V_3 = \frac{-I_1 R}{N} = \frac{R}{N} \frac{I_3}{N(k+1)} \tag{6.49}$$

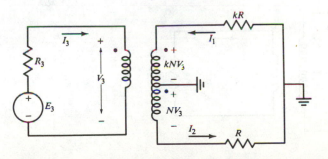

Figure 6.6 An asymmetrical three-winding transformer circuit in which ports 3 and 4 are conjugate.

Table 6.1 Resistance relationships in an asymmetrical hybrid transformer

$$R_1 = R$$

$$R_2 = \frac{R}{k}$$

$$R_3 = \frac{R}{N^2(1+k)}$$

$$R_4 = \frac{R}{1+k}$$

Thus

$$Z_{i_3} = \frac{E_3}{I_3} - R_3 = \frac{V_3}{I_3} = \frac{R}{N^2(k+1)} = \frac{R_4}{N^2} \qquad (6.50)$$

which agrees with the result previously obtained for the input impedance.

Table 6.1 summarizes the resistance relations required for the asymmetrical transformer to have isolated port pairs and maximum power transfer between ports.

Example 6.2 A transformer circuit is illustrated in Fig. 6.7. Determine the turns ratio N and R_4 so that the circuit is a hybrid transformer, and determine the output voltage E_o.

SOLUTION For the circuit to be a hybrid transformer, Table 6.1 shows that

$$N^2 = \frac{R_1}{2R_3} = \frac{1}{4}$$

and

$$R_4 = \frac{R}{2} = 25$$

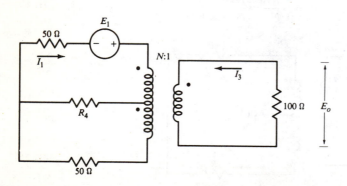

Figure 6.7 The transformer circuit discussed in Example 6.2.

The circuit will be a hybrid transformer provided $N = \frac{1}{2}$ and $R_4 = 25\,\Omega$. In this case the input impedance of port 1 is $50\,\Omega$, so

$$I_1 = \frac{E_1}{100}$$

The current I_2 will be zero, so

$$I_3 = NI_1 = -\frac{E_1}{200}$$

and

$$E_o = -I_3 R_L = \frac{E_1}{2}$$

Power Transfer in Hybrid Transformers

If power is applied at P_1, no power is transferred to P_2 because the ports are isolated. Since the transformer is lossless, the power must go to ports 4 and 3. That is,

$$P_1 = P_3 + P_4 \qquad (6.51)$$

Also, since the current through R_4 is equal to the current through R_1 and

$$R_4 = \frac{R_1}{1 + k} \qquad (6.52)$$

the power distribution for an applied power P_1 is

$$P_4 = \frac{P_1}{1 + k} \qquad (6.53)$$

and

$$P_3 = P_1 - P_4 = \frac{kP_1}{1 + k} \qquad (6.54)$$

or

$$\frac{P_3}{P_4} = \frac{N^2 k^2 (R_4/N^2 k)}{R_4} = k \qquad (6.55)$$

If the power is applied to port 2,

$$P_2 = P_3 + P_4 \qquad (6.56)$$

Since $I_4 = I_2$ ($I_1 = 0$),

$$P_4 = \frac{P_2 k}{1 + k} \qquad (6.57)$$

$$P_3 = P_2 \left(1 - \frac{k}{1 + k} \right) = \frac{P_2}{1 + k} \qquad (6.58)$$

and

$$\frac{P_3}{P_4} = k^{-1} \qquad (6.59)$$

Thus we see that the hybrid transformer can be used to split the power between two ports in a ratio determined by the transformer turns ratio. If the power applied to port 3 is divided between ports 1 and 2,

$$P_3 = P_1 + P_2 \tag{6.60}$$

Since $I_1 = I_2$ ($P_4 = 0$), the power is divided so that

$$P_1 = \frac{P_3}{1+k} \quad \text{and} \quad P_2 = \frac{kP_3}{1+k} \quad \text{or} \quad \frac{P_1}{P_2} = k \tag{6.61}$$

If the power is applied to port 4, $I_3 = 0$, so $I_2 = -kI_1$ and

$$\frac{P_1}{P_2} = \frac{R_1}{R_2 k^2} = k^{-1} \tag{6.62}$$

These power distribution results can be summarized as follows: If the power from port C is divided between ports B and D in the ratio k, then the power from port A, the conjugate port of C, will be divided between ports B and D in the ratio $1/k$.

Phase Distribution in Hybrid Transformers

One other property of hybrid transformers is concerned with the phase shift. While the phase shift between three of the four ports will be zero, the phase shift between the remaining port will be 180°. Various transformer configurations may be used, but in all of them the transmission path will have a phase shift differing 180° from the other three phase shifts. For example, consider the response I_3 to the applied voltages E_1 and E_2. The current I_3 is found using Eq. (6.14):

$$I_3 = \frac{E_1[-Z_4N - N(Z_1 + Z_4)] - E_2[-Z_4N - N(Z_1 + Z_4)]}{\det \mathbf{A}} \tag{6.63}$$

and

$$E_0 = -I_3 Z_3 = \frac{(E_1 - E_2)(Z_1 + 2Z_4)NZ_3}{\det \mathbf{A}}$$

$$= \frac{(E_1 - E_2)NZ_3}{Z_1 + 2Z_4} \tag{6.64}$$

The output is proportional to the difference of the input signals, and the device is so used in many communications circuits.

If we calculate I_4 in response to E_1 and E_2 we obtain

$$I_4 = I_1 - I_2 = \frac{E_1[(Z_1 + Z_4) + Z_4]}{\det \mathbf{A}} + \frac{E_2[(Z_1 + Z_4) + Z_4]}{\det \mathbf{A}}$$

$$= \frac{E_1 + E_2}{Z_1 + 2Z_4} \tag{6.65}$$

If the output is taken across Z_4, it is proportional to the sum of the two input signals. The hybrid transformer can therefore also be used to combine or subtract signals and may act as a signal combiner, such as in push-pull circuits. The usefulness of the hybrid transformer arises from its ability to isolate ports and, at the same time, to match impedances between the ports. The device can also be used to split a signal into two components, an operation that is not affected by a failure in the conjugate channel.

The Nonideal Three-Winding Transformer[6.1]

The previous discussion of the three-winding transformer assumed that the transformer was ideal, with no mutual coupling between the two sections of the primary. The derivations describe ideal circuit behavior, but they do not describe the limitations encountered in a practical circuit. The actual frequency characteristics of a three-winding transformer will now be described.

The equivalent circuit for a nonideal three-winding transformer with identical primary windings is shown in Fig. 6.8. L_b is the inductance of the secondary winding, L_a is the inductance of one primary winding, M is the mutual inductance between the secondary and one-half of the primary, and M_a is the mutual inductance between the two primary sections. With these definitions, the three loop equations are

$$E_1 - E_4 = (Z_1 + sL_a + Z_4)I_1 + (sM_a - Z_4)I_2 + sMI_3 \qquad (6.66)$$

$$-E_2 + E_4 = (sM_a - Z_4)I_1 + (Z_2 + Z_4 + sL_a)I_2 + sMI_3 \qquad (6.67)$$

$$E_3 = +sMI_1 + sMI_2 + (sL_b + Z_3)I_3 \qquad (6.68)$$

or $$E = AI \qquad (6.69)$$

If the current I_3 in response to an applied voltage E_4 is calculated, it is readily proved that if $Z_1 = Z_2$, ports 4 and 3 are isolated.

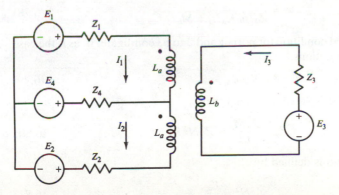

Figure 6.8 A circuit incorporating a nonideal three-winding transformer.

The current I_2 in response to E_1 is

$$I_2 = \frac{E_1[(sM)^2 + (Z_4 - sM_2)(sL_b + Z_3)]}{\det \mathbf{A}} \tag{6.70}$$

where the denominator is the determinant of the $\mathbf{A}$ matrix of Eq. (6.69).

In order for port 2 to be isolated from port 1, it is necessary that

$$(sM)^2 + (Z_4 - sM_a)(sL_b + Z_3) = 0 \tag{6.71}$$

or, in terms of the steady-state response,

$$-\omega^2 M^2 + Z_4 j\omega L_b + \omega^2 M_a L_b + Z_4 Z_3 - j\omega M_a Z_3 = 0 \tag{6.72}$$

The coefficient of coupling between the two primary sections can be defined as

$$K_a = \frac{M_a}{L_a} \tag{6.73}$$

and the coefficient of coupling between the primary and secondary can be defined as

$$K = \frac{M}{(L_a L_b)^{1/2}} \tag{6.74}$$

It follows that

$$M_a L_b = K_a L_a L_b = \frac{K_a M^2}{K^2} \tag{6.75}$$

If

$$K_a = K^2 \tag{6.76}$$

then Eq. (6.72) reduces to

$$Z_4 = \frac{Z_3 j\omega M_a}{Z_3 + j\omega L_b} \tag{6.77}$$

$$= \frac{Z_3}{Z_3/j\omega K_a L_a + M_b/M_a} \tag{6.78}$$

which is the required condition for ports 1 and 2 to be conjugate. If, as is the case in most practical situations,

$$K_a \omega L_a \gg Z_3 \tag{6.79}$$

ports 1 and 2 will be isolated, provided that

$$Z_4 \approx \frac{Z_3 M_a}{L_b} = Z_3 N^2 K_a \tag{6.80}$$

where the turns ratio is defined by

$$N^2 = \frac{L_a}{L_b} \tag{6.81}$$

This is identical to the requirement for the ideal transformer ($K_a = 1$). Hybrid transformers are normally designed with

$$Z_3 = \frac{Z_4}{N^2 K_a} \tag{6.82}$$

This requirement can be met over a broad range of frequencies, but the low-frequency performance of nonideal center-tapped transformers deteriorates when ωL_a is no longer much greater than Z_4; their high-frequency performance is limited by the stray and interwinding capacitances.

6.3 TRANSMISSION-LINE TRANSFORMERS

Transmission lines can also be used for impedance matching and power transfer. One of the principal functions of transformers in radio communication circuits is to match impedances between two networks. In applications where a narrowband circuit is required, the transformer can also serve as an integral part of the tuned circuit. However, unless special techniques are used, transformers are restricted to relatively narrowband applications, primarily because the interwinding capacitance resonates with the transformer inductance. Transformers built to minimize interwinding capacitances are usually bulky, especially if a great deal of power is to be transferred. Thus, engineers seek other means of matching impedance levels when large bandwidths are desired. A successful approach has been to construct transformers using transmission lines.[6.2]

The distinguishing feature of the transmission-line transformer is that the coils are so arranged that the interwinding capacitance combines with the inductance to form a transmission line that has no resonant frequencies that would limit the circuit bandwidth. For this reason the windings can be closely spaced while maintaining effective couplings. The net result of this construction of transmission-line transformers is a strong high-frequency response.

Consider the circuit shown in Fig. 6.9. It is assumed that the voltage travels from V_1 to V_2 via a transmission line. I_1 is the current at the sending end and I_2 is the current at the receiving end. The loop equations for the circuit can be written as

$$E_i = (I_1 + I_2)R_g + V_1 \tag{6.83}$$

$$= (I_1 + I_2)R_g - V_2 + I_2 R_L \tag{6.84}$$

The two transmission-line equations that relate the voltages and currents on a lossless line are[6.3]

$$V_1 = V_2 \cos \beta l + jI_2 Z_o \sin \beta l \tag{6.85}$$

and
$$I_1 = I_2 \cos \beta l + \frac{jV_2}{Z_o} \sin \beta l \tag{6.86}$$

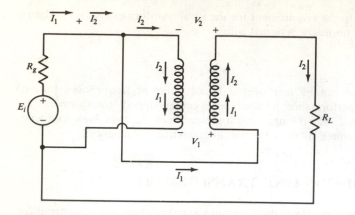

Figure 6.9 A transmission-line transformer circuit.

where l is the line length, β is the line phase constant, and Z_o is the characteristic impedance of the line. Both β and Z_o are determined by the line inductance L and capacitance C per unit length:

$$Z_o = \left(\frac{L}{C}\right)^{1/2} \tag{6.87}$$

and

$$\beta = (LC\omega)^{1/2} \tag{6.88}$$

The line wavelength is

$$\lambda = \frac{2\pi}{\beta} \tag{6.89}$$

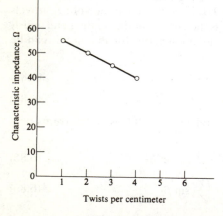

Figure 6.10 The characteristic impedance of a twisted-pair transmission line as a function of the number of twists.

The line inductance and capacitance depend upon the wire size and distance between conductors. Transmission-line transformers are normally made by twisting the two conductors together. The characteristic impedance decreases as the number of turns per unit length increases.[6.4] Figure 6.10 illustrates a typical plot of characteristic impedance as a function of the number of turns per centimeter.

If Eq. (6.85) is used to eliminate V_1, the equilibrium equations are

$$\begin{matrix} E_i = \\ E_i = \\ 0 = \end{matrix} \begin{bmatrix} R_g & R_g + jZ_o \sin \beta l & \cos \beta l \\ R_g & R_g + R_L & -1 \\ -1 & \cos \beta l & j \sin \beta l / Z_o \end{bmatrix} \begin{bmatrix} I_1 \\ I_2 \\ V_2 \end{bmatrix} \qquad (6.90)$$

Thus

$$I_2 = \frac{\begin{vmatrix} R_g & E_i & \cos \beta l \\ R_g & E_i & -1 \\ -1 & 0 & \sin \beta l / Z_o \end{vmatrix}}{\det \mathbf{A}}$$

$$= \frac{E_i(1 + \cos \beta l)}{2R_g(1 + \cos \beta l) + R_L \cos \beta l + j(R_g R_L + Z_o^2) \sin \beta l / Z_o} \qquad (6.91)$$

and the output power is

$$P_o = |I_2|^2 R_L = \frac{E_i^2(1 + \cos \beta l)^2 R_L}{[2R_g(1 + \cos \beta l) + R_L \cos \beta l]^2 + (R_g R_L + Z_o^2)^2 \sin^2 \beta l / Z_o^2} \qquad (6.92)$$

The value of characteristic impedance Z_o for which the output power is a maximum can be found by setting

$$\frac{dP_o}{dZ_o} = 0$$

The maximizing value of Z_o is found to be

$$Z_o = (R_g R_L)^{1/2} \qquad (6.93)$$

independent of the line length l.

Now with the value of characteristic impedance $Z_o = (R_g R_L)^{1/2}$,

$$P_o = |I_2^2| R_L = \frac{E_i^2(1 + \cos \beta l)^2 R_L}{[2R_g(1 + \cos \beta l) + R_L \cos \beta l]^2 + 4R_g R_L \sin^2 \beta l} \qquad (6.94)$$

To determine the value of R_L for which P_o is a maximum, the derivative dP_o/dR_L is set to zero. The maximizing value of R_L is

$$R_L = \frac{2R_g(1 + \cos \beta l)}{\cos \beta l} \qquad (6.95)$$

which is equal to $4R_g$ for $\beta l \approx 0$. For $Z_o = (R_g R_L)^{1/2}$ and $R_L = 4R_g$, Eq. (6.94) can be written

$$P_o = \frac{E_i(1 + \cos \beta l)^2 R_L}{[R_L/2(1 + \cos \beta l) + R_L \cos \beta l]^2 + R_L^2 \sin^2 \beta l}$$

$$= \frac{4E_i^2(1 + \cos \beta l)}{R_L(5 \cos \beta l + 1)} \tag{6.96}$$

Since the maximum available power from the source P_A is

$$P_A = \frac{E_i^2}{4R_g} = \frac{E_i^2}{R_L} \tag{6.97}$$

the ratio of output power to available power is

$$\frac{P_o}{P_A} = \frac{1 + \cos \beta l}{\frac{5}{4}(1 + \frac{6}{5} \cos \beta l + \cos^2 \beta l)} \tag{6.98}$$

Equation (6.98) shows that frequency response of this circuit starts to roll off when $\cos \beta l$ becomes significantly less than 1, and the power gain is zero when

$$\beta l = \pi$$

or

$$l = \frac{\pi}{\omega} = \frac{\lambda}{2} \tag{6.99}$$

The power gain is zero when the line is one-half wavelength long. Transmission-line transformers are designed so that the line length is normally less than a tenth of a wavelength at the highest frequency of interest. The shorter the physical length of the line, the higher the upper cutoff frequency of the device. Transmission-line transformers can be built with twisted-pair wires to have an upper cutoff frequency above 100 MHz. The upper frequency limit can be extended into the gigahertz region using thin-film fabrication techniques.

Equation (6.98) indicates that the low-frequency response extends down to zero frequency. Actually, the transmission-line model is not valid at the lower frequencies. This is seen when the circuit of Fig. 6.9 is redrawn as shown in Fig. 6.11 where the transformer is modeled as an autotransformer.

The step-down autotransformer can be replaced by the equivalent circuit shown in Fig. 6.12. It is readily shown that for the two circuits to be the same

$$n = \left(\frac{L_2}{L}\right)^{1/2} \tag{6.100}$$

where L is the total inductance shunting R_L and the inductance

$$L_2 = \omega^{-1} \left| \frac{V_2}{I_1} \right|_{I_2=0} \tag{6.101}$$

If the circuit in Fig. 6.12 is closely coupled with $k \approx 1$, the low-frequency

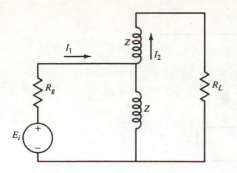

Figure 6.11 A low-frequency equivalent circuit of the circuit shown in Fig. 6.9.

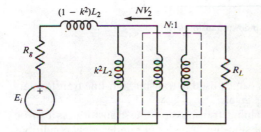

Figure 6.12 Another low-frequency equivalent circuit of the circuit shown in Fig. 6.9.

behavior is described by

$$V_2 = \frac{nE_i[sL_2 R'_L/(R_g + R'_L)]}{1 + sL_2(R_g + R'_L)/R_g R'_L} \tag{6.102}$$

Where

$$R'_L = n^2 R_L \tag{6.103}$$

the -3-dB frequency is

$$\omega_N = \frac{R_g R'_L}{L_2(R_g + R'_L)} \tag{6.104}$$

To reduce this cutoff frequency for a given R_g, L needs to be increased. This can be done by increasing the line length, but, as previously shown, this reduces the higher-frequency limit of the device. Another method for increasing the inductance L is to wind the wires on a magnetic core of high permeability material, since the inductance of a conductor is directly proportional to the permeability of the surrounding medium. Since equal and opposite currents flow in the two windings, the core does not influence the internal magnetic fields or the characteristic impedance of the line. The core's sole role is to minimize the shunting currents that determine the lower cutoff frequency. The core does not couple energy from input to output (except perhaps at the lowest frequencies); rather the power is coupled through the dielectric medium of the transmission

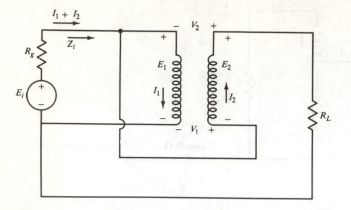

Figure 6.13 A four-to-one impedance step-down transformer.

line. Therefore, small ferrite cores can be used in transmission-line transformers operating at relatively high power levels.

The analysis of circuits containing transmission-line transformers can be simplified by analyzing the transmission-line transformer under ideal conditions (that is, the line is assumed lossless and matched so that $V_1 = V_2$). Consider the transformer in Fig. 6.13. Since the sum of the voltage drops around the loop ($V_1 + E_2 - V_2 - E_1 = 0$, if $V_1 = V_2$), then $E_1 = E_2$. Also, in the ideal transmission line, $I_1 = I_2$; thus under ideal conditions the circuit can easily be analyzed.

For the ideal case,

$$I_1 = I_2 \tag{6.105}$$

and

$$E_1 = E_2 \tag{6.106}$$

Thus

$$V_L = E_2 + E_1 = 2E_1 \tag{6.107}$$

and

$$I_L = I_1 \tag{6.108}$$

since

$$R_L = \frac{V_L}{I_L} = \frac{2E_1}{I_1} = 4Z_i \tag{6.109}$$

In this case the transmission line serves as a step-down transformer, thereby reducing the load impedance by a factor of 4. For maximum power transfer, R_L must satisfy the relation

$$R_L = 4R_g \tag{6.110}$$

and the line characteristic impedance must be

$$Z_o = (R_L R_g)^{1/2} = 2R_g \tag{6.111}$$

The Step-Up Transformer

The transmission-line transformer can also be used as a step-up transformer if used as shown in Fig. 6.14. If the line is ideal,

$$I_1 = I_2 = I \tag{6.112}$$

and
$$E_1 = E_2 = E \tag{6.113}$$

Therefore
$$I_L = 2I \tag{6.114}$$

$$V_L = E \tag{6.115}$$

and
$$R_L = \frac{V_L}{I_L} = \frac{E}{2I} \tag{6.116}$$

The input impedance

$$Z_i = \frac{V_i}{I} = \frac{2E}{I} = 4R_L \tag{6.117}$$

will be four times the load impedance.

Larger impedance ratios can be obtained using multiple cores, as shown in Fig. 6.15. The current I from the first core flows into the primary of the second core. This same current is also transferred to the secondary of the second core and hence through the third core, etc. It is evident that the total load current is

$$I_L = (N+1)I \tag{6.118}$$

where N is the number of cores. Also, if each line is matched, the load voltage $V_L = E$ will also be equal to the voltage across the output of each transmission-line transformer. The input voltage V_i must then be

$$V_i = (N+1)E \tag{6.119}$$

Hence
$$R_L = \frac{E}{(N+1)I} \tag{6.120}$$

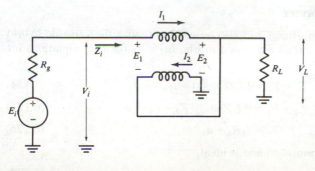

Figure 6.14 A four-to-one impedance step-up transformer.

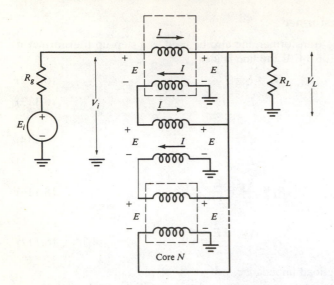

Figure 6.15 An $(N+1)^2:1$ impedance step-up transformer.

and

$$R_g = \frac{(N+1)E}{I} \qquad (6.121)$$

For maximum power transfer, the load resistance must be

$$R_L = \frac{R_g}{(N+1)^2} \qquad (6.122)$$

A similar method can be used to obtain the step-down transformation

$$R_L = (N+1)^2 R_g \qquad (6.123)$$

The Hybrid Transformer

Transmission-line transformers can also be used to realize the versatile hybrid transformer, as in the circuit illustrated in Fig. 6.16. Three loop equations for the circuit can be written as

$$E_1 - E_4 = I_1(Z_1 + Z_4) - I_2 Z_4 + a_2 \qquad (6.124)$$

$$-E_2 + E_4 = -I_1 Z_4 + I_2(Z_2 + Z_4) + a_1 \qquad (6.125)$$

$$E_3 = I_3 R_3 + a_1 \qquad (6.126)$$

In addition, if the transmission line is ideal,

$$a_1 = a_2 \qquad (6.127)$$

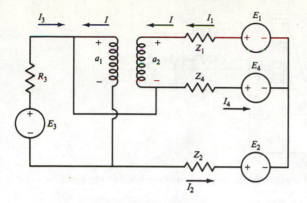

Figure 6.16 A transmission-line hybrid transformer.

and the following circuit identities hold:

$$I_1 = I \tag{6.128}$$

$$I_4 = I_1 - I_2 \tag{6.129}$$

$$I + I_3 + I = I_4 \tag{6.130}$$

If I and I_4 are eliminated from these three equations, it is found that

$$-I_3 = I_1 + I_2 \tag{6.131}$$

Equations (6.124) through (6.126) and (6.131) are identical to Eqs. (6.3) through (6.6) developed for the center-tapped transformer (with $N = 1$), so the results developed for that circuit apply here also. In particular, the ports of the center-tapped transformer will be biconjugate if

$$Z_1 = Z_2 \tag{6.132}$$

and

$$Z_3 = Z_4 \tag{6.133}$$

The transformer will be a hybrid transformer provided the additional relation

$$Z_3 = \frac{Z_1^*}{2} \tag{6.134}$$

is also satisfied.

A Power-Output Stage

Figure 6.17 illustrates a method for combining the outputs of a class B push-pull amplifier to drive an unbalanced load R_L. Two transmission-line transformers are used. The circuit operation can be easily understood by considering the transmission-line transformers one at a time. If the upper transistor Q_1 is on and the bottom transistor is off, the circuit can be drawn as shown in Fig. 6.18. (Ignore

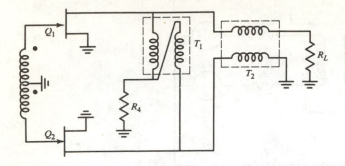

Figure 6.17 A class B push-pull amplifier incorporating transmission-line transformers in the output stage.

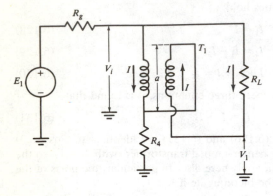

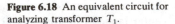

Figure 6.18 An equivalent circuit for analyzing transformer T_1.

for the moment that the load is not balanced.) If the current through R_L is I, then the current supplied by E_1 must be $2I$. If E_1 is off (open) and E_2 (Q_2) is on, then half the current supplied by E_2 will pass through R_L (but in the direction opposite to that supplied by E_1). The transformer T_1 serves to combine the two class B push-pull outputs.

The current in both transformer windings is I, so the current through R_L is also I, and the current supplied by the generator is $2I$. The input impedance of the transformer circuit (Fig. 6.18) is $Z_i = V_i/2I$. The voltage at the bottom of R_L is V_1, and the voltage drop across R_4 is $V_1 + a$, where a is the voltage across the transformer. Then $V_i = V_1 + 2a$ and

$$Z_i = \frac{V_i}{2I} = \frac{V_1 + a}{2I} + \frac{a}{2I} = R_4 + \frac{R_L}{4} \qquad (6.135)$$

since $R_L = 2a/I$.

This is the impedance seen by the amplifier with output impedance R_g. The

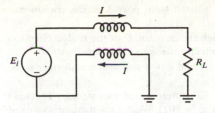

Figure 6.19 A balanced-to-unbalanced (BALUN) transformer.

remaining problem is that one side of the load in Fig. 6.18 is not grounded. This is solved by using a transmission-line transformer, as shown in Fig. 6.19 to convert from an unbalanced to a balanced load. It is easily shown that the impedance seen by the source E_i is R_L. Therefore, the transformer T_1 combines the two transistor outputs and serves as an impedance matcher. The transformer T_2 converts the balanced output to an unbalanced load.

Transmission-line transformers are relatively broadbanded and have very effective high-frequency performance (to several hundred megahertz). However, they do not have as good a low-frequency performance as conventional transformers and can only realize the turns ratio $(n + 1)$ where n is an integer.

PROBLEMS

6.1 Derive the relation between Z_1 and Z_2 for port 3 of Fig. 6.2 to be isolated from port 4.

6.2 Show that the equations

$$Z_1 = Z_2 = Z \qquad \text{and} \qquad Z_3 = \frac{Z_4}{N^2} = \frac{Z^*}{2N^2}$$

will result in maximum power transfer from ports 2 and 4 of the circuit shown in Fig. 6.2.

6.3 If the applied voltage $E_i = 1$ V rms in the circuit shown in Fig. P6.3, what is the output power?

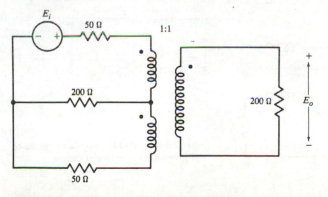

Figure P6.3 A three-winding transformer circuit.

6.4 Show that if $R_1 = kR_2$, then port 4 is isolated from port 3 in the transformer illustrated in Fig. 6.5.

6.5 Derive an equation for the current through the detector D of the bridge circuit in Fig. 6.3 (as a function of R_1 and R_2) if the detector has zero input impedance.

6.6 Derive the necessary conditions for ports 3 and 4 of a nonideal transformer to be conjugate.

6.7 Design a transformer circuit for taking the difference of two signals. The load impedance as well as each source impedance is to be 50 Ω. What is the minimum value of primary inductance which should be used at 20 MHz?

6.8 An autotransformer with a common ground for all ports is shown in Fig. P6.8. Determine the necessary relationships between the resistances in order for the transformer to be a hybrid transformer. Assume the transformer is ideal with no mutual coupling between turns.

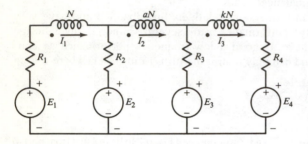

Figure P6.8 A three-winding autotransformer that can be used as a hybrid transformer.

6.9 Derive the relation required for the characteristic line impedance in order to obtain the maximum power transformer relation [Eq. (6.93)].

6.10 Derive Eq. (6.95).

6.11 Verify Eq. (6.100).

6.12 Figure P6.12 illustrates a circuit in which neither side of the load is grounded. Determine the value of R_L for maximum power transfer.

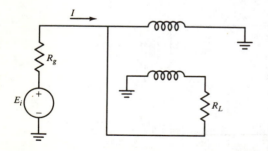

Figure P6.12 An unbalanced-to-balanced transformer circuit.

6.13 For the hybrid transformer illustrated in Fig. 6.5 ($k = 1$), determine the responses at ports 1 and 2 to signals amplified at ports 3 and 4. Compare the phases of the output signals with those obtained at ports 3 and 4 from signals applied at ports 1 and 2.

6.14 Figure P6.14 illustrates a transmission-line transformer circuit that can be used with balanced loads. What should R_L be for maximum power transfer?

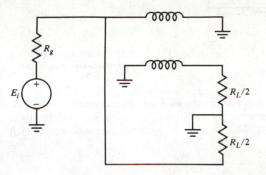

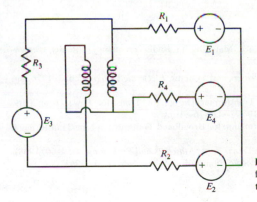

Figure P6.14 A transformer circuit with a balanced load.

6.15 Design a transmission-line transformer circuit with an input impedance

$$Z_i = \frac{R_L}{9}$$

6-16 Derive the relation between R_1, R_2, R_3, and R_4 required for the circuit shown in Fig. P6.16 to be a hybrid transformer. Assume the transmission line is ideal.

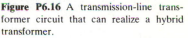

Figure P6.16 A transmission-line transformer circuit that can realize a hybrid transformer.

6.17 Design a transmission-line transformer circuit for combining the outputs of two class A amplifiers operating in push-pull. The load is 50 Ω unbalanced.

6.18 What is the input impedance of the transmission-line transformer circuit shown in Fig. P6.18?

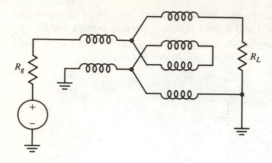

Figure P6.18 A circuit incorporating three transmission-line transformers.

6.19 A class B push-pull amplifier requires that the load seen by each transistor be 6 Ω. The load is 50 Ω unbalanced. Design a transmission-line coupling circuit which results in maximum output from the transistors. What should the characteristic impedance of each transmission line be?

REFERENCES

6.1 McIlwain, K., and J. G. Brainerd: *High-Frequency Alternating Currents*, 2d ed., Wiley, New York, 1939.

6.2 Ruthroff, C. L.: Some Broadband Transformers, *Proc. IRE*, **47**:1337–1342 (1959).

6.3 Skilling, H. H.: *Electric Transmission Lines*, McGraw-Hill, New York, 1951.

6.4 Krauss, H. L., and C. W. Allen: Designing Toroidal Transformers to Optimize Wideband Performance, *Electronics*, Aug. 16, 1973, pp. 113–116.

ADDITIONAL READINGS

Guanella, G.: New Method of Impedance Matching in Radio Frequency Circuits, *Brown-Boveri Rev.*, **31**:327–329 (1944).

Nagle, J. J.: Use of Wideband Autotransformers in rf Systems, *Electronic Design*, Feb. 2, 1976, pp. 64–70.

Pitzalis, O., and T. Couse: Broadband Transformer Design for Transistor Power Amplifiers, *USAECOM Tech. Rep.* 2989, July 1968 (AD 676-816).

———, and ———: Practical Design Information for Broadband Transmission Line Transformers, *Proc. IEEE*, **56**:738–739 (1968).

Satori, E. F.: Hybrid Transformers, *IEEE Trans. Parts, Materials and Packaging*, **4**:59–66 (1968).

Schlicke, H. M.: *Essentials of Dielectromagnetic Engineering*, Wiley, New York, 1961.

SEVEN
OSCILLATORS

7.1 INTRODUCTION

An *electronic oscillator* is a circuit with a periodic output signal but with no periodic input signal. It functions by converting dc power to a periodic output signal (ac power). A *harmonic oscillator* is one in which the output signal is approximately sinusoidal. If a crystal resonator is used in the circuit to closely control the oscillating frequency, the oscillator is referred to as a *crystal-controlled oscillator*. Modern communication systems often contain several oscillators, including a crystal-controlled reference oscillator, a voltage-controlled oscillator (VCO), and a voltage-controlled crystal oscillator (VCXO). There exist many integrated circuits that can be used for generating a periodic output signal, but oscillator design is one area in which discrete transistors have a marked advantage. Communication circuits require high-frequency, low-noise oscillators, and integrated-circuit components often cannot meet the noise and frequency specifications.

In this chapter, these various oscillators and three different methods of oscillator analysis are described. First, two mathematical approaches based on linear feedback theory and circuit analysis will be presented. Subsequently, a less rigorous approach will be presented, one which provides additional insight into the design process. After the conditions for oscillation are derived, we will discuss the crystal resonator. It will be shown that the crystal serves in oscillators as a very narrowband circuit and that the design of crystal-controlled oscillators is essentially the same as that of oscillators that do not contain a crystal.

7.2 CONDITIONS FOR OSCILLATION

Electronic oscillator circuits are feedback networks, and the extensive results of linear feedback analysis can be used for oscillator analysis and design. Oscillators are inherently a nonlinear circuit, but linear analysis techniques are the most useful for analysis and design. They provide accurate information for predicting the frequency of oscillation but have limited use for predicting the amplitude of oscillation. In this section we will consider the most widely developed method of linear feedback analysis, basing our discussion on interpretation of the block diagram model.

Nyquist Stability Criteria

Figure 7.1 shows, in block diagram form, the necessary components of an oscillator. It contains an amplifier with frequency-dependent forward loop gain $G(j\omega)$ and a frequency-dependent feedback network $H(j\omega)$. The output voltage is given by

$$V_o = \frac{V_i G(j\omega)}{1 + G(j\omega)H(j\omega)}$$

For an oscillator, the output V_o is nonzero even if the input signal V_i is zero. This can only be possible if the forward loop gain is infinite (which is not practical) or if the denominator is

$$1 + G(j\omega)H(j\omega) = 0$$

at some frequency ω_o. This leads to the well-known condition for oscillation—the Nyquist criterion: At some frequency ω_o

$$G(j\omega_o)H(j\omega_o) = -1 \tag{7.1}$$

That is, the magnitude of the open-loop transfer function is equal to 1, or

$$|G(j\omega_o)H(j\omega_o)| = 1 \tag{7.2}$$

and the phase shift is 180°, or

$$\arg G(j\omega_o)H(j\omega_o) = 180° \tag{7.3}$$

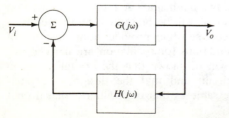

Figure 7.1 Block diagram representation of a linear negative-feedback system.

This can be expressed more simply as: If in a negative-feedback system, the open-loop gain has a total phase shift of 180° at some frequency ω_o, the system will oscillate at that frequency, providing the open-loop gain is unity. If the gain is less than unity at the frequency where the phase shift is 180°, the system will be stable; if the gain is greater than unity the system will be unstable. If positive feedback is used, the loop phase shift must be 0°. That is, arg $G(j\omega_o)H(j\omega_o) = 0°$.

The preceding expressions for oscillation are not precise for some complicated systems, but they are correct for those transfer functions normally encountered in oscillator design. The conditions for stability are also known as the *Barkhausen criteria*, which state that if the closed-loop transfer function is

$$\frac{V_o}{V_i} = \frac{\mu}{1 - \mu\beta} \tag{7.4}$$

then the system will oscillate, provided $\mu\beta = 1$. This is equivalent to the Nyquist criterion, the difference being that the transfer function is written for a loop with positive feedback. Both versions state that the total phase shift around the loop must be 360° at the frequency of oscillation, and the magnitude of the open-loop gain must be unity at that frequency.

If a single-stage common-emitter (or common-source) amplifier is used with feedback from collector to base, as illustrated in Fig. 7.2a, then the feedback network must supply 180° phase shift since there is 180° phase shift between the base and collector signals. If a common-base (or common-gate) amplifier (illustrated in Fig. 7.2b) is used, there is no phase shift between the emitter and collector signals; therefore, a necessary condition for oscillation is that there be no phase shift between the input and output of the feedback network. If a small phase shift occurs in the forward loop this must be compensated for by an equal and opposite phase shift in the feedback network.

An amplifier with feedback, but without a frequency-sensitive network,

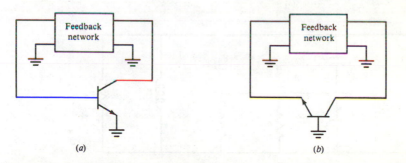

(a)   (b)

Figure 7.2 (a) A system in which the feedback network must provide 180° phase shift in order for oscillations to occur; (b) a system in which the feedback network must provide 0° phase shift in order for oscillations to occur.

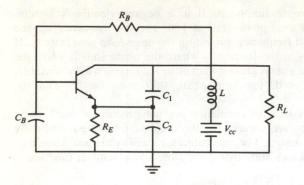

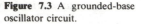

Figure 7.3 A grounded-base oscillator circuit.

can be made to oscillate. However, the frequency of oscillation will be difficult to control. Since the primary purpose of the feedback network is to control the frequency of oscillation, the network is designed so that the Nyquist criteria are satisfied at only a single frequency.

The following analysis of the relatively simple circuit shown in Fig. 7.3 illustrates the method of determining the conditions for oscillation. The linearized (and simplified) equivalent circuit of Fig. 7.3 is given in Fig. 7.4a in

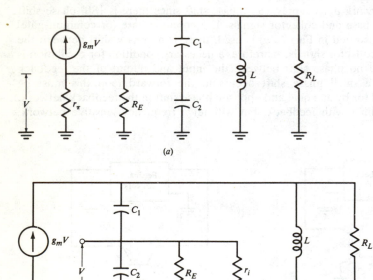

(a)

(b)

Figure 7.4 (a) A small-signal equivalent circuit of the grounded-base oscillator; (b) the circuit of Fig. 7.4a, but with the feedback path opened and terminated in the correct resistance.

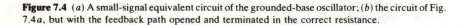

which the transistor output resistance r_o is ignored. Also, the capacitor connected to the base is assumed to be so large that the base is at ground potential for the small-signal analysis. Note that the transistor is connected in the common-base configuration and has no voltage phase inversion (i.e., the feedback is positive). The conditions for oscillation are

$$|G(j\omega_o)||H(j\omega_o)| = 1$$

and because the amplifier is noninverting, the phase shift of the network must also be $\arg G(j\omega_o)H(j\omega_o) = 0°$.

The loop gain is calculated by opening the feedback loop, applying a signal, and measuring the return difference. It is necessary that when the loop is opened the impedance seen at any point be the same as it is with the loop closed. In this case it is convenient to open the loop at the transistor emitter. The impedance shunting the capacitor C_2 is the resistor R_E in parallel with the input impedance r_i of the common-base amplifier. In the discussion of the common-base amplifier (Chap. 2) it was shown that the input resistance of the common-base circuit is

$$r_i = \frac{r_\pi}{\beta}$$

so the equivalent circuit of Fig. 7.4a can be redrawn as shown in Fig. 7.4b. The circuit analysis can be greatly simplified by assuming that

$$\{[\omega(C_2 + C_1)]^2\}^{-1} \ll \left(\frac{r_i R_E}{r_i + R_E}\right)^2$$

and also that the Q of the load impedance is high. In this case the circuit (see results for capacitive transformers in Chap. 4) reduces to that of Fig. 7.5, where the resistor shunting C_2 has been replaced by an equivalent resistance shunting C_1 and C_2. The circuit shown in Fig. 7.5 is not the small-signal equivalent of Fig. 7.3; rather it is a small-signal equivalent of the open-loop circuit with the loop or feedback network opened at the emitter and the circuit terminated in the correct impedance. The network could be opened at other

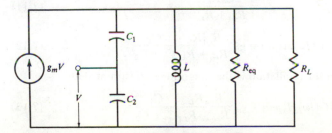

Figure 7.5 A circuit equivalent to Fig. 7.4b, provided $[\omega(C_1 + C_2)R]^2 \gg 1$.

points, but opening it at the emitter is particularly convenient because it is relatively easy to determine the terminating impedance at this point.

The feedback voltage is given by

$$V = \frac{V_o C_1}{C_1 + C_2}$$

and the equivalent resistance reflected across the coil is

$$R_{eq} = \frac{r_i R_E}{r_i + R_E} \left(\frac{C_1 + C_2}{C_1}\right)^2 \tag{7.5}$$

The forward loop gain is

$$\frac{V_o}{V} = G(j\omega) = g_m Z_L \tag{7.6}$$

where

$$Y_L = Z_L^{-1} = (j\omega L)^{-1} + R_{eq}^{-1} + R_L^{-1} + j\omega C \tag{7.7}$$

and

$$C = \frac{C_1 C_2}{C_1 + C_2}$$

The feedback ratio is

$$\frac{V}{V_o} = H(j\omega) = \frac{C_1}{C_1 + C_2} \tag{7.8}$$

A necessary condition for oscillation is that

$$\arg G(j\omega) H(j\omega) = 0° \tag{7.9}$$

Since H does not depend on frequency in this example, if arg GH is to be zero, the phase shift of the load impedance Z_L must be zero. This occurs only at the resonant frequency of the circuit where

$$\omega_o = \left[\left(L \frac{C_1 C_2}{C_1 + C_2}\right)^{1/2}\right]^{-1} \tag{7.10}$$

At this frequency

$$Z_L = \frac{R_{eq} R_L}{R_{eq} + R_L} \tag{7.11}$$

and

$$G(j\omega) H(j\omega) = g_m \frac{R_{eq} R_L}{R_{eq} + R_L} \frac{C_1}{C_1 + C_2} \tag{7.12}$$

The other condition for oscillation is the magnitude constraint that

$$|G(j\omega) H(j\omega)| = g_m \frac{R_{eq} R_L}{R_{eq} + R_L} \frac{C_1}{C_1 + C_2} = 1 \tag{7.13}$$

Example 7.1 The preceding results will now be used to design a 20-MHz common-base sinusoidal oscillator using a transistor with a minimum β of 100.

SOLUTION Oscillator design consists of a trial-and-error procedure using Eqs. (7.5) through (7.13). As a starting point we will assume a bias current $I_c = 1$ mA; then the common-base input resistance is

$$r_i = g_m^{-1} = \frac{V_T}{I_c} = 26\ \Omega$$

Since this is so small, we can safely assume that the emitter bias resistor R_E is much larger than r_i, so [from Eq. (7.5)]

$$R_{eq} \approx r_i \left(\frac{C_1 + C_2}{C_1}\right)^2$$

Equations (7.5) through (7.13) were based on the assumption that

$$\{[\omega(C_2 + C_1)]^2\}^{-1} \ll r_i^2$$

If $(\omega C_2)^{-1} = 8$ ($C_2 = 1000$ pF) so that

$$(\omega C_2)^{-1} < \frac{r_i}{3}$$

the above inequality is satisfied and the assumption is justified. In addition, for oscillations to occur the loop gain [Eq. (7.12)] must be at least 1. In oscillator design the loop gain is usually selected to be about 3, which allows for some error in the approximation. With a loop gain greater than 1 the system is unstable and the oscillations increase in amplitude until the transistor current begins to saturate. When this occurs, the β of the transistor is reduced, and thus g_m is also reduced. This reduces the loop gain and stabilizes the amplitude of oscillation. In this example the loop

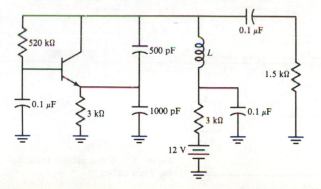

Figure 7.6 The oscillator circuit designed in Example 7.1.

gain is

$$|GH| \approx g_m r_i \frac{C_1 + C_2}{C_1} = \frac{C_1 + C_2}{C_1} = 3$$

So $C_1 = C_2/2 = 500 \, \text{pF}$. The value of the inductance is found from Eq. (7.10) to be $L = 1 \, \mu\text{H}$. Also, $R_{eq} = 234 \, \Omega$. A load resistor R_L can be shunted across the inductor without affecting the calculations if it is much bigger than R_{eq}. In this case a load resistance of $1500 \, \Omega$ could be safely added. The complete design would require selecting a supply voltage and bias resistors so that the quiescent collector current is 1 mA. A completed circuit schematic is given in Fig. 7.6.

The preceding oscillator analysis was based on a linear model for the circuit. In practice the analysis provides a good approximation for the frequency of oscillation and the minimum gain required, but it provides no information about the amplitude of oscillation. The system is initially unstable, and as the output amplitude increases the transistor begins to saturate, resulting in a reduction of loop gain and a stable oscillating amplitude. Such oscillators are referred to as *self-limiting* to distinguish them from oscillators that use an external means of regulating the oscillation amplitude.

Circuit Analysis

Identifying the open-loop gain is particularly useful for oscillator analysis and design. Further illustrations of this technique are given throughout this chapter. There are circuits in which it is difficult to identify the open-loop gain. For this reason two additional methods of oscillator design and analysis will be presented. A direct analysis of the circuit equations is frequently simpler and more informative than the block diagram interpretation (particularly for single-stage amplifiers). Figure 7.7 illustrates a transistor amplifier with three external impedances connected. Terminal 1 would represent the base terminal of

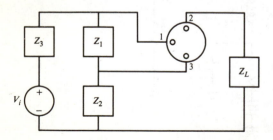

Figure 7.7 A generalized transistor oscillator circuit.

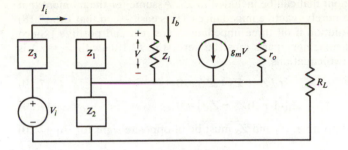

Figure 7.8 A small-signal equivalent circuit of Fig. 7.7.

a bipolar transistor or the gate terminal of a field-effect transistor, and terminal 3 represents the emitter or source terminal. The three external impedances, then, represent feedback connections between the transistor terminals. The small-signal equivalent circuit is given in Fig. 7.8. We will assume that the transistor output impedance is sufficiently large, and it will be neglected in the following analysis. Note that at this point the ground node for the circuit has not been identified. It will be subsequently shown that many different oscillator configurations can be realized for different ground points.

The loop equations are then

$$V_i = I_1(Z_2 + Z_3) + V + g_m V Z_2 \tag{7.14}$$

and

$$V = \frac{I_1 Z_1 Z_i}{Z_1 + Z_i} \tag{7.15}$$

For the amplifier to oscillate, the currents I_b and I_1 must be nonzero, even when $V_i = 0$. This is only possible if the system determinant Δ is

$$\Delta = \begin{vmatrix} Z_2 + Z_3 & 1 + g_m Z_2 \\ Z_1 & -\left(1 + \dfrac{Z_1}{Z_i}\right) \end{vmatrix} \tag{7.16}$$

is equal to 0. That is,

$$Z_1 + Z_2 + Z_3 + \frac{Z_1}{Z_i}(Z_2 + Z_3) + g_m Z_1 Z_2 = 0 \tag{7.17}$$

which reduces to

$$(Z_1 + Z_2 + Z_3)Z_i + Z_1 Z_2 \beta + Z_1(Z_2 + Z_3) = 0 \tag{7.18}$$

Only the case in which the transistor input impedance Z_i is real will be considered here ($Z_i = r_\pi$). The more complicated case in which Z_i is complex can be analyzed in the same manner. Normally Z_i will have a parallel capacitive component that can be included in Z_1. Assume for the moment that Z_1, Z_2, and Z_3 are purely reactive impedances. [It is easily seen that Eq. (7.18) does not have a solution if all three impedances are real and positive.] Since both the real and imaginary parts must be zero, Eq. (7.18) (with $Z_i = r_\pi$) is equivalent to the two equations

$$r_\pi(Z_1 + Z_2 + Z_3) = 0 \tag{7.19}$$

and

$$Z_1[(1+\beta)Z_2 + Z_3)] = 0 \tag{7.20}$$

Since β is real and positive, Z_2 and Z_3 must be of opposite signs for Eq. (7.20) to hold. That is,

$$(1+\beta)Z_2 = -Z_3 \tag{7.21}$$

Therefore, since r_π is nonzero, Eq. (7.19) reduces to

$$Z_1 + Z_2 - (1+\beta)Z_2 = 0 \tag{7.22}$$

or

$$Z_1 = \beta Z_2$$

Since β is positive, Z_1 and Z_2 will be reactances of the same kind. If Z_1 and Z_2 are capacitors, then Z_3 must be an inductor. An example of such a circuit with the ground located at the junction between Z_2 and Z_3 is shown in Fig. 7.9. This circuit is referred to as a *Colpitts oscillator*, named after the man that first published such a circuit for a vacuum tube oscillator. If the emitter is grounded, the oscillator will be as shown in Fig. 7.10. This circuit has become known as a *Pierce oscillator*. If Z_1 and Z_2 are inductors and Z_3 is a capacitor, the circuit is known as a *Hartley oscillator*. A common-base Hartley oscillator is illustrated in Fig. 7.11.

Equations (7.21) and (7.22) describe the necessary conditions for oscillation when the components external to the transistor are assumed to be ideal. These equations should be used with caution because they apply only to the ideal case. It appears that the conditions for oscillation are independent of the transistor input impedance. This is because the three external impedances are assumed to be purely reactive. In this case, the circuit will oscillate even

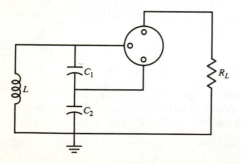

Figure 7.9 A Colpitts oscillator.

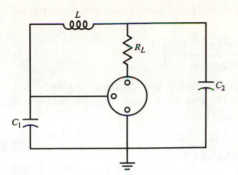

Figure 7.10 Another Colpitts oscillator, also known as a Pierce oscillator.

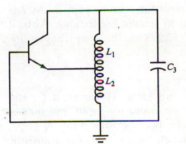

Figure 7.11 A Hartley oscillator.

without the transistor. The transistor is needed in the practical situation to supply the energy dissipated into a nonideal inductor and capacitor, to provide a means for amplitude limiting, and to furnish the energy that the oscillator must supply to the external circuitry. The preceding analysis will now be applied to the case of the nonideal component.

If Z_1, Z_2, Z_3, or Z_i is complex, the preceding analysis is more complicated, but the conditions for oscillation can still be obtained from Eq. (7.18). For example, if in the Colpitts circuit, there is a resistor r in series with L ($Z_3 = r + j\omega L$), Eq. (7.18) reduces to the two equations

$$r_\pi[\omega L - (\omega C_1)^{-1} - (\omega C_2)^{-1}] - \frac{r}{\omega C_1} = 0 \tag{7.23}$$

and

$$r_\pi r - \frac{1+\beta}{\omega^2 C_1 C_2} + (\omega C_1)^{-1}\omega L = 0 \tag{7.24}$$

If we define for notational simplicity,

$$C_1' = \frac{C_1}{1 + r/r_\pi} \tag{7.25}$$

then the reasonant frequency at which oscillations will occur is found from Eq. (7.23) to be

$$\omega_o = \left\{ \left[L \left(\frac{C_1' C_2}{C_1' + C_2} \right) \right]^{1/2} \right\}^{-1} \tag{7.26}$$

The resistance in series with the inductor changes the resonant frequency, but since r is normally much less than r_i, the change is small. For oscillations to occur, Eq. (7.24) must also be satisfied. That is,

$$rr_\pi = \frac{1 + \beta}{\omega_o^2 C_1 C_2} - \frac{L}{C_1} \tag{7.27}$$

It can be shown that if the product rr_π is greater than that given by Eq. (7.27) oscillations will die out. If rr_π is smaller than that given by Eq. (7.27), the system will be unstable and oscillations will increase in amplitude. Equation (7.27) illustrates the importance of the transistor-current gain β. As r is increased, the current gain must be increased to sustain oscillations. Also, it becomes increasingly difficult to satisfy Eq. (7.27) as the frequency is increased. In general, it is advantageous to have

$$X_{C_1} X_{C_2} = (\omega^2 C_1 C_2)^{-1}$$

as large as possible since then r can be sufficiently large. However, if C_1 and C_2 are too small (large X_{C_1} and X_{C_2}), then the transistor input and output capacitors which shunt C_1 and C_2, respectively, become important. A good, stable design will always have C_1 and C_2 much larger than the transistor capacitances they shunt.

Example 7.2 Design a 5-MHz Colpitts oscillator using a 10-μH inductor with an unloaded Q_u of 100. The transistor β is 100.

SOLUTION From Eq. (7.26) the equivalent capacitance is

$$\frac{C_1' C_2}{C_1' + C_2} = 100 \, \text{pF}$$

We will assume that C_1' and C_1 are equal; then the maximum r is found using Eq. (7.27). The series resistance of the inductor is found to be

$$r = 3.14 \, \Omega$$

One possible solution is to select $C_1 = C_2 = 200$ pF; then using Eq. (7.27),

$$3.14 r_\pi = (2.53 \times 10^6) - (5 \times 10^4)$$

or

$$r_\pi = 0.81 \times 10^6 \, \Omega$$

Since the transistor transconductance $g_m = 40I$, the transistor must be biased so that

$$I \geq 3.09 \times 10^{-6} \, \text{A}$$

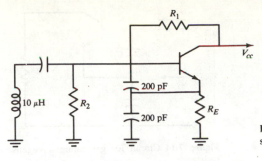

Figure 7.12 The oscillator circuit designed in Example 7.2.

A complete circuit is shown in Fig. 7.12. The resistors must be selected so that the bias current exceeds the minimum specified. Note that the assumption that r is much less than r_π is justified in this example.

Example 7.3 In the Colpitts oscillator circuit designed in Example 7.2 what will the frequency of oscillation be if the bias current is increased so that $r_\pi = 1000\ \Omega$?

SOLUTION From Eq. (7.25)

$$C_1' = 199\ \text{pF}$$

so the increased bias current will have a negligible effect and can be ignored.

Another Interpretation of the Oscillator Circuit

Although Eqs. (7.19) and (7.20) can be used to determine the exact expressions for oscillation, they are often difficult to use and add little insight to the design process. An alternative interpretation, originally presented by Gouriet[7.1] for vacuum tube oscillators, will now be presented. It is based on the fact that an ideal tuned circuit (infinite Q), once excited, will oscillate indefinitely because there is no resistance element present to dissipate the energy. In the actual case where the inductor Q is finite, the oscillations die out because energy is dissipated in the resistance.

It is the function of the amplifier to maintain oscillations by supplying an amount of energy equal to that dissipated. This source of energy can be interpreted as a negative resistor r_i in series with the tuned circuit, as shown in Fig. 7.13. If the total resistance is positive, the oscillations will die out, while the oscillation amplitude will increase if the total resistance is negative. To maintain oscillations the two resistors must be of equal magnitude. To see how a negative resistance is realized, the input impedance of the circuit in Fig. 7.14 will be derived.

If the transistor output impedance is sufficiently large, the equivalent

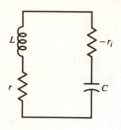

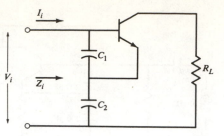

Figure 7.13 A resonant circuit including a negative resistor.

Figure 7.14 Circuit for generating a negative resistance.

circuit is as shown in Fig. 7.15. The steady-state loop equations are

$$V_i = I_i(X_{C_1} + X_{C_2}) - I_b(X_{C_1} - \beta X_{C_2}) \tag{7.28}$$

$$0 = -I_i(X_{C_1}) + I_b(X_{C_1} + r_\pi) \tag{7.29}$$

After I_b is eliminated from these two equations, Z_i is obtained as

$$Z_i = \frac{V_i}{I_i} = \frac{(1+\beta)X_{C_1}X_{C_2} + r_\pi(X_{C_1} + X_{C_2})}{X_{C1} + r_\pi} \tag{7.30}$$

If $X_{C_1} \ll r_\pi$, the input impedance is approximately equal to

$$Z_i \approx \frac{1+\beta}{r_\pi} X_{C_1} X_{C_2} + X_{C_1} + X_{C_2} \tag{7.31}$$

$$\approx \frac{-g_m}{\omega^2 C_1 C_2} + \left[j\omega \left(\frac{C_1 C_2}{C_1 + C_2} \right) \right]^{-1} \tag{7.32}$$

That is, the input impedance of the circuit shown in Fig. 7.13 is a negative resistor

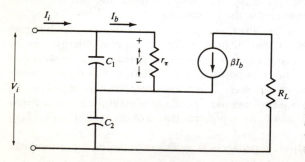

Figure 7.15 The small-signal equivalent circuit of Fig. 7.14 with the transistor output impedance neglected.

$$r_i = \frac{-g_m}{\omega^2 C_1 C_2} \tag{7.33}$$

in series with a capacitor

$$C_i = \frac{C_1 C_2}{C_1 + C_2} \tag{7.34}$$

which is the series combination of the two capacitors. With an inductor L (with series resistance r) connected across the input, it is clear that the condition for sustained oscillation is

$$r = \frac{g_m}{\omega^2 C_1 C_2} \tag{7.35}$$

and the frequency of oscillation is

$$f_o = \left[2\pi \left(L \frac{C_1 C_2}{C_1 + C_2} \right)^{1/2} \right]^{-1} \tag{7.36}$$

This interpretation of the oscillator readily provides several guidelines that can be used in the design. C_1 should be as large as possible so that $X_{C_1} \ll r_\pi$. Also, C_1 and C_2 should be much larger than the transistor output capacitances so that the transistor base-to-emitter and collector-to-emitter capacitances have a negligible effect on the circuit's performance. However, Eq. (7.35) limits the maximum value of the capacitances, since

$$r \le \frac{g_m}{\omega^2 C_1 C_2} \le \frac{G}{\omega^2 C_1 C_2} \tag{7.37}$$

where G is the maximum value of g_m. For a given series capacitance the product is a minimum when $C_1 = C_2 = C_m$. Then Eq. (7.37) can be written as

$$(\omega C_m)^{-1} > \left(\frac{r}{g_m} \right)^{1/2} \tag{7.38}$$

This equation states that for oscillations to be maintained, the minimum permissible resistance $(\omega C_m)^{-1}$ is a function of the resistance of the inductor and the transistor's mutual conductance g_m. If the two capacitors are not equal, then Eq. (7.37) must be satisfied.

The analytical approach of calculating the input impedance seen from the inductor (or capacitor in Hartley oscillators) can be used with all oscillators and is often the easiest way to analyze the circuit. The real part of the input impedance must be negative in order that the active device supply the energy dissipated in the inductor (or capacitor).

An oscillator circuit known as the *Clapp* or *Clapp-Gouriet circuit* is shown in Fig. 7.16. This oscillator is equivalent to the one just discussed, but it has the practical advantage of being able to provide another degree of design freedom by making C_o much smaller than C_1 and C_2. It is possible to use C_1 and C_2 to satisfy the condition of Eq. (7.37) and then adjust C_o for the desired

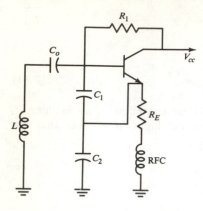

Figure 7.16 A Clapp-Gouriet oscillator.

frequency of oscillation ω_o, which is determined from

$$\omega_o L - (\omega_o C_o)^{-1} - (\omega_o C_1)^{-1} - (\omega_o C_2)^{-1} = 0 \qquad (7.39)$$

The following example illustrates the design procedure.

Example 7.4 Consider the Clapp-Gouriet oscillator shown in Fig. 7.16. The transistor is operated at a g_m of 6 mS. The coil used has an unloaded Q_u of 200 at 1 MHz and a reactive impedance of 800 Ω ($r = 4\ \Omega$). What are the required conditions for the circuit to oscillate?

SOLUTION In order to satisfy Eq. (7.38) we must have

$$(\omega C_m)^{-1} \geq \left(\frac{r}{g_m}\right)^{1/2} = 25.8\ \Omega$$

Therefore, at 1 MHz

$$C_m = 6200\ \text{pF}$$

C_m corresponds to the case of maximum series capacitance of the parallel combination of C_1 and C_2 and occurs for $C_1 = C_2 = C_m$. If both C_1 and C_2 equal 6200 pF, the reactance of the series combination of C_1 and C_2 is 51.6 Ω. C_o must then be selected so that $X_L = X_C$ at 1 MHz. If for example, $L = 82\ \mu$H, then $X_L = 515\ \Omega$ and C_o should be approximately 343 pF.

The Clapp-Gouriet oscillator is particularly effective at low frequencies. At higher frequencies C_1 and C_2 must be smaller in order that Eq. (7.37) be satisfied, and little fine-tuning advantage is obtained by adding C_o.

The gain requirement for oscillation [Eq. (7.35)] obtained using this model is somewhat different from that obtained with the matrix analysis [Eq. (7.27)]. However, it is readily shown that

$$\frac{1+\beta}{\omega_o^2 C_2} \gg L$$

so that Eq. (7.27) is approximated by

$$r \approx \frac{\beta}{r_\pi \omega_o^2 C_1 C_2} = \frac{g_m}{\omega_o^2 C_1 C_2}$$

Thus the two methods arrive at the same gain requirement for sustained oscillations. An exact expression for the required gain cannot be obtained with either model, since the actual oscillator analysis requires a nonlinear model, nor is an exact expression for the gain necessary. Oscillator design usually consists of realizing an open-loop gain three or four times larger than what the linear analysis predicts to be necessary for oscillations. This ensures that oscillations will begin. As the oscillation amplitude increases, the g_m of the active device begins to decrease until a stable oscillation amplitude that is self-limiting is reached.

The Pierce Oscillator

So far in our discusson we have considered two forms of the Colpitts oscillator. In the analysis of the generalized oscillator circuit no terminal was designated as ground. There are often practical reasons for grounding a particular terminal. Figure 7.3 describes a form of the Colpitts oscillator in which the base is grounded. If the emitter is grounded at the oscillating frequency the circuit appears as shown in Fig. 7.17 or 7.18. C_b and C_E serve to short-circuit the dc bias resistors R_b and R_E at the oscillating frequency. This circuit, often referred to as a *Pierce oscillator*, has the beneficial feature that the bias resistors do not

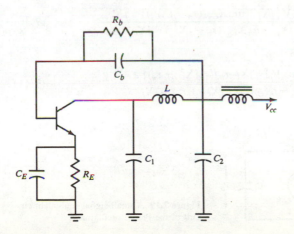

Figure 7.17 A Pierce oscillator.

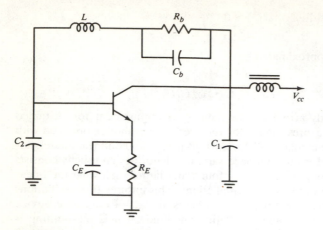

Figure 7.18 Another Pierce oscillator.

shunt the tuned circuit. Resistors connected across the tuned circuit have the effect of reducing the circuit Q, which reduces the frequency stability. The Pierce oscillator (grounded emitter) normally has the best frequency stability. (The common-base amplifier is most often used in high-frequency oscillators because the cutoff frequency of the common-base current gain is approximately β times greater than that of the common-emitter or common-collector configuration.)

The Pierce oscillator can be analyzed using the small-signal equivalent circuit shown in Fig. 7.19. It is assumed that the transistor output impedance is much larger than X_{C_1}. We will analyze the circuit by calculating the loop gain GH.

The voltage

$$V_o = \frac{-g_m V\left[jX_L - jr_\pi X_{C_2}/(r_\pi - jX_{C_2})\right] - jX_{C_1}}{jX_L - jr_\pi X_{C_2}/(r_\pi - jX_{C_2}) - jX_{C_1}} \tag{7.40}$$

and the voltage fed back is

$$V = \frac{V_o\left[-jX_{C_2}r_\pi/(-jX_{C_2} + r_\pi)\right]}{-jX_{C_2}r_\pi/(-jX_{C_2} + r_\pi) + jX_L} \tag{7.41}$$

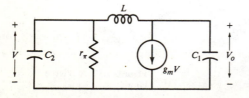

Figure 7.19 A small-signal equivalent circuit of the Pierce oscillator.

If V is eliminated from these two equations, we obtain

$$V_o = \frac{-g_m V_o r_\pi}{(j\omega L j\omega C_1)(j\omega r_\pi C_2 + 1) + j\omega r_\pi (C_2 + C_1) + 1}$$

$$= \frac{-g_m V_o r_\pi}{1 - \omega^2 L C_1 + j\omega r_\pi (C_1 + C_2 - \omega^2 L C_1 C_2)} \tag{7.42}$$

In order for the loop phase shift to be 360° the imaginary term must vanish. Therefore, the frequency of oscillation is

$$\omega_o = \left(\frac{C_1 + C_2}{L C_1 C_2}\right)^{1/2} \tag{7.43}$$

and it is also required that $\omega_o^2 L C_1 > 1$ in order for the phase shift to be 360°. That is [using Eq. (7.42)],

$$\frac{(C_1 + C_2) L C_1}{L C_1 C_2} > 1$$

which reduces to

$$\frac{C_1 + C_2}{C_2} > 1 \tag{7.44}$$

Since this is always the case, Eq. (7.43) determines the frequency of oscillation. For the open-loop gain GH to be greater than 1 at the resonant frequency, the magnitude of Eq. (7.42) must be greater than 1 at the resonant frequency. That is,

$$g_m r_\pi = \beta > \omega^2 L C_1 - 1 = \frac{C_1 + C_2}{C_2} - 1 \tag{7.45}$$

or

$$\beta C_2 > C_1 \tag{7.46}$$

This is the same requirement for oscillation as was obtained using the generalized circuit analysis [Eq. (7.22)].

The type of analysis to use depends on the circuit configuration. Sometimes calculating the loop gain is the most convenient. This is often so with FET oscillators, where the gate-to-source resistance can be neglected. The negative impedance interpretation is usually the easiest to use and provides the most insight for design.

7.3 AMPLITUDE STABILITY

Linearized analysis of oscillator circuits is convenient for determining the frequency of oscillation, but not for determining the amplitude of the oscillation. The Nyquist stability criterion states that the frequency of oscillation is the frequency at which the loop phase shift is 360°, but it says nothing about

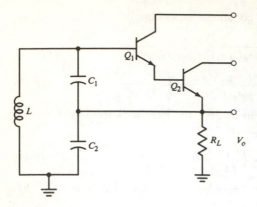

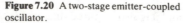

Figure 7.20 A two-stage emitter-coupled oscillator.

the oscillation amplitude. If no procedures are taken to control the oscillation amplitude it is susceptible to appreciable drift.

The two most frequently used methods for controlling the amplitude employ a self-limiting oscillator and an additional circuit or circuit element for amplitude regulation. The self-limiting oscillator is designed to be unstable; i.e., the loop gain is made greater than 1 at the frequency where the phase shift is 180°. (Usually the loop gain is designed to be two or three times that needed for oscillation.) As the amplitude increases, the transistor begins to saturate, causing the loop gain to decrease until the amplitude stabilizes—this is a self-limiting oscillator. There are nonlinear analysis techniques for predicting the amplitude of oscillation, but the results are only approximate, except in special idealized cases, and the designer must rely on an empirical approach to establish the oscillation amplitude.

An example of a two-stage emitter-coupled oscillator is shown in Fig. 7.20. In this circuit, amplitude stabilization occurs as a result of current limiting in the second stage.[7.2] This circuit has the additional advantage of output terminals that are isolated from the feedback path. The emitter of Q_2, which is rich in harmonics, is normally used for the output. Harmonics of the fundamental frequency can be obtained at the emitter of Q_2 by using an appropriately tuned circuit. Note that the collector of Q_2 is isolated from the feedback path.

7.4 PHASE STABILITY

An oscillator has a frequency or phase stability that can be considered in two separate parts. First there is the long-term stability in which the frequency changes over a period of minutes, hours, days, weeks, or even years. This frequency stability is normally limited by the circuit component's temperature coefficients and aging rates. The other part, short-term frequency stability, is

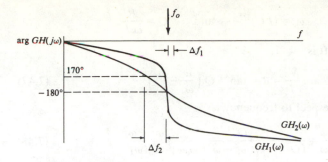

Figure 7.21 Phase plots of open-loop systems.

measured in seconds or even much shorter periods of time. One form of short-term instability is due to changes in phase of the system; here the term *phase stability* is used synonymously with *frequency stability*. It refers to how sensitive the frequency of oscillation is to small changes in phase shift of the open-loop system. It can be intuited that the system with the largest rate of change of phase as a function of frequency $(d\phi/df)$ will be the most stable in terms of frequency stability.

Figure 7.21 contains the phase plots of two open-loop systems used in oscillators. At the system crossover frequency the phase shift is $-180°$ (with negative feedback). If now some external influence causes a change in phase, say it adds $10°$ of phase lag, then the frequency will change so that the total phase shift is again $0°$. In this case the frequency will decrease to the point where the open-loop phase shift is $170°$. Figure 7.21 shows that Δf_2, the change in frequency associated with the $10°$ change in phase of GH_2, is greater than the change in frequency Δf_1, associated with open-loop system GH_1, whose phase is changing more rapidly near the open-loop crossover frequency.

This qualitative discussion illustrates that $d\phi/df|_{f=f_o}$ is a measure of an oscillator's phase stability. It provides a good means of quantitatively comparing the phase stability of two oscillators. Consider the simple parallel tuned circuit shown in Fig. 7.22. For the circuit the transimpedance is [see Eq. (4.20)]

$$\frac{V_o(j\omega)}{I(j\omega)} = \frac{R}{1 + jQ(\omega/\omega_o - \omega_o/\omega)}$$

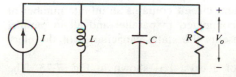

Figure 7.22 A parallel resonant circuit.

where
$$\omega_o = (LC)^{1/2} \quad \text{and} \quad Q = \frac{R}{\omega_o L}$$

The circuit phase shift is

$$\arg \frac{V_o}{I} = \phi = \tan^{-1} Q \left(\frac{\omega}{\omega_o} - \frac{\omega_o}{\omega} \right) \tag{7.47}$$

and derivative with respect to frequency is

$$\frac{d\phi}{d\omega} = \frac{Q^{-1}}{(Q^2)^{-1} + [(\omega_o^2 - \omega^2)/\omega_o\omega]^2} \frac{\omega^2 + \omega_o^2}{\omega_o(\omega)^2} \tag{7.48}$$

at the resonant frequency ω_o

$$\left. \frac{d\phi}{d\omega} \right|_{\omega=\omega_o} = \frac{2Q}{\omega_o} \tag{7.49}$$

The frequency stability factor S_F is defined as the change in phase divided by the normalized change in frequency $\Delta\omega/\omega_o$. That is,

$$S_F = 2Q \tag{7.50}$$

S_F is a measure of the short-term stability of an oscillator. Equation (7.50) indicates that the higher circuit Q, the higher the stability factor. This is one reason that high-Q circuits are widely used in oscillator circuits. Another reason is that the tuned circuit can be used to filter out undesired harmonics and noise. A piezoelectric or ceramic crystal can function in an electronic circuit as an inductor with a very high Q.

7.5 CRYSTAL OSCILLATOR CHARACTERISTICS

The previous discussion has shown that a high-Q circuit is desirable in an oscillator for short-term frequency stability. Piezoelectric (and ceramic) crystals are electromechanical devices that have very small dissipative losses and very high and stable electric circuit Q's. For these reasons they are usually employed when an oscillator with a very stable operating frequency is desired.

Crystals are three-dimensional, mechanically oscillating bodies with many modes of oscillation. These oscillations are excited through the piezoelectric properties and by the arrangement and shape of the electrodes and the crystal itself. Through the fabrication of the crystal, one has the capability of selecting certain oscillating modes and harmonics. At the electrical terminals of the crystal the observable equivalent circuit contains an infinite number of (lossy) series resonant circuits, all connected in parallel and all in parallel with a capacitance C_o, which represents the static capacitance of the electrode arrangement.

An equivalent electric circuit of a crystal is shown in Fig. 7.23. The

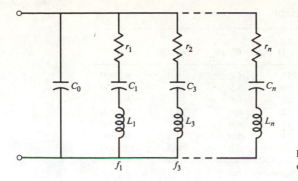

Figure 7.23 Electric circuit equivalent of a crystal.

circuit contains several series resonant circuits whose frequencies are all approximate (but not exact) odd harmonics of the fundamental frequency f_1; the higher resonant frequencies are referred to as *overtones of the fundamental frequency*. In a narrow frequency region around any resonant frequency f_i, the circuit can be simplified to that shown in Fig. 7.24. This simplified model can be considered sufficiently accurate for oscillator design, but precautions are often necessary to prevent the circuit from oscillating at an unwanted resonant frequency.

Quartz crystals have a Q ranging from about 10,000 to over a million. Ceramic resonators can be fabricated with a Q of several thousand. The inductance L_1 and capacitance C_1 of the equivalent circuit primarily depend on the mass and compliance of the quartz. The resistance r_1, which represents the losses of the circuit, is mainly attributable to damping resulting from the electrodes, the crystal mounting structure, the internal friction, and the lead resistance. The circuit components are referred to as the *crystal motional elements*, since they are the electrical equivalents of the vibratory (mechanical) motion of the crystal. Typical circuit parameters for fundamental, third, and fifth overtone crystals are given in Table 7.1.

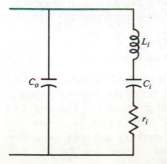

Figure 7.24 Electric circuit equivalent of a crystal, valid near the crystal's *i*th overtone frequency.

Table 7.1 Typical crystal data

f, MHz	Mode of oscillation	r_1, Ω	C_o, pF	C_1, fF	Q
1.0	Fundamental	250	4.0	9.0	65,000
2.0	Fundamental	70	3.5	10	110,000
5.0	Fundamental	15	6.0	24	85,000
10.0	Fundamental	12	6.0	24	50,000
20.0	Fundamental	12	6.0	24	25,000
45.0	3d overtone	25	5.0	1.5	90,000
100.0	5th overtone	40	5.0	0.3	130,000

Source: Courtesy of Savoy Electronics, Inc.

Table 7.1 indicates that the dissipation resistance r_1 of a crystal is relatively small. Figure 7.25 illustrates how the resistance r_1 varies as a function of fundamental frequency for three holder types. If r_1 is zero, the crystal input impedance is (in a narrow frequency region about f_o):

$$Z(j\omega) = \frac{(j\omega C_o)^{-1}[j\omega L_1 + (j\omega C_1)^{-1}]}{j\omega L_1 + (j\omega C_1)^{-1} + (j\omega C_o)^{-1}} = \frac{-[j/\omega(C_1 + C_o)](1 - \omega^2 L_1 C_1)}{1 - \omega^2 L_1[C_o C_1/(C_o + C_1)]} \quad (7.51)$$

The impedance $Z(j\omega)$ will be zero when the inductor L_1 and C_1 are in resonance [$\omega^2 = (LC)^{-1}$]. The frequency

$$f_s = [2\pi(L_1 C_1)^{1/2}]^{-1} \quad (7.52)$$

is referred to as the *series resonant frequency* of the crystal. The crystal impedance will be infinite at the frequency

$$f_a = \left\{ 2\pi \left[L_1 \left(\frac{C_o C_1}{C_o + C_1} \right) \right]^{1/2} \right\}^{-1} \quad (7.53)$$

f_a is referred to as the *antiresonant frequency* of the crystal. The ideal crystal ($r = 0$) behaves as both a series resonant and a parallel resonant circuit with infinite Q. The actual crystal also functions as both series and parallel resonant circuits, but with finite Q. The effect of a nonzero r on the circuit can be calculated from the equation

$$Z(j\omega) = \frac{(j\omega C_o)^{-1}[j\omega L_1 + r_1 + (j\omega C_1)^{-1}]}{j\omega L_1 + r_1 + (j\omega C_1)^{-1} + (j\omega C_o)^{-1}} \quad (7.54)$$

but it is usually not necessary, since the solution can be closely approximated with little difficulty. For all practical purposes the series resonant frequency is unchanged for a nonzero r_1 (using typical crystal parameters). The effect of r_1 is primarily a reduction of the circuit Q. The effect of r_1 on the antiresonant behavior is readily evaluated by making a series-to-parallel

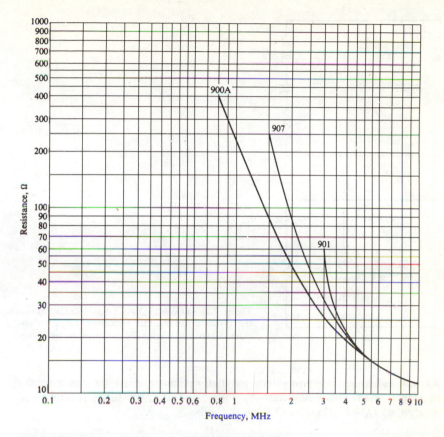

Figure 7.25 Crystal resistance r_1 as a function of fundamental frequency for three holder types. *(Courtesy of Savoy Electronics Inc.)*

transformation in the equivalent circuit, as shown in Fig. 7.26. The transformation, of course, is valid only at a single frequency. The equivalent parallel impedances are given by

$$R_p = r_1 \left(1 + \frac{X_s^2}{r_1^2}\right) = r_1(1 + Q_s^2) \tag{7.55}$$

and

$$X_p = X_s \left(1 + \frac{r_1^2}{X_s^2}\right) = X_s[1 + (Q_s^2)^{-1}] \tag{7.56}$$

where

$$X_s = \omega L_1 - (\omega C_1)^{-1} \quad \text{and} \quad Q_s = \frac{X_s}{r_1}$$

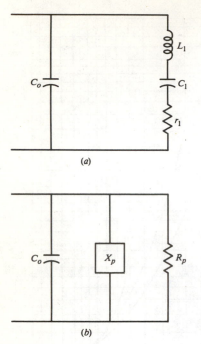

(a)

(b)

Figure 7.26 (a) Electric circuit equivalent, valid near the crystal's fundamental frequency; (b) a parallel equivalent of Fig. 7.26a, valid at a particular frequency.

At antiresonance, the equivalent parallel reactance must be equal to the reactance of the shunt capacitor $X_p = X_{C_o}$, and at f_a the series reactance X_s is large, so $Q_s \gg 1$. Therefore, $X_{C_o} \approx X_s$ and

$$R_p \approx \frac{X_{C_o}^2}{r_1} \tag{7.57}$$

Circuits containing crystals are frequently designed so that the frequency range of interest is between the series resonant and antiresonant frequencies of the crystal. In this frequency range the crystal impedance is reactive, as can be seen from an examination of Eq. (7.51). The ratio of the antiresonant to the resonant frequency is [from Eqs. (7.52) and (7.53)]

$$\frac{f_a}{f_s} = \frac{2\pi(L_1 C_1)^{1/2}}{2\pi\{L_1[C_1 C_o/(C_1 + C_o)]\}^{1/2}} = \left(1 + \frac{C_1}{C_o}\right)^{1/2} \tag{7.58}$$

The shunt capacitance C_o is normally much greater than C_1 so that

$$\frac{f_a}{f_s} = \left(1 + \frac{C_1}{C_o}\right)^{1/2} \approx 1 + \frac{C_1}{2C_o} = 1 + (2k)^{-1} \tag{7.59}$$

Typical values of k lie between 250 and 300. The crystal's antiresonant frequency is higher than its series resonant frequency.

Example 7.5 If the antiresonant frequency of the 5-MHz crystal in Table 7.1 is 5 MHz, what is the series resonant frequency f_s?

SOLUTION Since $k = C_o/C_1 = 6/0.024 = 250$ for this crystal,

$$f_s = 5 \times 10^6 \left(1 + \frac{1}{500}\right)^{-1} = 4.99 \text{ MHz}$$

The antiresonant frequency is 10 kHz higher than the series resonant frequency.

If Eq. (7.51) for the crystal impedance is rewritten as

$$Z(j\omega) = \frac{-(C_o + C_1)^{-1}(j/\omega)[1 - (\omega/\omega_s)^2]}{1 - (\omega/\omega_a)^2}$$

it is seen that the impedance is inductive for $\omega_s \le \omega \le \omega_a$, and it is capacitive for other frequency ranges. A plot of $Z(j\omega)$ is given in Fig. 7.27. This plot ignores the overtone circuits illustrated in Fig. 7.24. The actual crystal impedance will have multiple resonant and antiresonant frequencies, with the impedance inductive between each resonant and antiresonant frequency.

Example 7.6 Consider the 5-MHz crystal whose characteristics are given in Table 7.1. At the antiresonant frequency f_a the Q_s is large, so

$$X_s \approx X_{C_o}$$

If the total shunt capacitance is 6 pF, the magnitude of the shunt reac-

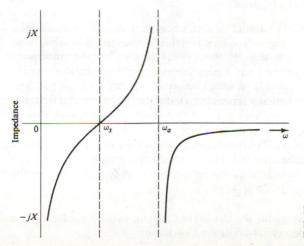

Figure 7.27 Crystal impedance as a function of frequency.

tance at 5 MHz is approximately 5305 Ω, which must be approximately the same as X_s. Therefore, since the series resistance $r = 15\,\Omega$, the circuit Q is

$$Q = \frac{5.3 \times 10^3}{15} = 3.53 \times 10^2$$

(Note that this is not the crystal Q, but solely an equivalent Q derived for easily expressing the series-to-parallel transformation.) X_s is composed of the series combination of inductive and capacitive reactances. That is,

$$X_L = X_s + X_{C_1} \approx X_{C_o} + X_{C_1}$$

Since $C_1 = 24 \times 10^{-3}\,\text{pF}$

$$X_{C_1} = [(24 \times 10^{-15})2\pi(5 \times 10^6)]^{-1} = 1.3 \times 10^6\,\Omega$$

C_1 is much smaller than the shunt capacitance, so the inductive reactance is

$$X_L \approx 1.3 \times 10^6\,\Omega$$

and the crystal Q is

$$Q = \frac{1.3 \times 10^6}{15} = 86,700$$

Actually X_L and X_{C_1} differ by a small amount (5305 Ω), but the calculation of Q is sufficiently accurate.

Parallel-Mode Crystal Oscillators

Crystals often serve as either parallel or series resonant circuits in oscillators. Their high Q provides greater frequency stability than is attainable with discrete inductors and capacitors. If the crystal is used in the antiresonant mode the circuit is referred to as a *parallel-mode crystal oscillator*. Series-mode oscillators use the crystal as a series resonant circuit, while in parallel-mode oscillators the crystal actually serves as an inductor. The oscillator design is the same as for noncrystal oscillators except that the biasing network can be different since crystals block dc voltages. A parallel-mode crystal is cut to be antiresonant at the desired oscillating frequency when the external capacitance across the crystal is a specified amount. The external capacitance is usually large enough that circuit stray capacitances can be neglected. A typical value for the specified crystal load capacitance is 32 pF.

Example 7.7 Convert again the 20-MHz Colpitts oscillator of Example 7.1 to a parallel-mode crystal-controlled oscillator.

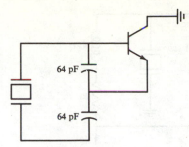

Figure 7.28 The crystal oscillator circuit designed in Example 7.7.

SOLUTION The oscillator can be converted to a crystal-controlled oscillator simply by replacing the inductor with a parallel-mode crystal antiresonant at 20 MHz. If the crystal load capacitance is specified to be 32 pF, then the parallel combination of C_1 and C_2 must be 32 pF. This could be satisfied by using 64 pF capacitors for both C_1 and C_2. In order for oscillations to occur the loop gain must still be greater than 1, or [from Eq. (7.35)]

$$\frac{g_m X_{C_1} X_{C_2}}{r_1} > 1$$

where r_1 is the series resistance of the crystal. The series resistance of crystals in this frequency range is approximately 12 Ω, so the inequality is easily satisfied. The complete circuit (except for biasing) would appear as shown in Fig. 7.28.

A difficulty in designing parallel-mode crystal oscillators is selecting the bias circuitry so that it does not reduce the circuit Q. Any resistance shunting a crystal will reduce the Q if the crystal is being used in the parallel mode. Figure 7.29 contains three parallel-mode oscillator circuits. A grounded-base oscillator is shown in Fig. 7.29a of the figure. C_b serves as a short circuit at the oscillating frequency. The crystal is shunted by a resistance R_s, consisting of the common-base input impedance r_i (a low impedance), in parallel with R_E increased by the turns ratio squared. That is,

$$R_s = \frac{r_i R_E}{r_i + R_E}\left(1 + \frac{C_2}{C_1}\right)^2 \tag{7.60}$$

In addition, the impedance of the RF choke must be high so that R_3 does not reduce the crystal Q. R_s is usually so small that the crystal Q is markedly degraded; this is not a good crystal oscillator circuit.

Figure 7.29b illustrates a grounded-collector oscillator. The crystal is shunted by the bias resistors R_1 and R_2. If these resistors are not sufficiently large they will significantly reduce the Q of the circuit. The bias resistors R_1 and R_2 do not shunt the crystal in the Pierce oscillator illustrated in Fig. 7.29c. C_E shunts the bias resistor R_E at the oscillating frequency. The Pierce

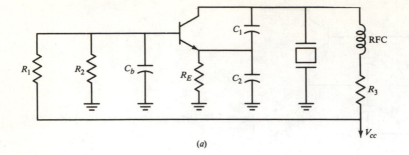

(a)

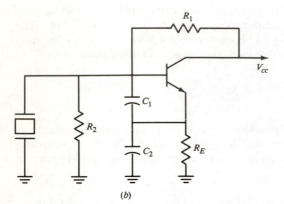

(b)

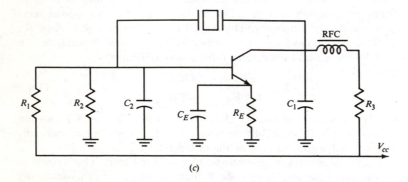

(c)

Figure 7.29 Three parallel-mode oscillator circuits: (a) a grounded-base circuit, (b) a grounded-collector circuit, and (c) a Pierce oscillator circuit.

circuit configuration is usually the best choice for a parallel-mode crystal oscillator—provided one side of the crystal does not have to be grounded. Since the bias resistors do not shunt the crystal, the Pierce oscillator normally has the highest Q and hence the best frequency stability.

Capacitor in parallel with the crystal. In many applications (such as in voltage-controlled oscillators) it is necessary to adjust the frequency of oscillation. Equation (7.53) describes how the parallel-mode (antiresonant) frequency can be varied by adding an external capacitor in parallel with the crystal, as shown in Fig. 7.30. The antiresonant frequency becomes

$$f_a = \left\{ 2\pi \left[L \, \frac{C_1(C_o + C_L)}{C_1 + C_o + C_L} \right]^{1/2} \right\}^{-1} \approx f_s \left[1 + \frac{C_1}{2(C_o + C_L)} \right] \qquad (7.61)$$

By increasing C_L, f_a can be decreased until $f_a \approx f_s$. It is left as an exercise to show that adding C_L does not significantly change the series resonant frequency. The frequency range

$$f_a - f_s = \frac{C_1}{2C_o} f_s \qquad (7.62)$$

is referred to as the *pulling range* of the crystal. Figure 7.31 illustrates how f_a varies as a function of the external load capacitance C_L. Although Eq. (7.61) indicates that the parallel antiresonant frequency can be "pulled" down to the series resonant frequency, in practice this results in poor performance for parallel-mode oscillators, since the crystal Q is simultaneously reduced.

Equation (7.55) shows that the equivalent parallel resistance at resonance is

$$R_p = r_1 \left(1 + \frac{X_s^2}{r_1^2} \right) \qquad (7.55)$$

For the antiresonant mode the parallel reactance

$$X_p = [\omega(C_o + C_L)]^{-1} \qquad (7.63)$$

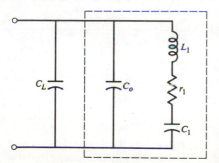

Figure 7.30 Adding a capacitor C_L in parallel with a crystal will reduce its antiresonant frequency.

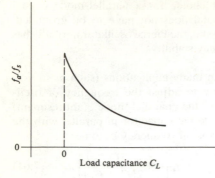

Figure 7.31 Variation in crystal antiresonant frequency as a function of load capacitance C_L.

must equal the series impedance X_s [Eq. (7.56)]. Therefore the equivalent parallel resistance is

$$R_p = r_1 \left[1 + \left(\frac{X_p}{r_1} \right)^2 \right] \tag{7.64}$$

As $C_o + C_L$ increases, X_p and hence R_p decrease, resulting in a reduction in the loop gain and eventually a cessation of oscillations. R_p versus C_L is plotted in Fig. 7.32 for a typical crystal. Equation (7.64) indicates that R_p will be small for high-frequency crystals, since X_p will be small. This is one reason that parallel-mode crystal oscillators are not used at frequencies above 20 MHz. The Q of the series resonant mode does not depend significantly upon the shunt capacitance $C_o + C_L$, so series-mode oscillators are used at the higher frequencies.

A practical rule of thumb is that the combination of the crystal capaci-

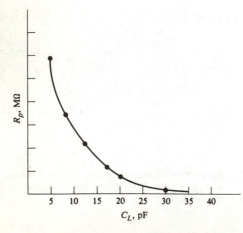

Figure 7.32 Equivalent parallel load resistance of a typical crystal as a function of additional load capacitance C_L.

tance C_o and C_L can be used in an antiresonant circuit as long as

$$[\omega(C_o + C_L)r_1]^{-1} > 4 \qquad (7.65)$$

This relation is based upon the fact that for values less than 4, the slope of the phase shift of the crystal near the frequency f_a is not sufficiently steep to provide good phase-frequency stability.[7.3]

Example 7.8 Consider again the 5-MHz crystal of Table 7.1. Assume that the specified load capacitance is 32 pF. If it is desired to decrease the antiresonant frequency, an additional capacitor C_L must be added in parallel with the 32-pF-load capacitor.

For example, if an additional 22 pF is added in parallel, the new frequency is

$$f'_a = \frac{f_a[1 + 0.024/(38 + 22)]}{1 + 0.024/2(38)} = \frac{f_a(1.0002)}{1.0003} = 4.9995 \text{ MHz}$$

That is, an additional 22 pF in parallel with C_o will reduce the antiresonant frequency by 500 Hz. The equivalent parallel resistance is found [using Eq. (7.55)] to be

$$R_p = 15\{1 + [(15 \times 2\pi \times 4.9995 \times 10^6 \times 60 \times 10^{-12})^{-1}]^2\} = 18.8 \times 10^3 \ \Omega$$

The additional 22 pF reduces the antiresonant frequency by 500 Hz and decreases R_p from 46.8 kΩ to 18.8 kΩ.

Series-Mode Crystal Oscillators

The results of the analysis of the Clapp-Gouriet oscillator can be used to show that, for crystal oscillators operating in the parallel mode, the g_m of the active device must satisfy the relation

$$g_m \geq C_1 C_2 \omega^2 r_1$$

As the frequency is increased, C_1 and C_2 must be reduced in order for this relation to hold. Once the value of the capacitors approaches the size of the transistor terminal capacitances, the oscillator stability is seriously degraded, since the transistor capacitances cannot be precisely controlled. To overcome this problem, high-frequency oscillators are operated with the crystals in the series resonant mode. (Most oscillators operating above 20 MHz are, in fact, in the series resonant mode.)

Another characteristic of crystals is that their fundamental frequency is inversely proportional to the crystal thickness (for most crystal cuts). High-frequency crystals require thin plates that are very fragile and sensitive to contamination. For this reason, high-frequency crystals usually operate on an overtone of the fundamental frequency, which allows for thicker, less fragile crystals. Overtone crystals are almost always used in the series mode. As a

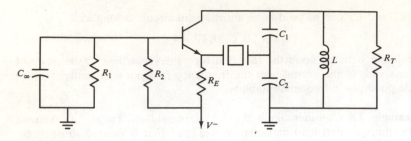

Figure 7.33 A series-mode crystal oscillator.

general rule, third-overtone crystals are used from 20 to 60 MHz, and fifth-overtone crystals from 60 to 125 MHz.

A crystal operating in the series mode functions as a short circuit at the oscillating frequency and as a large impedance at other frequencies. An example of a series-mode crystal oscillator is shown in Fig. 7.33. At frequencies other than its series resonant frequency the crystal impedance is large enough to prevent current from being fed back to the emitter. The tank circuit is designed to be antiresonant at the series resonant frequency of the crystal. It is far from obvious that the circuit Q is determined by the crystal and not the tank circuit, but the following analysis shows that it is. The small-signal equivalent circuit is shown in Fig. 7.34. Z_x is the crystal impedance, and r_i is the transistor common-base input resistance, which is assumed to be much smaller than the emitter bias resistor R_E. The voltage V is

$$V = \frac{V_o C_1}{C_1 + C_2} \frac{r_i}{r_i + Z_x}$$

provided $\omega C_2 \ll |Z_x + r_i|$, which will be the case near the series resonant

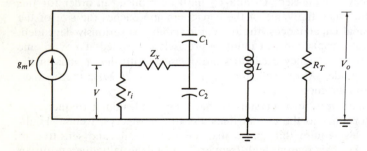

Figure 7.34 A small-signal equivalent circuit of the oscillator shown in Fig. 7.33, with the feedback path opened and terminated.

frequency in a well-designed oscillator. The output voltage is

$$V_o = g_m V Z_L$$

where Z_L is the equivalent load seen by the collector. At the series resonant frequency, Z_L will consist of the crystal resistance r_x and r_i, transformed by the turns ratio squared, in parallel with R_T. That is,

$$Z_L(j\omega_o) = R_T \| (r_x + r_i) \left(\frac{C_2 + C_1}{C_1} \right)^2$$

The open-loop gain is

$$A_o = \frac{Z_L C_1}{C_1 + C_2} (Z_x + r_i)^{-1} \qquad (7.66)$$

It was explained earlier in the chapter that a figure of merit for an oscillator is the rate of change of the phase shift of the open-loop gain, evaluated at the resonant frequency. That is,

$$S_F = \frac{d \arg A_o}{d\omega} \bigg|_{\omega = \omega_o} = \frac{d \arg Z_L}{d\omega} + \frac{d \arg (Z_x + r_i)^{-1}}{d\omega} \qquad (7.67)$$

The impedance Z_L includes the shunting effects of the crystal circuit, but it is easily shown that near the resonant frequency

$$Z_L \approx \frac{R_p}{1 + jQ(\omega/\omega_o - \omega_o/\omega)}$$

where

$$R_p = Z_L(j\omega_o)$$

Likewise

$$(Z_x + r_i)^{-1} = \left[1 + jQ_x \left(\frac{\omega}{\omega_o} - \frac{\omega_o}{\omega} \right) \right]^{-1}$$

where

$$Q_x = \frac{\omega_o L_i}{r_x + r_i} = \frac{Q_1}{1 + r_i/r_x}$$

and Q_1 is the crystal Q. Since the phase shift of a resonant circuit at the resonant frequency is [Eq. (7.50)]

$$\frac{d \arg Z_L}{d\omega} = \frac{2Q_L}{\omega_o}$$

the stability factor is

$$S_F = \omega_o \frac{d \arg A_o}{d\omega} = 2(Q_L - Q_x) \qquad (7.68)$$

Q_L is the Q of the parallel tuned circuit which is much less than that of the crystal Q_x; that is, $Q_L \ll Q_x$. Any parallel loading of the tank circuit will reduce Q_L, which justifies the assumption of ignoring the effect of crystal loading on

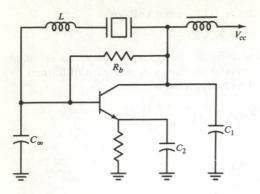

Figure 7.35 An impedance-inverting Pierce oscillator.

the tank circuit. Therefore,

$$S_F \approx -2Q_x$$

The stability factor is approximately that of $2Q_x$. Q_x is proportional to the crystal Q. The transistor common-base input resistance r_i appears in series with the crystal, but since r_i is of the same order of magnitude as the crystal resistance r_x, the circuit is controlled by the high Q of the crystal. The crystal could also be inserted in the base circuit, but then the crystal is in series with the common-emitter input resistance, which is much larger, and the phase stability factor is reduced proportionally.

Another series-mode oscillator is shown in Fig. 7.35. This circuit is often referred to as the *impedance-inverting Pierce oscillator*. A simple test to determine whether a crystal operates in the series or parallel mode in a circuit is to replace the crystal by a short circuit. If the circuit will not oscillate with the crystal short-circuited it is a parallel-mode oscillator; otherwise it usually is a series-mode oscillator.

Capacitor in series with the crystal. The frequency of series-mode crystal oscillators cannot be adjusted by adding a capacitor in parallel with the crystal, since such a capacitor has a negligible effect on the series resonant frequency f_s.

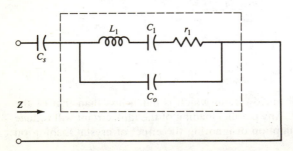

Figure 7.36 A capacitor placed in series with the crystal will increase the series resonant frequency.

The series resonant frequency can be altered by adding a capacitor in series with the crystal, as shown in Fig. 7.36. For this circuit the input impedance is

$$Z(j\omega) = -jX_o + \frac{-jX_o(jX_1 - jX_{C_1} + r_1)}{jX_1 - jX_{C_1} - jX_o + r_1} \tag{7.69}$$

where $\qquad X_s = (\omega C_s)^{-1} \qquad X_{C_o} = (\omega C_o)^{-1}$

and $\qquad X_1 = \omega L_1 \qquad X_{C_1} = (\omega C_1)^{-1}$

The input impedance can be written as

$$\begin{aligned}
Z(j\omega) &= \frac{-jX_s[jX_1 - j(X_{C_1} + X_o) + r_1] - jX_o(jX_1 - jX_{C_1} + r_1)}{jX_1 - jX_{C_1} - jX_o + r_1} \\
&= \frac{(jX_1 + r_1)(-jX_s - jX_o) + j(X_o + X_s)jX_{C_1} + j^2 X_o X_s}{jX_1 + r_1 - j(X_o + X_{C_1})} \\
&= \frac{-j(X_o + X_s)\{jX_1 + r_1 - j[X_{C_1} + X_o X_s/(X_o + X_s)]\}}{jX_1 + r_1 - j(X_o + X_{C_1})} \tag{7.70}
\end{aligned}$$

Let $\qquad X' = \left[\omega \left(\dfrac{C_o C_s}{C_o + C_s} \right) \right]^{-1} = nX_o \tag{7.71}$

where $\qquad n = \dfrac{C_o + C_s}{C_s} = \dfrac{X_o^{-1} + X_s^{-1}}{X_s^{-1}} = \dfrac{X_o + X_s}{X_o} \tag{7.72}$

and define

$$X'_{C_1} = X_{C_1} + \frac{X_o X_s}{X_o + X_s} \tag{7.73}$$

Then

$$\begin{aligned}
X_o + X_{C_1} &= X'_{C_1} + X_o - \frac{X_o X_s}{X_o + X_s} \\
&= X'_{C_1} + \frac{X_o^2}{X_o + X_s} \tag{7.73a}
\end{aligned}$$

and

$$\begin{aligned}
Z(j\omega) &= \frac{-j(X_o + X_s)[(jX_1 + r_1) - jX'_{C_1}]}{jX_1 + r_1 - j[X'_{C_1} + X_o^2/(X_o + X_s)]} \\
&= \frac{-jnX_o(jX_1 + r_1 - jX'_{C_1})}{jX_1 + r_1 - j(X'_{C_1} + X_o/n)} \\
&= \frac{-jnX_o(jn^2 X_1 + n^2 r_1 - jn^2 X'_{C_1})}{jn^2 X_1 + n^2 r_1 - jn^2 X'_{C_1} - jX_o n} \tag{7.74}
\end{aligned}$$

That is, the crystal-plus-series capacitor can be represented by the equivalent circuit shown in Fig. 7.37, where

$$(C'_1)^{-1} = C_1^{-1} + (C_o + C_s)^{-1}$$

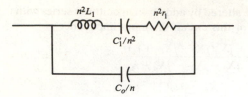

Figure 7.37 A convenient equivalent circuit of a capacitor in series with a crystal.

or

$$C_1' = \frac{C_1(C_o + C_s)}{C_1 + C_o + C_s} \tag{7.75}$$

This equivalent circuit is of the same form as that of the crystal (minus the series capacitor), so the same expressions can be used as were used in deriving the crystal characteristics. For example, the antiresonant frequency of the circuit with the capacitor in series with the crystal is [Eq. (7.53)]

$$\omega_a = \left\{ \left[L_1 n^2 \frac{(C_1'/n^2)(C_o/n)}{C_1'/n^2 + C_o/n} \right]^{1/2} \right\}^{-1}$$

This expression is readily simplified to

$$\omega_a = \left[\left(L \frac{C_1 C_o}{C_1 + C_o} \right)^{1/2} \right]^{-1}$$

which is the same as the antiresonant frequency of the crystal without a series

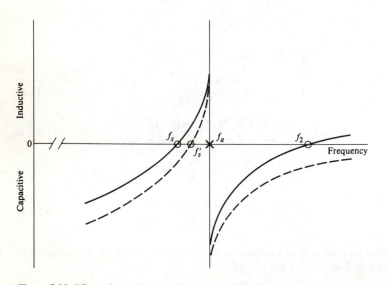

Figure 7.38 Effect of a series capacitor on crystal impedance (dashed curve).

capacitance. The series capacitance does not change the antiresonant frequency.

The new series resonant frequency is

$$\omega'_s = [(n^2 L_1 C'_1/n^2)^{1/2}]^{-1} = \left\{ \left[L_1 \frac{C_1(C_o + C_s)}{C_1 + C_o + C_s} \right]^{1/2} \right\}^{-1} \qquad (7.76)$$

The addition of the series capacitor moves the series resonant frequency toward the antiresonant frequency. The crystal-plus-series capacitor reactance plots are illustrated in Fig. 7.38.

Note that without the series capacitance, C_o has a negligible effect on the series resonant frequency, but once C_s is added, ω_s can be changed by changing C_o. Changing C_o will, however, also change ω_a. The ability to change the oscillating frequency by adding a capacitor in series or parallel is used in the design of voltage-controlled crystal oscillators.

7.6 VOLTAGE-CONTROLLED OSCILLATORS AND VOLTAGE-CONTROLLED CRYSTAL OSCILLATORS

The preceding sections have shown how the frequency of oscillation can be varied by the addition of a capacitor. Diodes exist (referred to as *varicaps* or *varactors*) which function as voltage-variable capacitors. If the varicap is included in the oscillator circuit and the frequency of oscillation is varied by changing the dc bias voltage across the varicap, the oscillator is referred to as a *voltage-controlled oscillator* (*VCO*). If the VCO is crystal-controlled, the oscillator is referred to as a *voltage-controlled crystal oscillator* (*VCXO*). These devices find many applications, such as in frequency modulators, telemetry, Doppler radar, spectrum analyzers, television tuners, and frequency synthesizers.

A frequently used VCO is shown in Fig. 7.39a. The modulating voltage f_m changes the varicap voltage and thus the capacitance shunting the inductor, thereby changing the frequency of oscillation. If the varicap is added in series with the inductor, as shown in Fig. 7.39b, the VCO is a Clapp-Gouriet oscillator. In this configuration a smaller value of capacitance can be used to change the oscillating frequency. A main difficulty in VCO design is to achieve a linear voltage-frequency transfer characteristic. The idealized voltage-frequency transfer characteristic is illustrated in Fig. 7.40.

A simplified model of a VCXO is illustrated in Fig. 7.41. The transistor amplifier, operating near the series resonant frequency of the feedback circuit, is represented by a voltage amplifier with frequency-independent gain A. The crystal is operating in the series mode (if the crystal can be replaced by a short circuit without stopping the oscillation, the crystal is being used in the series mode). The frequency is controlled by the bias voltage applied to the voltage-variable capacitance C_s. Since the addition of C_s increases the series resonant frequency, the reactance of L_s is chosen equal in magnitude to that of

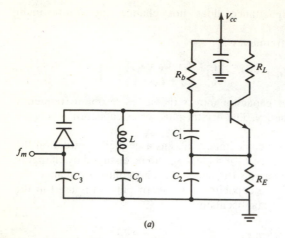

(a)

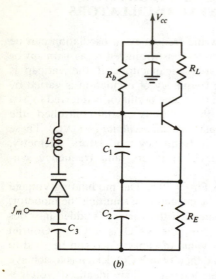

(b)

Figure 7.39 (a) A voltage-controlled oscillator; (b) another voltage-controlled oscillator.

C_s, making the series resonant frequency again that of the crystal. The resonant frequency can be increased or decreased with the control voltage.

The overall Q will be close to that of the crystal, so the circuit frequency stability is close to that of the fixed-frequency crystal. To verify that the Q remains high, consider a 20-MHz series-mode crystal-plus-series capacitance and inductor, as shown in Fig. 7.42. At the resonant frequency the magnitude of the crystal inductance is 2 MΩ and the series resistance of the crystal is 25 Ω ($Q_x = 80,000$). The magnitude of the series inductance is 750 Ω and the coil

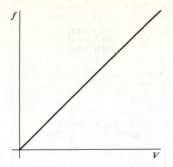

Figure 7.40 Ideal voltage-frequency transfer characteristics.

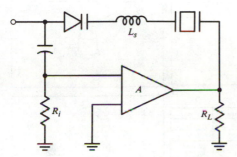

Figure 7.41 A voltage-controlled crystal oscillator (VCXO).

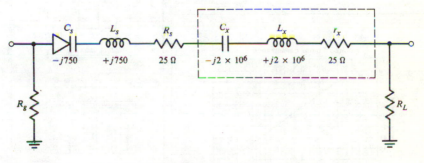

Figure 7.42 A small-signal equivalent circuit of the voltage-controlled oscillator.

resistance is 25 Ω ($Q_u = 30$). Therefore, the total series inductance reactance is approximately 2 MΩ and the total series resistance is 50 Ω, the circuit $Q_u = 40,000$. The addition of the series inductor has reduced the circuit Q by a factor of 2, but it is still a very high Q circuit.

Figure 7.43a illustrates the capacitance of a tuning diode as a function of the reverse-bias voltage across the diode. The capacitance is an approximately

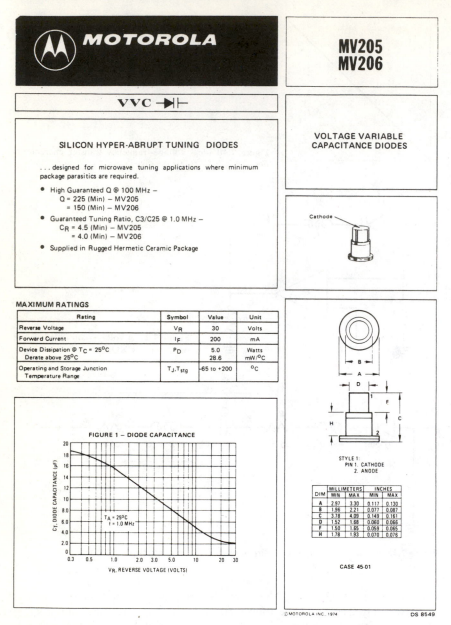

MOTOROLA

MV205
MV206

VVC →▶︎|—

SILICON HYPER-ABRUPT TUNING DIODES

... designed for microwave tuning applications where minimum package parasitics are required.

- High Guaranteed Q @ 100 MHz —
  Q = 225 (Min) — MV205
  = 150 (Min) — MV206
- Guaranteed Tuning Ratio, C3/C25 @ 1.0 MHz —
  C_R = 4.5 (Min) — MV205
  = 4.0 (Min) — MV206
- Supplied in Rugged Hermetic Ceramic Package

VOLTAGE VARIABLE CAPACITANCE DIODES

Cathode

MAXIMUM RATINGS

Rating	Symbol	Value	Unit
Reverse Voltage	V_R	30	Volts
Forward Current	I_F	200	mA
Device Dissipation @ T_C = 25°C Derate above 25°C	P_D	5.0 28.6	Watts mW/°C
Operating and Storage Junction Temperature Range	T_J, T_{stg}	-65 to +200	°C

FIGURE 1 — DIODE CAPACITANCE

T_A = 25°C
f = 1.0 MHz

C_T, DIODE CAPACITANCE (pF)

V_R, REVERSE VOLTAGE (VOLTS)

STYLE 1:
PIN 1. CATHODE
2. ANODE

DIM	MILLIMETERS		INCHES	
	MIN	MAX	MIN	MAX
A	2.97	3.30	0.117	0.130
B	1.96	2.21	0.077	0.087
C	3.78	4.09	0.149	0.161
D	1.52	1.68	0.060	0.066
F	1.50	1.65	0.059	0.065
H	1.78	1.93	0.070	0.076

CASE 45-01

© MOTOROLA INC., 1974   DS 8549

Figure 7.43 (a) Tuning diode characteristics; (b) additional tuning diode characteristics. *(Courtesy of Motorola Inc.)*

MV205 ● MV206

ELECTRICAL CHARACTERISTICS (T_A = 25°C unless otherwise noted.)

Characteristic — All Types	Symbol	Min	Typ	Max	Unit
Reverse Breakdown Voltage (I_R = 10 μAdc)	BV_R	30	–	–	Vdc
Reverse Voltage Leakage Current (V_R = 28 V) (V_R = 28, T_A = 60°C)	I_R	– –	– –	50 0.5	nAdc μAdc
Series Inductance (1) (f = self resonant frequency)	L_S	–	0.8	–	nH
Case Capacitance (2) (f = 1.0 MHz)	C_C	–	0.15	–	pF
Diode Capacitance Temperature Coefficient (6) (V_R = 3.0 Vdc, f = 1.0 MHz, -55°C to +125°C)	TC_c	–	–	400	ppm/°C

Device Type	C_T (2),(3) V_R = 25 Vdc pF		Q (5) f = 100 MHz C_T = 9 pF	C_R (4) f = 1.0 MHz C_3/C_{25}	
	Min	Max	Min	Min	Max
MV205	2.0	2.3	225	4.5	6.0
MV206	1.8	2.8	150	4.0	6.0

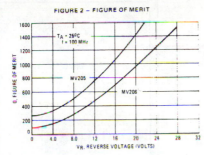

FIGURE 2 – FIGURE OF MERIT

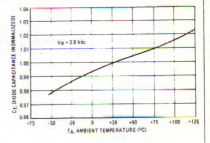

FIGURE 3 – DIODE CAPACITANCE

PARAMETER TEST METHODS

1. L_S, SERIES INDUCTANCE

L_S is determined from the self resonant frequency and the junction capacity of the device.

$$L_S = \frac{1}{\omega^2_{res} C_J}$$

2. C_C, CASE CAPACITANCE

C_C is measured on an open package at 1.0 MHz using a capacitance bridge (Boonton Electronics Model 75A or equivalent.)

3. C_T, DIODE CAPACITANCE

(C_T = C_C + C_J). C_T is measured at 1.0 MHz using a capacitance bridge (Boonton Electronics Model 75A or equivalent.)

4. C_R, CAPACITANCE RATIO

C_R is the ratio of C_T measured at 3.0 Vdc divided by C_T measured at 25 Vdc.

5. Q, FIGURE OF MERIT

Q is calculated by taking the G and C readings of an admittance bridge at the specified frequency and substituting in the following equations:

$$Q = \frac{2\pi f C}{G}$$

(Boonton Electronics Model 33AS8 or equivalent).

6. TC_c, DIODE CAPACITANCE TEMPERATURE COEFFICIENT

TC_c is guaranteed by comparing C_T at V_R = 3.0 Vdc, f = 1.0 MHz, T_A = -55°C with C_T at V_R = 3.0 Vdc, f = 1.0 MHz, T_A = +125°C in the following equation, which defines TC_c:

$$TC_c = \frac{|C_{T(+125°C)} - C_{T(-55°C)}| \times 10^6}{(55 + 125) \, C_{T(25°C)}}$$

Accuracy limited by C_T measurement, ±0.1 pF.

MOTOROLA *Semiconductor Products Inc.*

BOX 20912 ● PHOENIX, ARIZONA 85036 ● A SUBSIDIARY OF MOTOROLA INC

Figure 7.43a *Continued.*

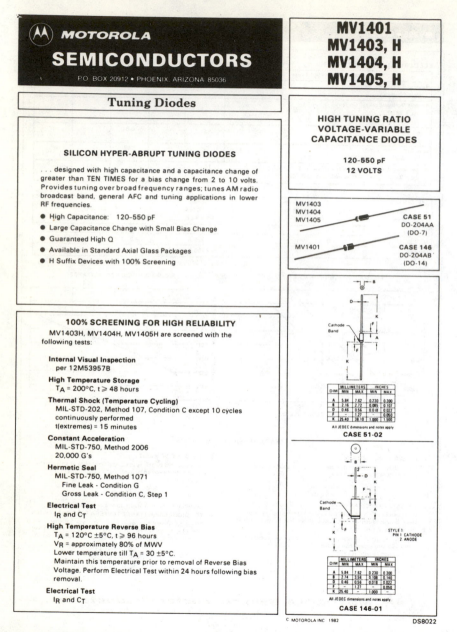

MOTOROLA SEMICONDUCTORS

P.O. BOX 20912 • PHOENIX, ARIZONA 85036

Tuning Diodes

MV1401
MV1403, H
MV1404, H
MV1405, H

**HIGH TUNING RATIO
VOLTAGE-VARIABLE
CAPACITANCE DIODES**

120–550 pF
12 VOLTS

SILICON HYPER-ABRUPT TUNING DIODES

. . . designed with high capacitance and a capacitance change of greater than TEN TIMES for a bias change from 2 to 10 volts. Provides tuning over broad frequency ranges; tunes AM radio broadcast band, general AFC and tuning applications in lower RF frequencies.

● High Capacitance: 120–550 pF
● Large Capacitance Change with Small Bias Change
● Guaranteed High Q
● Available in Standard Axial Glass Packages
● H Suffix Devices with 100% Screening

MV1403
MV1404
MV1405

CASE 51
DO-204AA
(DO-7)

MV1401

CASE 146
DO-204AB
(DO-14)

100% SCREENING FOR HIGH RELIABILITY

MV1403H, MV1404H, MV1405H are screened with the following tests:

Internal Visual Inspection
per 12M53957B

High Temperature Storage
T_A = 200°C, t ≥ 48 hours

Thermal Shock (Temperature Cycling)
MIL-STD-202, Method 107, Condition C except 10 cycles continuously performed
t(extremes) = 15 minutes

Constant Acceleration
MIL-STD-750, Method 2006
20,000 G's

Hermetic Seal
MIL-STD-750, Method 1071
    Fine Leak - Condition G
    Gross Leak - Condition C, Step 1

Electrical Test
I_R and C_T

High Temperature Reverse Bias
T_A = 120°C ±5°C, t ≥ 96 hours
V_R = approximately 80% of MWV
Lower temperature till T_A = 30 ±5°C.
Maintain this temperature prior to removal of Reverse Bias Voltage. Perform Electrical Test within 24 hours following bias removal.

Electrical Test
I_R and C_T

Cathode
Band

DIM	MILLIMETERS		INCHES	
	MIN	MAX	MIN	MAX
A	5.84	7.62	0.230	0.300
B	2.16	2.72	0.085	0.107
D	0.46	0.56	0.018	0.022
F	–	1.27	–	0.050
K	25.40	38.10	1.000	1.500

All JEDEC dimensions and notes apply

CASE 51-02

Cathode
Band

STYLE 1
PIN 1 CATHODE
2 ANODE

DIM	MILLIMETERS		INCHES	
	MIN	MAX	MIN	MAX
A	5.84	7.62	0.230	0.300
B	2.74	3.56	0.108	0.140
D	0.46	0.56	0.018	0.022
F	–	1.27	–	0.050
K	25.40	–	1.000	–

All JEDEC dimensions and notes apply.

CASE 146-01

© MOTOROLA INC. 1982

DS8022

Figure 7.43b *Continued.*

MV1401 ● MV1403, H ● MV1404, H ● MV1405, H

MAXIMUM RATINGS

Rating	Symbol	Value	Unit
Reverse Voltage	V_R	12	Volts
Forward Current	I_F	250	mA
Device Dissipation @ T_A = 25°C Derate above 25°C	P_D	400 2.67	mW mW/°C
Junction Temperature	T_J	+175	°C
Storage Temperature Range	T_{stg}	–65 to +200	°C

ELECTRICAL CHARACTERISTICS (T_A = 25°C unless otherwise noted)

Characteristic — All Types	Symbol	Min	Typ	Max	Unit
Reverse Breakdown Voltage (I_R = 10 μAdc)	$V_{(BR)R}$	12	—	—	Vdc
Leakage Current at Reverse Voltage (V_R = 10 Vdc, T_A = 25°C)	I_R	—	—	0.10	μAdc
Series Inductance (f = 250 MHz, Lead Length ≈ 1/16″)	L_S	—	5.0	—	nH
Case Capacitance (f = 1.0 MHz, Lead Length ≈ 1/16″)	C_C	—	0.25	—	pF

	C_T, Diode Capacitance						Q, Figure of Merit	TR, Tuning Ratio	
	V_R = 1.0 Vdc, f = 1.0 MHz pF			V_R = 2.0 Vdc, f = 1.0 MHz pF			V_R = 2.0 Vdc, f = 1.0 MHz	C_1/C_{10} f = 1.0 MHz	C_2/C_{10} f = 1.0 MHz
Device	Min	Nom	Max	Min	Nom	Max	Min	Min	Min
MV1401	468	550	633	—	—	—	200	14	—
MV1403, H	—	—	—	140	175	210	200	—	10
MV1404, H	—	—	—	96	120	144	200	—	10
MV1405, H	—	—	—	200	250	300	200	—	10

PARAMETER TEST METHODS

1. **L_S, SERIES INDUCTANCE**
   L_S is measured on a shorted package at 250 MHz using an impedance bridge (Boonton Radio Model 250A RX Meter).

2. **C_C, CASE CAPACITANCE**
   C_C is measured on an open package at 1.0 MHz using a capacitance bridge (Boonton Electronics Model 75A or equivalent).

3. **C_T, DIODE CAPACITANCE**
   ($C_T = C_C + C_J$) C_T is measured at 1.0 MHz using a capacitance bridge (Boonton Electronics Model 75A or equivalent).

4. **TR, TUNING RATIO**
   TR is the ratio of C_T measured at 2.0 Vdc (1.0 Vdc for MV1401) divided by C_T measured at 10 Vdc.

5. **Q, FIGURE OF MERIT**
   Q is calculated by taking the G and C readings of an admittance bridge at the specified frequency and substituting in the following equation:

$$Q = \frac{2\pi f C}{G}$$

   (Boonton Electronics Model 33AS8). Use Lead Length ≈ 1/16″.

FIGURE 1 — DIODE CAPACITANCE versus REVERSE VOLTAGE

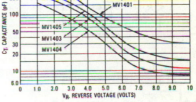

Ⓜ MOTOROLA *Semiconductor Products Inc.*

BOX 20912 ● PHOENIX, ARIZONA 85036 ● A SUBSIDIARY OF MOTOROLA INC.

Figure 7.43b *Continued.*

linear function of voltage for reverse voltages of 1 to 10 V. (The capacitance varies from 16 to 5 pF.) If this diode is used as the tuning capacitor in Fig. 7.42 (a capacitive reactance of 750 Ω corresponds to 10.6 pF at 20 MHz), the control voltage should be approximately 2.5 V. If the voltage is reduced to 1 V, the series capacitance will be 16 pF (497 Ω), and the net series reactance will be equal to 253 (750 − 497) Ω inductive, which will cause a decrease in the series resonant frequency. If the control voltage is increased to 10 V the series capacitance will be 5 pF (1591 Ω), and the net series reactance will be equal to 841 (1591 − 750) Ω capacitive, resulting in an increase in the series resonant frequency. If it is desired to use this circuit to frequency-modulate an audio signal, the diode should be biased approximately in the middle of its linear region ($V_R = 4.5$ V), and the audio voltage should be approximately 9 V peak to peak if the entire linear region of the tuning diode is to be utilized. Figure 7.43b describes the voltage-versus-capacitance characteristics of another family of tuning diodes.

7.7 FIELD-EFFECT TRANSISTOR OSCILLATORS

FETs are used extensively in oscillator circuits because they have several inherent advantages over bipolar transistors. Their high input impedance

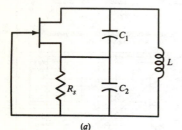

(a)

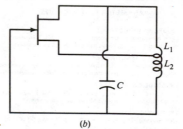

(b)

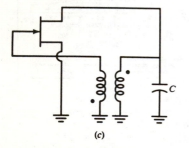

(c)

Figure 7.44 (a) A FET Colpitts oscillator; (b) a FET Hartley oscillator; (c) a FET oscillator with transformer feedback.

permits operation at lower current levels with less power dissipation, and hence the thermal problems introduced by the power dissipation are reduced. Also, when operated in the square-law region, the transconductance of the device is not a function of signal level. Figure 7.44 illustrates (with the biasing circuitry removed) three configurations commonly used in FET oscillators. Figure 7.44*a* contains a Colpitts oscillator, and Fig. 7.44*b* illustrates a Hartley common-gate oscillator. The oscillator shown in Fig. 7.44*c* utilizes the phase inversion possible with a transformer to obtain a 360° loop phase shift (180° via the transformer, plus the 180° phase shift present in the common-source configuration). Example 7.9 will illustrate the main points of FET oscillator design.

FET Pierce Oscillator

Figure 7.45 illustrates a Pierce oscillator employing a FET. The equivalent circuit is shown in Fig. 7.46. If r_d is neglected, the small-signal loop gain is

$$\frac{g_m V_o (X_{C_1} X_{C_2})}{-jX_{C_1} - jX_{C_2} + jX_L + R_s} = V_o \qquad (7.77)$$

or

$$\frac{g_m (\omega^2 C_1 C_2)^{-1} V_o}{R_s + j(X_L - X_{C_1} - X_{C_2})} = V_o$$

For the loop phase shift to be 360° the reactances must cancel so that

$$X_L = X_{C_1} + X_{C_2}$$

If the phase shift is 360°, the circuit will oscillate, provided

$$\frac{g_m X_{C_1} X_{C_2}}{R_s} \geq 1 \qquad (7.78)$$

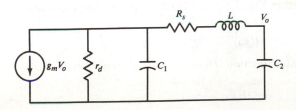

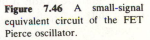

Figure 7.45 A FET Pierce oscillator.

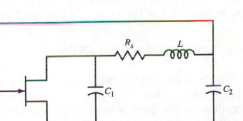

Figure 7.46 A small-signal equivalent circuit of the FET Pierce oscillator.

In the preceding analysis r_d was assumed large enough to be neglected. In the design X_{C_1} should be selected so that it is much less than r_d.

Example 7.9 Design a FET oscillator with a transistor whose parameters are $r_d = 50 \text{ k}\Omega$ and $g_m = 5 \times 10^{-3}$. The frequency of oscillation is to be 10^8 rad/s.

SOLUTION The Pierce oscillator shown in Fig. 7.45 can be used. X_{C_1} can be about 1 kΩ; r_d will then have a negligible loading effect. Also for oscillations to occur it is necessary that $g_m X_{C_1} X_{C_2} > R_s$. If $R_s = 15$ and $X_{C_1} = 1 \text{ k}\Omega$, then

$$X_{C_2} > \frac{R_s}{g_m X_{C_1}} = \frac{R_s}{g_m} 10^{-3}$$

or

$$X_{C_2} > 10^{-3} \frac{15}{g_m} = \frac{15}{10^3 \times 5 \times 10^{-3}} = 3$$

The corresponding capacitance values are

$$C_2 = 3300 \text{ pF}$$

and

$$C_1 = 10 \text{ pF}$$

This value of C_1 may be so small that the transistor output capacitance has an effect. Therefore it is desirable to increase C_1. If C_1 is increased by a factor of 10 so that, $X_{C_1} = 100$, then C_2 must also be increased:

$$X_{C_2} > \frac{R_s}{g_m \times 100} = \frac{15}{0.5} = 30$$

That is, C_2 must be less than 330 pF for oscillations to occur. The inductance L is found from $X_L = X_{C_1} + X_{C_2} = 130 \, \Omega$ or $L = 1.3 \times 10^{-6}$ H.

7.8 OSCILLATOR CONTROL USING DELAY LINES

The criteria for oscillation are (1) that the magnitude of the loop gain be unity and (2) that the loop output signal be fed back so that it is in phase with the input at the frequency at which the magnitude of the gain is unity. The oscillators described up to this point have used an LC resonator to obtain the desired phase shift. Another method that can be used is to incorporate a delay line in the feedback path, as illustrated in Fig. 7.47. An ideal delay line has the transfer function

$$H(j\omega) = e^{-j\omega T}$$

The magnitude of this transfer function $|H(j\omega)|$ is 1 at all frequencies, and the phase shift

$$\arg H(j\omega) = -\omega T$$

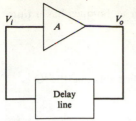

Figure 7.47 A delay-line oscillator.

is a linearly decreasing function of frequency. The phase shift can be described as shown in Fig. 7.48. Recent advances in acoustic surface-wave technology now make the design of delay lines a practical task in the frequency region above 10 MHz. Surface-wave delay lines can be designed to have circuit Q's between those of crystals and LC resonators. The delay line will find applications in designs where it is desired to have a greater frequency stability than is provided by an LC oscillator and greater "pullability," or frequency deviation, than can be obtained with a crystal oscillator. With delay-line oscillators, wideband frequency modulation can be achieved without the use of a complex frequency-multiplier chain.

The Q of a delay line is defined by comparing the slope of its phase shift to the slope of an RLC network at its resonant frequency. The slope of the phase shift of a parallel RLC network at its resonant frequency is [Eq. (7.49)]

$$\frac{d\phi}{d\omega} = \frac{-2Q}{\omega_o}$$

The slope of the phase shift of a delay line $d\phi/d\omega = -T$ is equal to the line time delay. The delay line Q is defined as

$$Q = \frac{\omega_o T}{2}$$

The longer the delay of the line, the higher the Q of the line, and the greater will be the frequency stability of the line.

Acoustic surface-wave delay lines provide a good approximation to the ideal delay line.[7.4] The magnitude and phase characteristics depend upon the

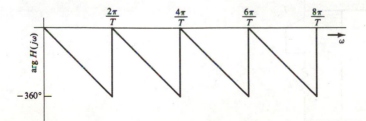

Figure 7.48 Delay-line phase shift as a function of frequency.

fabrication and crystal cut. One of the best fabrication techniques has a time delay of

$$T = \frac{N}{f_o}$$

where N is the number of wavelengths of line length. For this line the Q is

$$Q = \pi N$$

The magnitude of the line frequency response is given by

$$|H(j\omega)| = \left[\frac{\sin (f - f_o)2\pi f_o}{(f - f_o)2\pi/f_o} \right]^2$$

7.9 INTEGRATED-CIRCUIT OSCILLATORS

Oscillators can also be realized using integrated circuits, but their performance is not as good as a well-designed and constructed discrete-component oscillator.

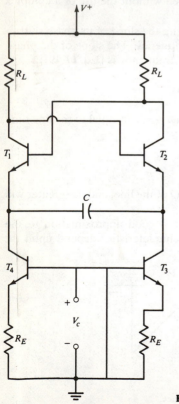

Figure 7.49 An IC oscillator.

Discrete-component oscillators can be built with a better noise performance and are capable of operating at higher frequencies than the IC oscillators. Many IC oscillators which do use external components are of the type shown in Fig. 7.49. This circuit contains a free-running multivibrator using transistors T_1 and T_2. Transistors T_3 and T_4 function as constant current sources. The emitter resistors R_E linearize the voltage-current relationship so that the current is proportional to the control voltage V_c. The frequency of oscillation is

$$f = \frac{I_o}{4CV}$$

where the voltage V is the voltage required to turn the multivibrator off (approximately 0.6 V). Since

$$I_o \approx \frac{V_c}{R_E}$$

the oscillating frequency is directly proportional to the control voltage. The oscillator output can be either a square of a triangular waveform, but additional wave-shaping circuitry is required to obtain a sinusoidal output waveform.

Figure 7.50 shows a commercial crystal-controlled oscillator that can operate between 0.1 and 20 MHz. The circuit uses an external series crystal for precise frequency regulation. The device includes wave-shaping circuitry which generates a sinusoidal output waveform. Outputs are also available for directly interfacing with emitter-coupled logic and with transistor-transistor-logic (TTL) circuitry.

MC12060 • MC12560
MC12061 • MC12561

The MC12060/12560 and MC12061/12561 are designed for use with an external crystal to form a crystal controlled oscillator. In addition to the fundamental series mode crystal, two bypass capacitors are required (plus usual power supply pin bypass capacitors). Translators are provided internally for MECL and MTTL outputs.

- Frequency Range = 100 kHz to 2.0 MHz for MC12060/12560
  = 2.0 MHz to 20 MHz for MC12061/12561
- Temperature Range = -55°C to +125°C for MC12560, 61
  = 0°C to +70°C for MC12060, 61
- Single Supply Operation: +5.0 Vdc or -5.2 Vdc
- Three Outputs Available:
  1. Complementary Sine Wave (600 mVp-p typ)
  2. Complementary MECL
  3. Single Ended MTTL

L SUFFIX
CERAMIC PACKAGE
CASE 620

P SUFFIX
PLASTIC PACKAGE
CASE 648
MC12060/MC12061 only.

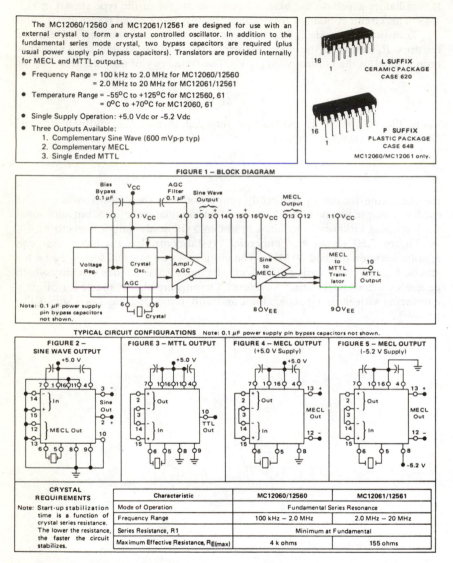

FIGURE 1 – BLOCK DIAGRAM

TYPICAL CIRCUIT CONFIGURATIONS Note: 0.1 µF power supply pin bypass capacitors not shown.

FIGURE 2 – SINE WAVE OUTPUT

FIGURE 3 – MTTL OUTPUT

FIGURE 4 – MECL OUTPUT (+5.0 V Supply)

FIGURE 5 – MECL OUTPUT (-5.2 V Supply)

CRYSTAL REQUIREMENTS

Note: Start-up stabilization time is a function of crystal series resistance. The lower the resistance, the faster the circuit stabilizes.

Characteristic	MC12060/12560	MC12061/12561
Mode of Operation	Fundamental Series Resonance	
Frequency Range	100 kHz – 2.0 MHz	2.0 MHz – 20 MHz
Series Resistance, R1	Minimum at Fundamental	
Maximum Effective Resistance, $R_{E(max)}$	4 k ohms	155 ohms

Figure 7.50 An IC oscillator. (*Courtesy of* Motorola Inc.)

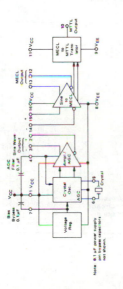

Note: 0.1 μF power supply pin bypass capacitors not shown.

				MC12560, MC12561										MC12060, MC12061									
		Pin Under		@ Test Temperature	−55°C		+25°C			+125°C			0°C		+25°C			+75°C					
Characteristic	Symbol	Test	Unit	Min	Max	Min	Typ	Max	Min	Max	Min	Max	Min	Typ	Max	Min	Max	Min	Max			
Power Supply Drain Current — MC12060/12560 — MC12061/12561	I_{CC}	1 1 11 16	mAdc	1 1 13 18		1 1 16 23		19 28	1 1 13 18	19 28	13 18	19 28	13 18	16 23	19 28	13	19	13 16	19			
						3.0 16		4.0 19		4.0 19		4.0 19		3.0 16	4.0 19							
Input Current	I_{INH}	14 15	μAdc					250 250		250 250		250 250			250 250			14 15				
	I_{INL}	14 15	μAdc					1.0 1.0		1.0 1.0		1.0 1.0			1.0 1.0			15 14				
Differential Offset Voltage MC12060/12560 MC12061/12561	ΔV	4 to 7 2 to 3 2 to 3	mVdc	40 −220 −100	325 +220 +100		0 0	325 +220 +100	40 −300 −200	325 +300 +200	40 −300 −200	325 +300 +200		8 0 0	325 +300 +200			8 14,15 8,15				
Output Voltage Level	V_{out}	2 3	Vdc	3.5 3.5		3.5 3.5					3.5 3.5							8				
Logic "1" Output Voltage	V_{OH1}*	12 13	Vdc	3.92 3.92	4.07 4.07	4.04 4.04		4.19 4.19	4.12 4.12	4.37 4.37	4.00 4.00	4.16 4.16	4.04 4.04		4.19 4.19	4.10 4.10	4.28 4.28	14 15	15 14	14 15	16 16	1 1	8 8
	V_{OH2}	10	Vdc	2.4		2.4			2.4		2.4		2.4			2.4					10		8,9
Logic "0" Output Voltage	V_{OL1}*	12 13	Vdc	2.97 2.97	3.39 3.39	3.00 3.00		3.44 3.44	3.04 3.04	3.50 3.50	2.98 2.98	3.43 3.43	3.00 3.00		3.44 3.44	3.02 3.02	3.47 3.47	15 14	14 15	16 16		12 13	8 8
	V_{OL2}	10	Vdc	0.5 0.5		0.5 0.5			0.5 0.5		0.5 0.5		0.5 0.5			0.5 0.5		15 15	14 14	1,16	10 10		8,9 8,9
Logic "1" Threshold Voltage	V_{OHA}	12 13	Vdc	3.90 3.90		4.02 4.02			4.15 4.15		3.98 3.98		4.02 4.02			4.08 4.08			16 16	16 16		12 13	8 8
Logic "0" Threshold Voltage	V_{OLA}	12 13	Vdc	3.41 3.41		3.41 3.41			3.52 3.52		3.45 3.45		3.41 3.41			3.46 3.46		15 14	14 15	16 16		12 13	8 8
Output Short-Circuit Current	I_{OS}	10	mAdc	20	60	20		60	20	60	20	60	20		60	20	60	15		11,16			89,10

TEST VOLTAGE/CURRENT VALUES

	Volts										mA		
	V_{IHmax}	V_{ILmin}	V_{IHAmin}	V_{ILAmax}	V_{IHT}	V_{CCL}	V_{CC}	V_{CCH}	I_{OL}	I_{OH}	I_L	Gnd
@ Test Temperature −55°C +25°C +125°C 0°C +25°C +75°C	4.07 4.19 4.37 4.16 4.19 4.28	3.18 3.21 3.25 3.19 3.21 3.23	3.72 3.90 4.03 3.86 3.90 3.96	3.49 3.52 3.60 3.51 3.52 3.55	4.0 4.0 4.0 4.0 4.0 4.0	4.5 4.5 4.5 4.75 4.75 4.75	5.0 5.0 5.0 5.0 5.0 5.0	5.5 5.5 5.5 5.25 5.25 5.25	16 16 16 16 16 16	−0.4 −0.4 −0.4 −0.4 −0.4 −0.4	−2.5 −2.5 −2.5 −2.5 −2.5 −2.5			

TEST VOLTAGE/CURRENT APPLIED TO PINS LISTED BELOW

*Devices will meet standard MECL logic levels using $V_{EE} = -5.2$ Vdc and $V_{CC} = 0$.

Figure 7.50 *Continued.*

283

FIGURE 6 – AC CHARACTERISTICS – MECL AND MTTL OUTPUTS

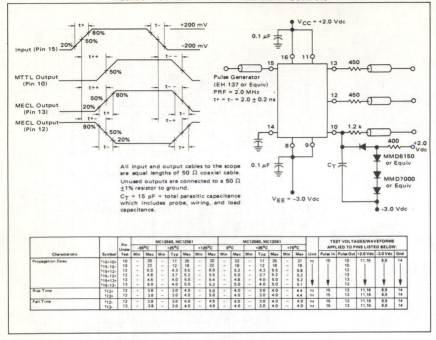

Characteristic	Symbol	Pin Under Test	MC12560, MC12561									MC12060, MC12061									Unit	TEST VOLTAGES/WAVEFORMS APPLIED TO PINS LISTED BELOW:					
			−55°C		+25°C			+125°C		0°C		+25°C			+75°C						Pulse In	Pulse Out	+2.0 Vdc	−3.0 Vdc	Gnd
			Min	Max	Min	Typ	Max	Min	Max	Min	Max	Min	Typ	Max	Min	Max									
Propagation Delay	t15+10+	10	−	30	−	17	25	−	30	−	22	−	17	25	−	27	ns	15	10	11,16	8,9	14
	t15−10−	10	−	22	−	12	18	−	22	−	19	−	12	18	−	18		15	10			
	t15+12+	12	−	5.0	−	4.3	5.5	−	6.0	−	5.2	−	4.3	5.5	−	5.8			12			
	t15−12+	12	−	4.8	−	3.7	5.2	−	5.5	−	5.0	−	3.7	5.2	−	5.2			12			
	t15+13+	13	−	4.6	−	4.0	5.0	−	5.4	−	4.8	−	4.0	5.0	−	5.2			13			
	t15−13−	13	−	5.0	−	4.0	5.0	−	5.2	−	5.0	−	4.0	5.0	−	5.1			13			
Rise Time	t12+	12	−	3.8	−	3.0	4.0	−	5.0	−	4.0	−	3.0	4.0	−	4.4	ns	15	12	11,16	8,9	14
	t13+	13	−	3.8	−	3.0	4.0	−	5.0	−	4.0	−	3.0	4.0	−	4.4	ns	15	13	11,16	8,9	14
Fall Time	t12−	12	−	3.8	−	3.0	4.0	−	4.5	−	4.0	−	3.0	4.0	−	4.0	ns	15	12	11,16	8,9	14
	t13−	13	−	3.8	−	3.0	4.0	−	4.5	−	4.0	−	3.0	4.0	−	4.0	ns	15	13	11,16	8,9	14

OUTLINE DIMENSIONS

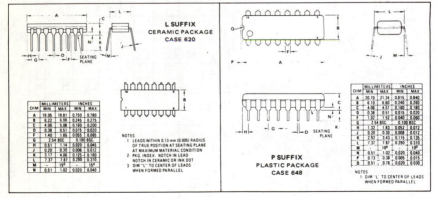

L SUFFIX
CERAMIC PACKAGE
CASE 620

DIM	MILLIMETERS		INCHES	
	MIN	MAX	MIN	MAX
A	19.05	19.81	0.750	0.780
B	6.22	6.98	0.245	0.275
C	4.06	5.08	0.160	0.200
D	0.38	0.51	0.015	0.020
F	1.40	1.65	0.055	0.065
G	2.54 BSC		0.100 BSC	
H	0.51	1.14	0.020	0.045
J	0.20	0.30	0.008	0.012
K	3.17	4.06	0.125	0.160
L	7.37	7.87	0.290	0.310
M	−	15°	−	15°
N	0.51	1.02	0.020	0.040

NOTES
1 LEADS WITHIN 0.13 mm (0.005) RADIUS OF TRUE POSITION AT SEATING PLANE AT MAXIMUM MATERIAL CONDITION
2 PKG INDEX NOTCH IN LEAD NOTCH IN CERAMIC OR INK DOT
3 DIM "L" TO CENTER OF LEADS WHEN FORMED PARALLEL

P SUFFIX
PLASTIC PACKAGE
CASE 648

DIM	MILLIMETERS		INCHES	
	MIN	MAX	MIN	MAX
A	20.70	21.34	0.815	0.840
B	6.10	6.60	0.240	0.260
C	4.06	4.57	0.160	0.180
D	0.38	0.51	0.015	0.020
F	1.02	1.52	0.040	0.060
G	2.54 BSC		0.100 BSC	
H	1.32	1.83	0.052	0.072
J	0.20	0.30	0.008	0.012
K	2.92	3.43	0.115	0.135
L	7.37	7.87	0.290	0.310
M	−	10°	−	10°
N	0.51	1.02	0.020	0.040
P	0.13	0.38	0.005	0.015
Q	0.51	0.76	0.020	0.030

NOTES
1 DIM "L" TO CENTER OF LEADS WHEN FORMED PARALLEL

Figure 7.50 *Continued*.

FIGURE 7 – AC TEST CIRCUIT – SINE WAVE OUTPUT

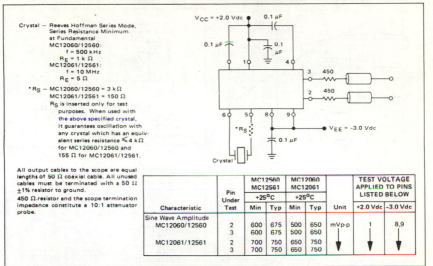

Crystal – Reeves Hoffman Series Mode,
Series Resistance Minimum
at Fundamental
MC12060/12560:
f = 500 kHz
$R_E = 1 k \Omega$
MC12061/12561:
f = 10 MHz
$R_E = 5 \Omega$

* R_S – MC12060/12560 = 3 kΩ
MC12061/12561 = 150 Ω
R_S is inserted only for test
purposes. When used with
the above specified crystal,
it guarantees oscillation with
any crystal which has an equiv-
alent series resistance $\leqslant 4$ kΩ
for MC12060/12560 and
155 Ω for MC12061/12561.

All output cables to the scope are equal
lengths of 50 Ω coaxial cable. All unused
cables must be terminated with a 50 Ω
±1% resistor to ground.
450 Ω resistor and the scope termination
impedance constitute a 10:1 attenuator
probe.

Characteristic	Pin Under Test	MC12560 MC12561 +25°C		MC12060 MC12061 +25°C		Unit	TEST VOLTAGE APPLIED TO PINS LISTED BELOW	
		Min	Typ	Min	Typ		+2.0 Vdc	-3.0 Vdc
Sine Wave Amplitude								
MC12060/12560	2	600	675	500	650	mVp-p	1	8,9
	3	600	675	500	650			
MC12061/12561	2	700	750	650	750			
	3	700	750	650	750			

OPERATING CHARACTERISTICS

The MC12060/12560 and MC12061/12561 consist of three basic sections: an oscillator with AGC and two translators (Figure 1). Buffered complementary sine wave outputs are available from the oscillator section. The translators convert these sine wave outputs to levels compatible with MECL and/or MTTL.

Series mode crystals should be used with the oscillator. If it is necessary or desirable to adjust the crystal frequency, a reactive element can be inserted in series with the crystal – an inductor to lower the frequency or a capacitor to raise it. When such an adjustment is necessary, it is recommended that the crystal be specified slightly lower in frequency and a series trimmer capacitor be added to bring the oscillator back on frequency. As the oscillator frequency is changed from the natural resonance of the crystal, more and more dependence is placed on the external reactance, and temperature drift of the trimming components then affects overall oscillator performance.

The MC12060/12560 and MC12061/12561 are designed to operate from a single supply – either +5.0 Vdc or -5.2 Vdc. Although each translator has separate V_{CC} and V_{EE} supply pins, the circuit is NOT designed to operate from both voltage levels at the same time. The separate V_{EE} pin from the MTTL translator helps minimize transient disturbance. If neither translator is being used, all unused pins (9 thru 16) should be connected to V_{EE} (pin 8). With the translators not powered, supply current drain is typically reduced from 35 mA to 16 mA for the MC12060/12560, and from 42 mA to 23 mA for the MC12061/12561.

Frequency Stability

Output frequency of different oscillator circuits (of a given device type number) will vary somewhat when used with a given test setup, however the variation should be within approximately ±0.001% from unit to unit.

Frequency variations with temperature (independent of the crystal, which is held at 25°C) are small – about –0.08 ppm/°C

for MC12061/12561 operating at 8.0 MHz, and about –0.16 ppm/°C for MC12060/12560 operating at 1.0 MHz (see Figure 8).

Signal Characteristics

The sine wave outputs at either pin 2 or pin 3 will typically range from 800 mVp-p (no load) to 500 mVp-p (120 ohm ac load). Approximately 500 mVp-p can be provided across 50 ohms by slightly increasing the dc current in the output buffer by the addition of an external resistor (680 ohms) from pin 2 or 3 to ground, as shown in Figure 9. Frequency drift is typically less than 0.0003% when going from a high-impedance load (1 megohm, 15 pF) to the 50-ohm load of Figure 9. The dc voltage level at pin 2 or 3 is nominally 3.5 Vdc with $V_{CC} = +5.0$ Vdc.

Harmonic distortion content in the sine wave outputs is crystal as well as circuit dependent. The largest harmonic (third) will usually be at least 15 dB down from the fundamental. The harmonic content is approximately load independent except that the higher harmonic levels (greater than the fifth) are increased when the MECL translator is being driven.

Typically, the MECL outputs (pins 12 and 13) will drive up to five gates, as defined in Figure 10, and the MTTL output (pin 10) will drive up to ten gates, as defined in Figure 11.

Noise Characteristics

Noise level evaluation of the sine wave outputs using the circuit of Figure 12, with operation at 1.0 MHz for MC12060/12560 and 9.0 MHz for MC12061/12561, indicates the following characteristics:

1. Noise floor (200 kHz from oscillator center frequency) is approximately –122 dB when referenced to a 1.0 Hz bandwidth. Noise floor is not sensitive to load conditions and/or translator operation.
2. Close-in noise (100 Hz from oscillator center frequency) is approximately –88 dB when referenced to a 1.0 Hz bandwidth.

Figure 7.50 *Continued.*

FIGURE 8 — FREQUENCY SHIFT versus TEMPERATURE

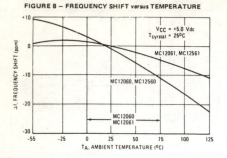

FIGURE 9 — DRIVING LOW-IMPEDANCE LOADS

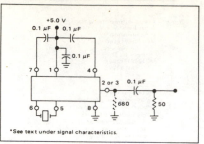

*See text under signal characteristics.

FIGURE 10 — MECL TRANSLATOR LOAD CAPABILITY

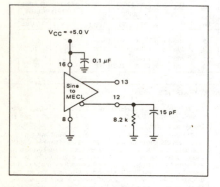

FIGURE 11 — MTTL TRANSLATOR LOAD CAPABILITY

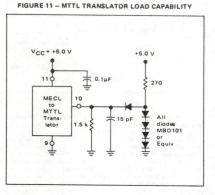

FIGURE 12 — NOISE MEASUREMENT TEST CIRCUIT

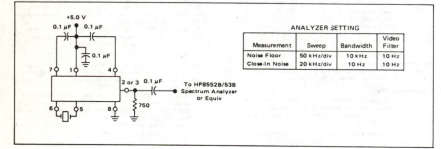

ANALYZER SETTING			
Measurement	Sweep	Bandwidth	Video Filter
Noise Floor	50 kHz/div	10 kHz	10 Hz
Close-In Noise	20 kHz/div	10 Hz	10 Hz

Figure 7.50 *Continued.*

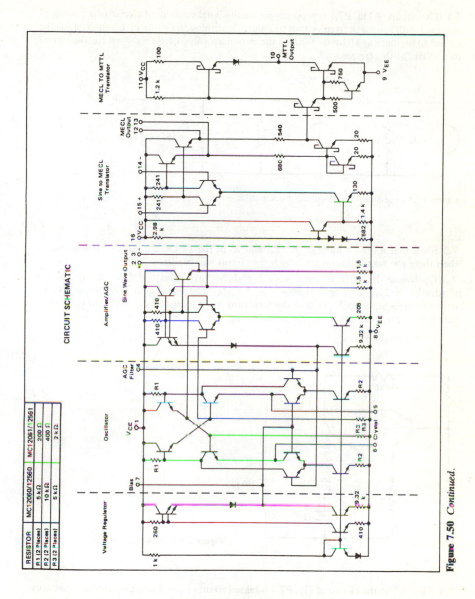

Figure 7.50 Continued.

PROBLEMS

7.1 The circuit of Fig. P7.1 represents the small-signal equivalent circuit of a two-stage amplifier with feedback from output to input. Determine the value of L required for the circuit to oscillate at 10 MHz. What is the minimum value of g_m required for the circuit to oscillate at this frequency?

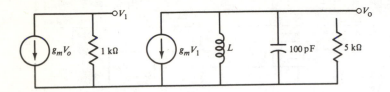

Figure P7.1 A second-order feedback circuit.

7.2 Show that if one or more of Z_1, Z_2, and Z_3 of Eq. (7.18) are positive real resistors, then there are no conditions for which the circuit will oscillate.

7.3 Determine the value of inductance L and the turns ratio N_1/N_2 so that the circuit illustrated in Fig. P7.3 will oscillate at 5 MHz. The loop gain should initially be approximately equal to 3. Assume the transistor input impedance is sufficiently large so that it does not load down the autotransformer.

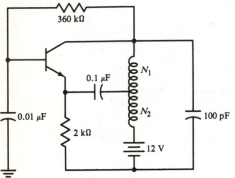

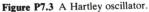

Figure P7.3 A Hartley oscillator.

7.4 The FET in the circuit of Fig. P7.4 is biased so that $g_m = 5$ mS. Determine capacitors C_1 and C_2 so that the circuit will oscillate at 10 MHz. The open-loop gain should be at least 2.5 to ensure that oscillations begin.

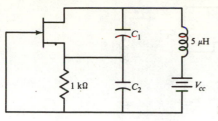

Figure P7.4 A common-gate Colpitts oscillator.

7.5 The FET in the circuit illustrated in Fig. P7.5 is biased so that $g_m = 5$ mS. Determine the inductance L and N_1/N_2 so that the circuit will oscillate at 10 MHz.

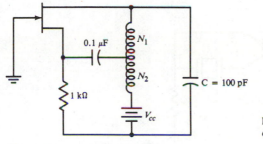

Figure P7.5 A common-gate Hartley oscillator.

7.6 Design a Colpitts 3.5-MHz oscillator using a 1.5-μH coil with a Q_μ of 150. The load resistance is 4 kΩ and the transistor has a minimum β of 100. The supply voltage is 12 V. Specify the complete circuit, including bias resistors.

7.7 Repeat the oscillator design of Prob. 7.6 using a FET with a g_m of 6×10^{-3} S. Estimate the power delivered to the 4 kΩ.

7.8 Design a 20-MHz oscillator using a 2N3904 transistor ($\beta_{min} = 100$) and a 12-V supply. Specify the complete circuit, including bias resistors.

7.9 If a capacitor is added in series with a crystal does the antiresonant frequency of the composite circuit change?

7.10 Design a 15-MHz crystal-controlled oscillator using a crystal that is antiresonant at 15 MHz, provided a 32-pF load is connected across it.

7.11 A crystal has $C_o = 3$ pF, $C_1 = 0.01$ pF, $L_1 = 0.1$ H, and $r = 15$ Ω. Calculate the series and parallel resonant frequencies of the crystal. How much capacitance must be added to change the antiresonant frequency by 0.01 percent? How much capacitance must be added to change the series resonant frequency by 0.01 percent?

7.12 Calculate the output impedance of the common-base amplifier shown in Fig. P7.12. Under what conditions will this circuit oscillate when the inductor with a finite Q_μ is connected across it?

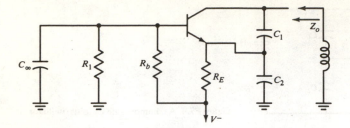

Figure P7.12 A common-base oscillator.

7.13 Derive an expression for the loop gain of the grounded-base crystal oscillator illustrated in Fig. P7.13. Show that the loop gain is always less than 1.

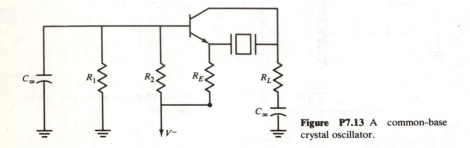

Figure P7.13 A common-base crystal oscillator.

7.14 Figure P7.14 illustrates a FET Pierce oscillator. Assume the 10-MHz crystal of Table 7.1 is antiresonant at 10 MHz if the external load capacitance is 32 pF. Select C_1 and C_2 and determine the minimum g_m for this circuit to oscillate at 10 MHz. The transistor input impedance is $10\,\text{M}\Omega$ shunted by $3\,\text{pF}$, and the transistor output impedance is $15\,\text{k}\Omega$.

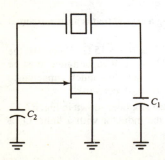

Figure P7.14 A FET Pierce crystal oscillator.

7.15 Design a series resonant 40-MHz crystal oscillator. The supply voltage is 20 V. Show all component values, including the bias network.

7.16 The circuit shown in Fig. P7.16 is frequently used as a tuned-input, tuned-output amplifier. Show that the collector-to-base capacitance C_f can cause the circuit to oscillate.

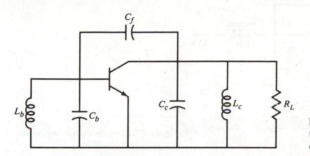

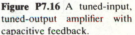

Figure P7.16 A tuned-input, tuned-output amplifier with capacitive feedback.

7.17 The 45-MHz third-overtone crystal described in Table 7.1 is used in a series-mode oscillator. An inductor with a Q of 100 is placed across the crystal to resonate out C_o at 45 MHz. Estimate the resulting Q of the combination at 45 MHz.

7.18 Derive an expression for the change in the series resonant frequency if an inductor is added in series with a crystal.

7.19 Derive an expression for the change in the antiresonant frequency if an inductor is added in parallel with a crystal.

7.20 Given a crystal that is antiresonant at 10 MHz with a 32-pF load and that has $r_i = 60\ \Omega$ plus an NPN transistor with a minimum β of 100 and a 12-V supply, design a 10-MHz oscillator. Show all circuit details, including the bias network. Any additional components used can be assumed to be ideal.

7.21 The circuit shown in Fig. P7.21 uses a time-delay network in the feedback path. What value of T will be required for the circuit to oscillate at 10 MHz?

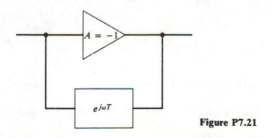

Figure P7.21

7.22 The FET of the circuit illustrated in Fig. P7.22 is biased so that $g_m = 4$ mS. If C_1 and C_2 are each 64 pF, what is the maximum value of crystal resistance r_1 for which the circuit will oscillate? Show how you would modify the circuit so that you can use a 20-MHz series-mode crystal with one terminal connected to the transistor gate.

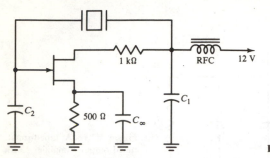

Figure P7.22 A delay-line oscillator.

7.23 Select values for C_1 and C_2 that will enable the circuit shown in Fig. P7.23 to oscillate at 40 MHz. The inductor $Q_u = 100$. The input impedance looking into the source of the grounded-gate amplifier is approximately 100 Ω. What will be the minimum g_m of the transistor necessary for the circuit to oscillate?

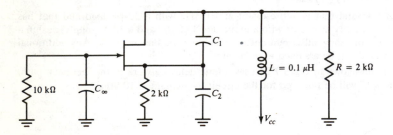

Figure P7.23 A grounded-gate oscillator.

7.24 Consider the series-mode oscillator circuit of Fig. P7.24. It uses the 45-MHz crystal described in Table 7.1.

(a) What value of capacitance must be added to increase the oscillating frequency by 0.1 percent?

(b) Where should the capacitor be added?

(c) Estimate the power dissipated in the crystal if the peak-load voltage is 10 V.

(d) What value of L is required? Assume the inductor Q is infinite.

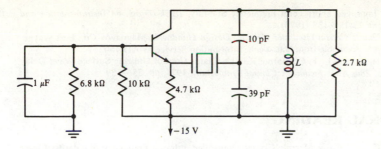

Figure P7.24 A grounded-base oscillator.

7.25 Figure P7.25 illustrates a FET VCO with a buffered output stage. Explain the operation of the circuit, including the purpose of each transistor and estimate the turns ratio N_1/N_2 required for an open-loop gain of 3.

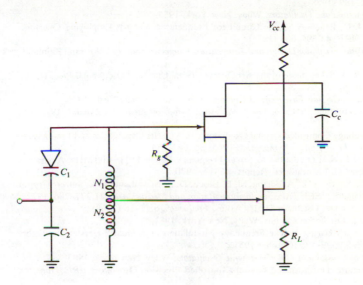

Figure P7.25 A voltage-controlled oscillator.

REFERENCES

7.1 Gouriet, G. G.: High Stability Oscillator, *Wireless Engineer*, April 1950, pp. 105–112.
7.2 Driscoll, M. M.: Two-Stage Self-Limiting Series Mode Type Quartz Crystal Oscillator

Exhibiting Improved Short-Term Frequency Stability, *IEEE Trans. on Instrumentation and Measurement*, **22**:130–138 (1973).

7.3 Firth, D.: *Quartz Crystal Oscillator Circuits*, *Design Handbook*, Magnavox Co., Fort Wayne, Indiana, 1965. Available from Nat. Tech. Information Service as AD 460377.

7.4 Vollmers, H. G., and L. T. Claiborne: R.F. Oscillator Control Utilizing Surface Wave Delay Lines, *Proc. 28th Ann. Frequency Control Symposium*, 1974, pp. 256–259.

ADDITIONAL READING

Anderson, T. C., and F. G. Merrill: Crystal Controlled Primary Frequency Standards: Latest Advances for Long Term Stability, *IRE Trans. on Instrumentation*, **9**:136–140 (1960).

Barnes, J. A., et al.: Characterization of Frequency Stability, U.S. Dept. of Comm. Nat. Bureau of Standards *NBS Technical Note 394*, 1970.

Baxandall, P. J.: Transistor Crystal Oscillators and the Design of a 1 Mc/s Oscillator Circuit Capable of Good Frequency Stability, *Radio and Electronic Engineer*, April 1965.

Clapp, J. K.: An Inductance Capacitance Oscillator of Unusual Frequency Stability, *Proc. IRE*, **36**:356–358 (1949).

Cote, A. J., Jr.: Matrix Analysis of Oscillators and Transistor Applications, *IRE Trans. on Circuit Theory*, **5**:181–189 (1958).

Edson, W. A.: *Vacuum Tube Oscillators*, Wiley, New York, 1953.

Felch, E. O., and J. O. Israel: A Simple Circuit for Frequency Standards Employing Overtone Crystals, *Proc. IRE*, **43**:596–603 (1955).

Frerking, M. R.: *Crystal Oscillator Design and Temperature Compensation*, Van Nostrand Reinhold, New York, 1978.

Gerber, E. A., and R. A. Sykes: State of the Art Quartz Crystal Units and Oscillators, *Proc. IEEE*, **54**:103–116 (1966).

Heising, R. A.: *Quartz Crystals for Electrical Circuits*, Van Nostrand, New York, 1946.

Helle, J.: VCXO Theory and Practice, *Proc. 29th Ann. Symp. on Frequency Control*, 1975, pp. 300–307.

Kent, R. L.: The Voltage Controlled Crystal Oscillator (VCXO), Its Capabilities and Limitations, *Proc. 19th Ann. Freq. Control Symposium*, 1965, pp. 642–654.

Lane, M.: Transistor Crystal Oscillators to Cover Frequency Range 1 kHz–100 MHz, Australian Post Office Research Laboratories, Report #6513, 1970.

Layden, O. P., W. L. Smith, A. E. Anderson, M. B. Bloch, D. E. Newell, and P. C. Sulzer: Crystal Controlled Oscillator, *IEEE Trans. on Instrumentation and Measurement*, **21**:277–286 (1972).

Marker, Thomas F.: Crystal Oscillator Design Notes, *Frequency*, **6**:12–16 (1968).

Mathys, R. J.: *Crystal Oscillation Circuits*, Wiley, New York, 1983.

Mortley, W. S.: Circuit Giving Linear Frequency Modulation of a Quartz Crystal Oscillator, *Wireless World*, **57**:399–403 (October 1951).

Parzen, B.: *Design of Crystal and Other Harmonic Oscillators*, Wiley, New York, 1983.

Smith, W. L.: Miniature Transistorized Crystal Controlled Precision Oscillators, *IRE Trans. on Instrumentation*, **9**:141–148 (1960).

Warner, A. W.: High Frequency Crystal Units for Primary Frequency Standards, *Proc. IRE*, **40**:1030–1033 (1952).

Ziegler, R. R.: Know Your Oscillators, *Microwave J.*, June 1976, pp. 44–47.

EIGHT

PHASE-LOCKED LOOPS

8.1 INTRODUCTION

A *phase-locked loop* (*PLL*) is a feedback system in which the feedback signal is used to lock the output frequency and phase to the frequency and phase of an input signal. The input waveform can be of many different types, including sinusoidal or digital. The first known application of the phase-locked technique was in 1932 for the synchronous detection of radio signals.[8.1,8.2] These early applications were all concerned with the detection of a transmitted signal.

Starting in the 1960s, the NASA satellite programs used the phase-locked technique to determine the frequency of the signals transmitted by satellites. Although the transmission was designed to take place at 108 MHz, oscillator drift and Doppler shift resulted in an uncertainty of several kilohertz in the received signal. The transmitted signal was of very narrow bandwidth, but because of the frequency drifts it was necessary that the receiver bandwidth be much wider, with a resultant increase in noise power. (It was demonstrated in Chap. 3 that the receiver noise power is proportional to the bandwidth.) However, the satellite communication system was improved by using a phase-locked loop to lock onto the transmitted frequency, and thus permit a much narrower receiver bandwidth with much less output noise power.

The phase-locked loop has been used for filtering, frequency synthesis, motor-speed control, frequency modulation, demodulation, signal detection, and a variety of other applications. The realization of the phase-locked loop as a relatively inexpensive integrated circuit has made it one of the most frequently used communication circuits. Phase-locked loops can be analog or digital, but the majority are composed of both analog and digital components. Some authors apply the term "PPL" to a digital phase-locked loop that contains one or more digital components. But since virtually all PLLs contain digital

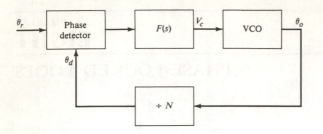

Figure 8.1 Block diagram of a phase-locked loop.

components, in this textbook the notation "digital phase-locked loop" (DPLL) will be reserved for PLLs in which all of the components are digital.

Figure 8.1 illustrates the basic architecture of the phase-locked loop. The phase detector generates an output signal that is a function of the difference between the phases of the two input signals. The detector output is filtered (and perhaps amplified), and the dc component of the error signal is applied to the voltage-controlled oscillator. The signal fed back to the phase detector is the VCO output frequency divided by N. The VCO control voltage $V_c(t)$ forces the VCO to change frequency in the direction that reduces the difference between the input frequency and the divider output frequency. If the two frequencies are sufficiently close, the PLL feedback mechanism forces the two-phase detector input frequencies to be equal, and the VCO is "locked" with the incoming frequency. That is,

$$f_r = f_d$$

and the divider output frequency is

$$f_d = \frac{f_o}{N}$$

The output frequency

$$f_o = Nf_r$$

is an integral multiple of the input frequency. If a divider is not used, N equals 1. Once the loop is in lock there will be a small phase difference between the two phase-detector input signals. This phase difference results in a dc voltage at the phase-detector output which is required to shift the VCO from its free-running frequency and keep the loop in lock. (This is not true for type II PLLs, which are described in the next chapter.)

The self-correcting ability of the PLL allows it to track frequency changes in the input signal once it is locked. The range of frequencies over which the PLL can remain locked to an input signal is known as its *lock range*. *Capture range* is the range of frequencies over which the loop can acquire lock, and this range is less than the lock range.

Since the PLL output frequency is an integral multiple of the reference frequency, it can be changed simply by changing the divide ratio N. Integrated circuitry has made the digital programmable divider an inexpensive circuit component. This provides a means of easily generating multiple frequencies from a single input frequency. Frequency synthesis is a major application of PLLs, and an entire chapter is devoted to it in this text. Before examining various applications of the phase-locked loop, we will first develop a mathematical model for the system and examine the characteristics of the various types of phase detectors.

8.2 LINEAR MODEL OF THE PHASE-LOCKED LOOP

Although the PLL is nonlinear because the phase detector is nonlinear, it can be accurately modeled as a linear device when the phase difference between the phase-detector input signals is small. For the linear analysis, it is assumed that the phase-detector output is a voltage which is a linear function of the difference in phase between its inputs; that is,

$$V_a = K_d(\theta_r - \theta_d) \tag{8.1}$$

where θ_r and θ_d are the phases of the input and feedback signals, respectively. K_d is the *phase-detector gain factor* and has dimensions of volts per radian. The characteristics of several types of phase detectors are discussed in detail later in the chapter. It will also be assumed that the VCO can be modeled as a linear device whose output frequency deviates from its free-running frequency by an increment of frequency:

$$\Delta\omega = K_o V_c \tag{8.2}$$

where V_c is the VCO input voltage and K_o is the *VCO gain factor* with dimensions of radians per second per volt. The output frequency is

$$\omega_o = \omega_c + \Delta\omega = \omega_c + K_o V_c$$

where ω_c is the free-running frequency of the VCO. Since frequency is the time derivative of phase, the VCO operation can be described as

$$\Delta\omega = \frac{d\theta_o}{dt} = K_o V_c \tag{8.3}$$

The output of the frequency divider f_d is the divider input frequency divided by N. That is,

$$f_d = \frac{f_o}{N}$$

or, since phase is the time integral of frequency,

$$\theta_d = \frac{\theta_o}{N}$$

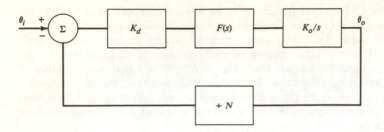

Figure 8.2 A small-signal linear model of a phase-locked loop.

For the PLL model, the divide-by-N circuit can be replaced by a frequency-independent scalar equal to $1/N$. With these assumptions, the PLL can be represented by the linear model shown in Fig. 8.2. $F(s)$ is the transfer function of the low-pass filter. The linear transfer function relating the output phase $\theta_o(s)$ and the input phase $\theta_r(s)$ is

$$\frac{\theta_o(s)}{\theta_r(s)} = \frac{K_d K_o F(s)/s}{1 + K_d K_o F(s)/Ns} = \frac{G(s)}{1 + G(s)/N} \tag{8.4}$$

The same transfer function relates the input and output frequencies $f_r(s)$ and $f_o(s)$.

If no low-pass filter is used, the transfer function is

$$\frac{\theta_o}{\theta_r} = \frac{K_d K_o}{s + K_d K_o/N} = \frac{NK_v}{s + K_v}$$

which is equivalent to the transfer function of a simple low-pass filter with a dc gain of N and a bandwidth equal to K_v, where

$$K_v = \frac{K_d K_o}{N}$$

is defined to simplify the notation.

This PLL is referred to as a *first-order loop* since it can be described by a first-order differential equation.

With the mathematical model in use here the phase-locked loop appears to be a low-pass filter, but the output phase and frequency represent deviations from the free-running frequency ω_c. The PLL is actually a bandpass filter centered at the frequency of the input waveform. The phase-detector output is a low-frequency signal that is filtered by a low-pass filter. It is much easier to build narrow bandwidth, low-pass filters than the high-Q filters that would otherwise be required. This is one of the principal advantages of the PLL.

Example 8.1 A frequency synthesizer uses a PLL to synthesize a 1-MHz signal from a 25-kHz reference frequency. In order to realize an output frequency of 1 MHz a division of

$$N = \frac{10^6}{25 \times 10^3} = 40$$

must be included in the feedback path. If no filtering is included, the closed-loop transfer function will be

$$\frac{\theta_o}{\theta_r} = \frac{K_d K_o / s}{1 + K_d K_o / sN} = \frac{K_d K_o}{s + K_d K_o / N}$$

A typical value for K_d is 2 V/rad, and a typical value for the VCO gain factor K_o (for a 1-MHz VCO) is 100 Hz/V. With these values the closed-loop transfer function is

$$\frac{\theta_o}{\theta_r} = \frac{(2 \times 100) 2\pi}{s + (2 \times 100 \times 2\pi)/40}$$

The synthesizer bandwidth will be $(2 \times 100)/40$, or 5 Hz.

Normally the loop will also contain a filter to filter out undesirable components from the phase detector and to provide further control over the loop's frequency response. If $F(s)$ is a simple low-pass filter, then

$$F(s) = \left(\frac{s}{\omega_L} + 1\right)^{-1}$$

and the closed-loop transfer function is

$$\frac{\theta_o(s)}{\theta_r(s)} = \frac{NK_v}{s(s/\omega_L + 1) + K_v} = \frac{N}{(s^2/\omega_n^2) + (2\zeta/\omega_n)s + 1} \tag{8.5}$$

where

$$K_v = \frac{K_d K_o}{N} \tag{8.6}$$

$$\omega_n^2 = K_v \omega_L \tag{8.7}$$

$$2\zeta = \frac{\omega_n}{K_v} = \left(\frac{\omega_L}{K_v}\right)^{1/2} \tag{8.8}$$

Equation (8.5) is the general form of the second-order low-pass transfer function. It occurs so frequently in PLL analysis that its characteristics are described in detail here. The magnitude of the steady-state frequency response is

$$\left|\frac{\theta_o}{\theta_r}(j\omega)\right| = \frac{N}{[(1 - \omega^2/\omega_n^2)^2 + (2\zeta\omega/\omega_n)^2]^{1/2}} \tag{8.9}$$

and the phase shift is

$$\arg \frac{\theta_o}{\theta_r}(j\omega) = -\tan^{-1} \frac{2\zeta\omega}{\omega_n(1 - \omega^2/\omega_n^2)} \tag{8.10}$$

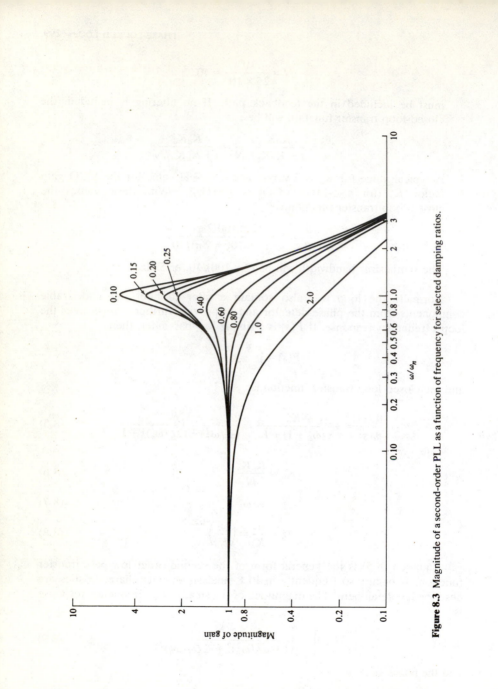

Figure 8.3 Magnitude of a second-order PLL as a function of frequency for selected damping ratios.

The magnitude of the frequency response [Eq. (8.9)] of this second-order transfer function is plotted in Fig. 8.3 for selected values of ζ. For $\zeta = 0.707$, the transfer function becomes the second-order "maximally flat" Butterworth response. For values of $\zeta < 0.707$, the gain exhibits peaking in the frequency domain. The maximum value of the frequency response M_p as a function of the damping ratio can be found by setting the derivative of Eq. (8.9)—with respect to frequency—equal to zero. M_p is found to be

$$M_p = \frac{N}{2\zeta(1 - \zeta^2)^{1/2}} \tag{8.11}$$

and the frequency ω_p at which the maximum occurs is

$$\omega_p = \omega_n(1 - 2\zeta^2)^{1/2} \tag{8.12}$$

The 3-dB bandwidth ω_n can be derived by solving for the frequency ω_n at which the value of Eq. (8.9) is equal to 0.707 for the dc again (0.707 N). ω_n is found to be (provided $\zeta < 1$)

$$\omega_h = \omega_n[1 - 2\zeta^2 + (2 - 4\zeta^2 + 4\zeta^4)^{1/2}]^{1/2} \tag{8.13}$$

The time it takes for the output to rise from 10 to 90 percent of its final value is called the *rise time* t_r. Rise time is approximately related to the system bandwidth by the relation

$$t_r = \frac{2.2}{\omega_h} \tag{8.14}$$

which is exact for first-order systems.

Normally the designer would like to have the bandwidth narrow for maximum filtering and have the rise time as short as possible so that the loop can follow changes in the input waveform. Equation (8.14) shows that this is not possible; rather the designer must make a trade-off between the system speed of response and system bandwidth.

Example 8.2 Example 8.1 described a frequency synthesizer with $K_v = 10\pi$ rad/s. The closed-loop bandwidth is 10π rad/s. What value of low-pass filter should be used so that the closed-loop system approximates a second-order Butterworth filter?

SOLUTION For a Butterworth filter the damping ratio $\zeta = 0.707$. From Eq. (8.8)

$$2\zeta = 1.414 = \left(\frac{\omega_L}{K_v}\right)^{1/2} = \left(\frac{\omega_L}{10\pi}\right)^{1/2}$$

so the required low-pass filter bandwidth is

$$\omega_L = 20\pi \text{ rad/s}$$

The bandwidth of the closed-loop system is [from Eq. (8.13)]

$$\omega_h = \omega_n \qquad (\zeta = 0.707)$$
$$= (K_v \omega_L)^{1/2} = 14.14\,\pi\,\text{rad/s}$$

The corresponding system rise time is estimated to be [Eq. (8.14)]

$$t_r = \frac{2.2}{\omega_n} = 49.4 \times 10^{-3}\,\text{s}$$

The system characteristics can be changed by changing the loop gain, the filter bandwidth, or adding a higher-order filter. A detailed analysis of phase-locked loops is provided in the next chapter.

PLL performance characteristics vary depending upon the type of phase detector used. The three most frequently used forms of phase detector are the digital detector in which the output signal is restricted to two or three possible levels, the analog mixer or multiplier, and the sampling phase detector. These three types of phase detectors will now be described.

8.3 PHASE DETECTORS (PDs)

Digital Phase Deectors

Logic circuits now serve as the most frequently used phase detectors because they are readily available as small inexpensive integrated circuits. The output of logic circuit PDs is a constant amplitude pulse whose width is proportional to the phase difference between the two input signals (which can be either analog or digital).

Exclusive-OR Phase Detectors

The exclusive-OR circuit shown in Fig. 8.4 often serves as one of the simplest types of PDs. The output of the exclusive-OR circuit is high if, and only if, one of the two input signals is high. In digital PDs phase error is defined as

$$\phi_\epsilon = \frac{\tau}{T} 2\pi \tag{8.15}$$

where T is the period of the input signals and τ is the time difference between the leading edges of the two signals. (If the two inputs are not of the same frequency, phase error is ambiguous.) The average value of the exclusive-OR

Figure 8.4 An exclusive-OR phase detector.

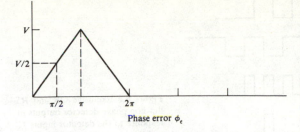

Figure 8.5 Average voltage output as a function of phase error for the exclusive-OR phase detector.

gate output as a function of phase error is plotted in Fig. 8.5. It is assumed that both input signals have a 50 percent duty cycle. The output is a maximum (the gate output is high at all times) when the two signals are 180° out of phase. There are two values of phase error for each value of output voltage, but one value will correspond to a negative loop gain and the other value to a positive loop gain value. For a positive value of loop gain the closed-loop system is unstable, and the error will adjust itself to the phase error corresponding to a negative-feedback loop. One disadvantage of the exclusive-OR phase detector is that the output depends on the duty cycle of the input waveforms.

Flip-Flop Detectors

The simple set-reset flip-flop illustrated in Fig. 8.6 can also be used as a phase detector. The signals f_1 and f_2, consisting of narrow pulses, are connected to the set and reset inputs. The average value of the Q output will be proportional to the phase difference between the two signals. The average-voltage-versus-phase transfer characteristic will be as shown in Fig. 8.7. This flip-flop phase

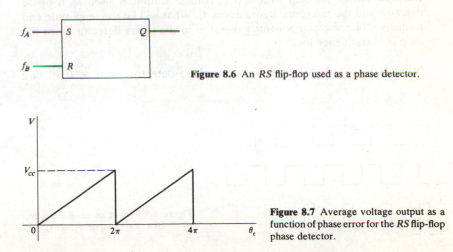

Figure 8.6 An *RS* flip-flop used as a phase detector.

Figure 8.7 Average voltage output as a function of phase error for the *RS* flip-flop phase detector.

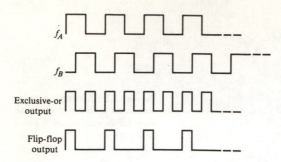

Figure 8.8 Exclusive-OR and *RS* flip-flop phase-detector outputs in response to the detector inputs f_A and f_B.

detector has an advantage over the exclusive-OR circuit in that it has twice the phase range (0 to 2π). That is, the output is V volts only when the phase error reaches 2π rad. A disadvantage of this phase detector is that the output requires more filtering than the exclusive-OR phase-detector output. Consider the timing diagram of Fig. 8.8. (It is assumed here that some means was used to convert the input signals A and B to digital pulses.) The exclusive-OR circuit output is at twice the frequency of the input signals, whereas the flip-flop output frequency is the same as the input frequency. This implies that the first ac component of the exclusive-OR output is twice as fast as that of the flip-flop output, and therefore the low-pass filter requirements will be less stringent if an exclusive-OR phase detector is used. The *RS* flip-flop works best with low duty-cycle input waveforms. The output will have a flat spot of width corresponding to the width of the input waveform, which will have a negative effect on PLL performance.

Example 8.3 A flip-flop with a 0 V voltage output is used as a phase detector and the reference frequency is f_r. What will be the amplitude and frequency of noise components generated in the phase detector when the loop is in frequency lock?

SOLUTION   When the loop is in lock the phase-detector output $\theta_\epsilon(t)$ will be a

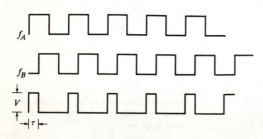

Figure 8.9 Output of an *RS* flip-flop phase detector to input signals f_A and f_B.

rectangular pulse train, as shown in Fig. 8.9. The error signal is

$$\theta_\epsilon(t) = \sum_{n=0}^{\infty} p(t - nT)$$

where
$$p(t) = V \qquad 0 \leq t \leq \tau$$
$$= 0 \qquad T \geq t > \tau$$

and
$$T = f_r^{-1}$$

τ is the time delay between the reference pulse and the divider output. If the time origin is shifted by $\tau/2$ (which does not alter the amplitude of the harmonics), $\theta_\epsilon(t)$ can be expanded in a Fourier series as

$$\theta_\epsilon(t) = \sum_{n=0}^{\infty} C_n \cos n\omega_0 t$$

where
$$C_o = \frac{V}{T}\tau$$

and
$$C_n = \frac{2V}{T} \frac{\sin(n\omega_r\tau/2)}{n\omega_r} \qquad n \neq 0$$

Each C_n is a maximum when $\sin n\omega_r\tau/2 = 1$. That is,

$$\frac{n\omega_r\tau}{2} = \frac{\pi}{2}$$

which can be written as

$$\tau = \frac{2\pi}{2n\omega_r} = \frac{T}{2n}$$

The amplitude of the component at the reference frequency ($n = 1$) will have a maximum value of

$$C_1 = \frac{2V}{T\omega_r} = \frac{V}{\pi}$$

when the two input signals are 180° out of phase. The maximum amplitudes of the other harmonics will occur at different time delays between the two input signals; the maximum amplitude of the nth harmonic is

$$(C_n)_{max} = \frac{V}{2\pi} = \frac{C_1}{n}$$

Dual-D Flip-Flops

The phase-voltage characteristics of the preceding set-reset flip-flop are sensitive to the width of the input signals. If they are of finite width, non-linearities will occur in the characteristics. The dual-D flip-flop shown in Fig.

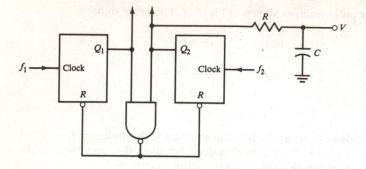

Figure 8.10 A dual-D flip-flop phase detector, including a low-pass filter.

8.10 is less sensitive to the duty cycle of the waveforms. D flip-flops go high on the leading edge of the input waveform and remain high until they are reset. The reset signal occurs when both inputs are high. When both signals are in phase and of the same frequency, both outputs will remain low and no pump signals will be applied to the low-pass filter. When the two signal frequencies are the same, but not necessarily in phase, the dc output-voltage transfer characteristic will be the same as shown in Fig. 8.7 for the RS flip-flop. If the two signal frequencies are not the same, the output voltage will depend on both the relative frequency and phase differences. The timing diagram of Fig. 8.11

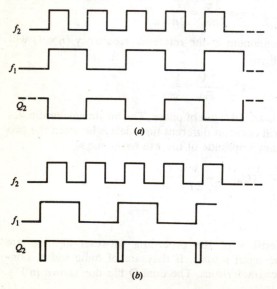

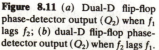

Figure 8.11 (a) Dual-D flip-flop phase-detector output (Q_2) when f_1 lags f_2; (b) dual-D flip-flop phase-detector output (Q_2) when f_2 lags f_1.

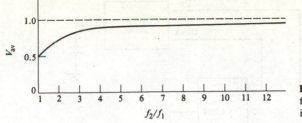

Figure 8.12 Average dual-D flip-flop output as a function of input frequency difference.

illustrates the case in which $f_2 = 2f_1$. In Fig. 8.11a the leading edge of f_1 occurs just after that of f_2, so Q_2 (which goes high when f_2 goes high and then resets when f_1 goes high) is high 50 percent of the time, and the average value of the PD output is 0.5 V. In Fig. 8.11b the leading edge of f_1 occurs just before that of f_2, so Q_2 is high almost all of the time, and the average output voltage is approximately V. The output voltage averaged over all of the phase differences is then 0.75 V for $f_2 = 2f_1$. In general, it can be induced that the average output (averaged over all phase differences) is given by

$$V_{av} = \left(1 - \frac{f_1}{2f_2}\right) V$$

provided f_2 is greater than f_1. This expression is plotted in Fig. 8.12.

Phase-Frequency Detectors

The exclusive-OR and flip-flop phase detectors, although simple, have several limitations. One limitation is that the output requires substantial filtering to extract the dc value. Also, the loop can be slow to respond if the two input signals are not of the same frequency. The *phase-frequency*, or *three-state*, phase detector is designed to reduce these limitations. The phase-frequency detector acts as a phase detector during lock and provides a frequency-sensitive signal to aid acquisition when the loop is out of lock. Phase-frequency detectors are available in integrated-circuit form and usually contain a charge pump as an integral part of the device. The essential idea of a charge pump is illustrated in Fig. 8.13. The charge pump consists of a voltage-controlled current source that

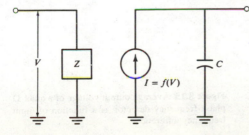

Figure 8.13 A charge-pump circuit.

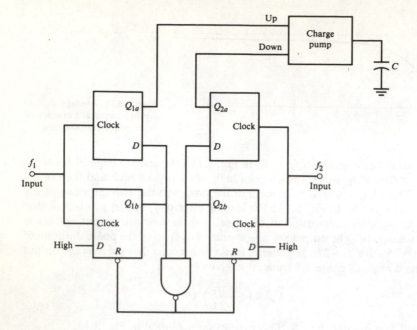

Figure 8.14 A quad-D phase-frequency detector.

outputs a current of plus or minus 1 depending on the value of control voltage. For other values of control voltage the current is zero (i.e., open-circuited). If the capacitor is part of an integrator, another pole at the origin is added to the transfer function, and the loop becomes a type II loop. If C is shunted by a resistor, the loop remains a type I loop.

Many manufacturers now produce a quad-D phase-frequency detector (as shown in Fig. 8.14). If the two input frequencies are the same, flip-flops Q_{1a} and Q_{2a} are never set, and the circuit functions like the dual-D flip-flop. If the frequencies are not equal, then either Q_{1a} or Q_{2a} will be set. These two

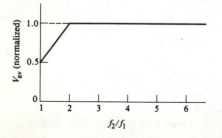

Figure 8.15 Average output voltage of a quad-D phase-frequency detector as a function of input frequency differences.

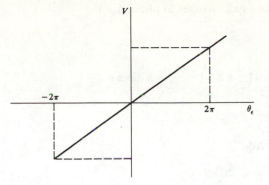

Figure 8.16 Average output voltage as a function of phase error for the quad-D phase-frequency detector.

flip-flops then serve as an out-of-frequency detector. For example, if f_1 is at least twice as fast as f_2, then Q_{1a} or Q_{1b} will be high all of the time.

The average-voltage-versus-relative-frequency characteristic plotted in Fig. 8.15 applies for $f_1 \geq f_2$. Note that, if the frequencies are not the same, the average output voltage is greater than that of the dual-D flip-flop. Therefore, a larger voltage is applied to the VCO, and the loop is quicker to respond. Once the loop reaches frequency lock, then the phase error can be obtained from Q_{1b} and Q_{2b} just as in the dual-D flip-flop. If the loop is in lock, the average-voltage-versus-phase characteristic is as shown in Fig. 8.16. The phase range for the phase-frequency detector is 720°.

Mixers

Mixers (and multipliers) are often used as phase detectors in analog PLLs. If the input signal is $\theta_i = A_i \sin \omega_o t$ and the reference signal is $\theta_r = A_r \sin(\omega_o t + \phi)$, where ϕ is the phase difference between the two signals, then the output signal θ_ϵ is

$$\theta_\epsilon = \theta_i \theta_r = \frac{A_i A_r}{2} K \cos \phi - \frac{A_i A_r}{2} K \cos(2\omega_o t + \phi) \qquad (8.16)$$

where K is the mixer gain. One of the primary functions of the loop's low-pass filter is to eliminate the second harmonic term before it reaches the VCO. The second harmonic will be assumed to be filtered out, and only the first term will be considered. Therefore,

$$\theta_\epsilon = \frac{A_i A_r}{2} K \cos(\phi) \qquad (8.17)$$

When the error signal is zero, $\phi = \pi/2$. The error signal is proportional to phase differences about 90°. For small changes in phase $\Delta\phi$,

$$\phi \cong \frac{\pi}{2} + \Delta\phi$$

$$\theta_\epsilon = \frac{A_i A_r}{2} K \cos\left(\frac{\pi}{2} + \Delta\phi\right) = \frac{A_i A_r}{2} K \sin \Delta\phi$$

$$\simeq \frac{A_i A_r}{2} K \Delta\phi$$

For a small phase perturbation $\Delta\phi$

$$\theta_\epsilon \simeq \frac{A_i A_r K}{2} \Delta\phi \tag{8.18}$$

since the phase-detector output was assumed to be $\theta_\epsilon = K_d(\theta_i - \theta_o)$. The phase-detector scale factor K_d is given by

$$K_d = \frac{A_i A_r K}{2} \tag{8.19}$$

The phase-detector scale factor K_d depends on the input signal amplitudes; the device can only be considered linear for constant amplitude input signals and for small deviations in phase. For larger deviations in phase

$$\theta_\epsilon = K_d \sin \Delta\phi \tag{8.20}$$

which describes the nonlinear relation between θ_ϵ and ϕ.

Sampling Detectors

Phase detection can also be accomplished with a linear time-varying switch that is closed periodically. Mathematically, the switch can be described as a pulse modulator, as shown in Fig. 8.17. If the operation of the sampling switch is time-periodic, that is, if the sampler closes for a short interval P at instants $T = 0, T, 2T, \ldots, nT$, the sampling is uniform. The waveshapes of the input and output signals of a uniform-rate-sampling device are shown in Fig. 8.18. The output can be considered to be

$$\theta_\epsilon(t) = \theta_i(t)\,\theta_r(t) \tag{8.21}$$

Figure 8.17 A switch modeled as a phase modulator.

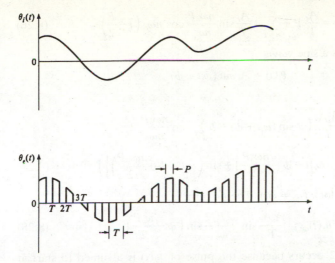

Figure 8.18 An example of the input and output waveforms of a uniform-rate-sampling device.

where $\theta_r(t)$ can be assumed to be a periodic train of constant amplitude pulses of amplitude A_r, width P, and period T. Since $\theta_r(t)$, illustrated in Fig. 8.19, is periodic, it can be expanded in a Fourier series as

$$\theta_r(t) = \sum_{n=-\infty}^{\infty} C_n e^{jn\omega_o t} \tag{8.22}$$

where

$$C_n = T^{-1} \int_0^P A_n e^{-jn\omega_o t} \, dt \tag{8.23}$$

$$= 4 \frac{A_r}{T} \sin \frac{n\omega_o P}{2} e^{\frac{-jn\omega_o P}{2}} \qquad n \neq 0 \tag{8.24}$$

$$= \frac{A_r}{T} P \qquad n = 0 \tag{8.25}$$

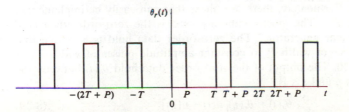

Figure 8.19 Pulse modulation is used to model the uniform-rate sampler.

Thus,
$$\theta_r(t) = \frac{A_r}{T} P + \sum_{n=1}^{\infty} 4 \frac{A_r}{T} \sin \frac{n\omega_o P}{n\omega_o} \cos n\omega_o \left(t - \frac{P}{2} \right) \tag{8.26}$$

If the input signal is a sine wave
$$\theta_i(t) = A_i \sin (\omega_i t + \phi)$$

then

$$\theta_\epsilon(t) = \theta_r(t)\theta_i(t) = \frac{A_i A_r}{T} \left(P \sin (\omega_i t + \phi) + 2 \sum_{n=1}^{\infty} \sin \frac{n\omega_o P}{n\omega_o} \right.$$

$$\times \left\{ \sin \left[(n\omega_o + \omega_i)t + \phi - \frac{n\omega_o P}{2} \right] + \sin \left(\omega_i t + \phi + \frac{n\omega_o P}{2} \right) \right\} \right) \tag{8.27}$$

when the loop is in lock ($\omega_i = \omega_o$). The dc term is

$$\theta_\epsilon(t)_{dc} = \frac{A_i A_r}{T} \sin \frac{\omega_o P/2}{\omega_o} \sin \left(\phi - \frac{\omega_o P}{2} \right) \tag{8.28}$$

The term $\omega_o P/2$ occurs because the pulse of $\theta_r(t)$ is assumed to start at $t = 0$. If the time origin is shifted to the middle of the pulse, this term does not appear. For small phase differences ϕ, the error signal is proportional to the phase difference. Therefore, the linear time-varying switch is able to serve as a phase detector. It differs from the mixer in that the dc output is zero when $\phi = \omega_o P/2$. That is, the output is zero when the oscillator and reference signal are in phase. This differs from the mixer type of phase detector, which is nulled when the two signals are in phase quadrature. As with the mixer, the sampling-phase-detector gain constant K_d is proportional to the amplitude of the applied signals. The necessary conditions for both types of loops to exhibit linear characteristics are that the input signal amplitudes be constant and the phase error be sufficiently small that

$$\sin \left(\phi - \frac{\omega_o P}{2} \right) \approx \phi - \frac{\omega_o P}{2}$$

When the loop is in lock ($\omega_i = \omega_o$), the mixer output contains a dc term and the second harmonic, whereas the sampled output contains a dc term plus all harmonics of the input frequency. Therefore the low-pass filter requirements for the sampling type of phase detector are more stringent than those for the sinusoidal mixer. Fortunately, there are filters that can easily be implemented for the sampling PD. The most commonly used is the zero-order data hold (ZODH) or "boxcar generator." The zero-order data hold is a device that converts the pulses of width P to constant amplitude pulses of width T, as shown in Fig. 8.20. The output of the zero-order data hold $\theta_o(t)$ between the sampling instants t_i and t_{i+1} is

$$\theta_o(t) = \theta_\epsilon(t_i)[u(t) - u(t_i)] \tag{8.29}$$

where $\theta_\epsilon(t_i)$ is the value of $\theta_\epsilon(t)$ at the sampling time t_i. Although the exact

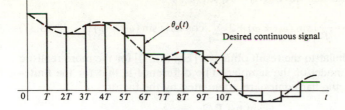

Figure 8.20 Output of a zero-order data hold compared with the ideal output (dashed line).

analysis of the finite-pulse-width sampler and ZODH combination is complex, the frequency response can be closely approximated if the sampling process is replaced by an "ideal sampler" whose output is a train of impulses. That is, the sampled signal $\theta^*(t)$ is a train of amplitude-modulated impulses

$$\theta^*(t) = \theta_i(t)\delta_T(t) \tag{8.30}$$

where $\delta_T(t)$ is a unit-impulse train of period T,

$$\delta_T(t) = \sum_{n=-\infty}^{\infty} \delta(t - nT) \tag{8.31}$$

$\delta(t - nT)$ represents an impulse of unit area occurring at time $t = nT$. Since $\delta_T(t)$ is periodic, it can be expressed by the Fourier series

$$\delta_T(t) = \sum_{n=-\infty}^{\infty} C_n e^{-jn\omega_o t} \tag{8.32}$$

where $\omega_o = 2\pi/T$. The constants C_n are determined from

$$C_n = T^{-1} \int_{-T/2}^{T/2} \delta_T(t) e^{-jn\omega_o t}\, dt = T^{-1} \tag{8.33}$$

$\delta_T(t)$ can be expanded in a Fourier series

$$\delta_T(t) = T^{-1} \sum_{n=-\infty}^{\infty} e^{jn\omega_o t} \qquad \omega_o = \frac{2\pi}{T} \tag{8.34}$$

and since $e^{jn\omega_o t} + e^{-jn\omega_o t} = 2\cos n\omega_o t$,

$$\delta_T(t) = T^{-1} + \frac{2}{T}\sum_{n=1}^{\infty} \cos n\omega_o t \tag{8.35}$$

That is, the frequency spectrum of an impulse train of period T contains a dc term plus the fundamental frequency and all harmonics with an amplitude of $1/T$. Therefore, Eq. (8.30) can be written

$$\theta^*(t) = \theta_i(t)\left(T^{-1} + \frac{2}{T}\sum_{n=1}^{\infty} \cos n\omega_o t\right) \tag{8.36}$$

If the input $\theta_i(t)$ is a sine wave $\theta_i(t) = A_i \sin(\omega_i t + \phi)$,

$$\theta^*(t) = \frac{A_i}{T} \left[\sin(\omega_i t + \phi) + 2 \sum_{n=1}^{\infty} \cos n\omega_o t \sin(\omega_i t + \phi) \right] \tag{8.37}$$

This equation is similar to the result obtained [Eq. (8.27)] for the more realistic finite-pulse-width model of the sampler. The difference is that for the finite-pulse-width model the harmonics are attenuated by the factor

$$\frac{\sin n\omega_o P/2}{n\omega_o}$$

With the impulse sampler all harmonics are attenuated by $2/T$.

The impulse response of the ZODH is a pulse T s wide, or

$$\theta_o(t) = u(t) - u(t - T)$$

so its frequency-dependent transfer function is

$$G_z(s) = \frac{1 - e^{-sT}}{s} \tag{8.38}$$

and the ZODH frequency response is

$$G_z(j\omega) = \frac{1 - e^{-j\omega T}}{j\omega} = \frac{T}{2} e^{-j\omega T/2} \frac{e^{j\omega T/2} - e^{-\omega T/2}}{j\omega T/2}$$

$$= T e^{-j\omega T/2} \frac{\sin \omega T/2}{\omega T/2} \tag{8.39}$$

which is a low-pass filter with a linear phase shift, as illustrated in Fig. 8.21. An important feature of this filter is that it has zero gain at the sampling frequency and at all harmonics of the sampling frequency. As Eq. (8.37)—or Eq. (8.27)—shows, when the input and sampling frequencies are equal, the output of the sampler contains a dc term and all harmonics of the sampling frequency.

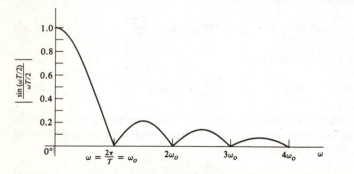

Figure 8.21 Frequency response of a zero-order data hold.

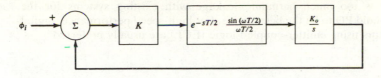

Figure 8.22 A simplified model of a phase-locked loop containing a sample-and-hold module.

Since the ZODH has zero gain at these nonzero frequencies, the unwanted harmonics are completely removed by the filter. This is one of the primary reasons for the widespread application of samplers in phase-locked loops. The ZODH also has a phase lag that increases linearly with frequency. The effect of this negative phase shift on loop stability is discussed in the next chapter.

Although a PLL containing a sampling detector is often analyzed using Z-transform techniques, an equally accurate approximation to loop performance can be obtained using continuous techniques. The inaccuracy inherent in the Z-transform analysis of PLLs is further considered in the section on large-signal behavior in the following chapter. When the input and feedback frequencies are equal, Eq. (8.37) can be written

$$\theta^*(t) = A_i \frac{\sin \phi}{T} + \text{high-frequency terms} \tag{8.40}$$

Since the high-frequency terms are filtered out by the zero-order data hold,

$$\theta^*(t) \approx A_i \frac{\phi}{T} \tag{8.41}$$

for small ϕ. The frequency response of the loop can then be estimated using the model shown in Fig. 8.22, provided the VCO is the only other frequency-dependent component in the loop. In this model, K is composed of the phase-detector factor A_i times any additional gain in the loop. This model is used to analyze loop stability and frequency response characteristics in the next chapter.

Phase-Detector Comparisons

Which type of phase comparator to select for a particular application depends on many factors—including cost, size, speed, and noise performance. The double-balanced mixer has the best noise performance of all the PDs, but it is only capable of producing approximately 0.5 V output. Since most VCOs require 2 to 10-V input, a preamplifier will be required with this type of PD, but the additional noise contributed may be so large that it is no longer the best choice. The double-balanced mixer finds application primarily in loops where little VCO pulling range is necessary, as when VCXOs are used. The sample-and-hold discriminator works well from 20 to 100 kHz, but above this range there is too much harmonic leakage with existing systems for the sample-and-hold PD to be the best choice. For high-speed performance, digital phase detectors using emitter-coupled logic (ECL) are usually preferred.

8.4 VOLTAGE-CONTROLLED OSCILLATORS (VCOs)

VCOs are described in detail in Chap. 7 and the effect of their noise on the performance of PLLs used as frequency synthesizers is discussed in Chap. 10, here we will only be concerned with the main points of their importance when used in PLLs. The main properties of a VCO used in a PLL are discussed below.

1. *Frequency deviation*. The maximum PLL capture range is equal to the open-loop gain, provided the VCO frequency-deviation capability is at least this great. If it is less, then the PLL capture range is limited by the maximum VCO frequency-deviation capability.
2. *Frequency stability*. If high-frequency stability is required, VCXOs are normally employed. Frequency stability is of the utmost importance in frequency synthesizers. However, as mentioned previously, the VCXO has a small frequency deviation and is not able to follow signals with a large frequency deviation.
3. *Modulation sensitivity*. The modulation sensitivity K_o should be high. A small change in dc voltage should produce a relatively large change in VCO frequency.
4. *Response*. The VCO should respond quickly enough that it does not affect the loop stability characteristics. Normally, the VCO poles should lie outside the dominant poles of the system.
5. *Frequency-voltage characteristics*. The VCO frequency/voltage-characteristics must be linear. The tolerance on linearity depends on the particular application. PLLs that include a microprocessor in the loop can use the microprocessor plus a D/A converter to compensate for VCO non-linearities.
6. *Spectral purity*. In some applications, such as analog frequency synthesizers, the VCO output should be as pure a sine wave as possible. In other applications, the VCO output can be a rectangular wave train.

8.5 LOOP FILTERS

The loop filter is a low-pass filter, usually of the first-order, but higher-order filters are used when additional suppression of the ac components of the phase detector output is desired. In some instances a notch network is included in the filter for suppression of a particular frequency. The network configuration depends on whether the phase-detector output can be modeled as a voltage

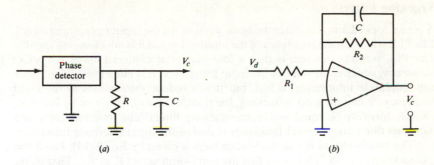

Figure 8.23 A first-order filter realized with a charge-pump output phase detector; (*b*) a first-order filter realized with a voltage output phase detector.

source (low output impedance) or a current source (high output impedance). Figure 8.23*a* illustrates a first-order filter that can be used with a charge-pump output. The voltage is given by

$$V_c = \frac{I(V)R}{sRC + 1}$$

where $I(V)$ is the amplitude of the phase-detector output. Figure 8.23*b* illustrates a first-order active filter, which can be used with a phase detector with a low output impedance. The transfer function

$$\frac{V_c}{V_d} = -\frac{R_2}{R_1}(sR_2C + 1)^{-1}$$

is a first-order filter with a dc gain of

$$A_o = -\frac{R_2}{R_1}$$

which can be adjusted to modify loop performance. The selection of loop filters for loop stability, transient performance, and noise suppression is described in more detail in the following chapter.

8.6 PHASE-LOCKED LOOP APPLICATIONS

The PLL is an exceptionally versatile device suitable for a variety of frequency selective modulation, demodulation detection, tracking, and synthesis applications. A few of the basic applications are described here. The use of the PLL for frequency synthesis is described in detail in Chap. 10, and additional applications of the PLL for modulation and detection are given in Chap. 12.

Tracking Filters

A phase-locked loop can filter the noise present on the input (reference) signal. The PLL will track the frequency of the input as long as it is not changing rapidly. The PLL transfer function is then a low-pass filter centered about the VCO frequency, which is the same as the input frequency. It is not a linear filter, since any amplitude information is lost, but it will reduce fluctuations in the input frequency. When the loop is tracking, the transfer function is given by Eq. (8.4) (N will normally be equal to 1 in the tracking filter). The loop functions as a bandpass filter whose center frequency is that of the input reference frequency.

The bandwidth of the second-order loop is given by Eq. (8.13). From this equation and Eq. (8.7) it is seen that the bandwidth $\omega_n \propto (K_v \omega_L)^{1/2}$. That is, the closed-loop bandwidth (from the center frequency to one 3-dB point) is proportional to the square root of the loop filter bandwidth times the loop gain constant. It is possible for the loop to lock on to harmonics or subharmonics of the loop, depending on what type of phase detector and VCO are used. For example, with a mixing-type phase detector, the PLL can lock onto subharmonics of the input if the VCO output is a square wave.

The ability of the PLL to automatically tune its center frequency to that of the input signal makes it an attractive solution to modulation and demodulation problems.

Angle Modulation

A phase-locked loop provides a ready means of phase modulation and indirect frequency modulation. Figure 8.24 illustrates a phase-locked loop with the modulating signal added before the low-pass filter. If a linear model is assumed for the loop then superposition can be used to find the output phase:

$$\theta_o(s) = \frac{\theta_r(s)[K_o K_d F(s)/s]}{1 + K_o K_d F(s)/sN} + \frac{M(s)[K_o K_d F(s)/s]}{1 + K_o K_d F(s)/sN} \tag{8.42}$$

where $M(s)$ is the Laplace transform of the modulating signal $m(t)$. At low frequencies

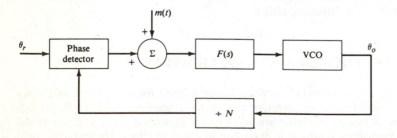

Figure 8.24 A phase-locked loop with a phase-modulating signal $m(t)$.

$$\left| \frac{K_o K_d F(j\omega)}{j\omega N} \right| \gg 1 \qquad (8.43)$$

and

$$\theta_o(j\omega) = N[\theta_r(j\omega) + M(j\omega)]$$

The phase-locked loop is a low-pass filter, and the inequality [Eq. (8.43)] is satisfied at frequencies within the loop bandwidth. In this frequency region the output phase is modulated by $m(t)$, and

$$f_o(t) = \frac{d\theta_o(t)}{dt} = N\left[\frac{d\theta_r(t)}{dt} + \frac{dm(t)}{dt} \right]$$

If the input phase is constant, the output frequency is proportional to the derivative of the modulating signal. This is referred to as *indirect frequency modulation* to distinguish it from direct FM, in which the frequency is directly proportional to the modulating signal.

Frequency Demodulation

If the PLL is locked on an input frequency, the VCO control voltage is proportional to the VCO's frequency shift from its free-running frequency. If the input frequency shifts, the control voltage shifts accordingly (provided the frequency shift is within the loop's tracking page). If the input signal is frequency-modulated, then the VCO control voltage will be the demodulated output. The PLL can be used for detecting either narrowband or wideband (high deviation) FM signals with a higher degree of linearity than can be obtained by other FM detection methods. If the maximum phase-detector output voltage is V volts, then the maximum control voltage applied to the VCO is KV volts, where K is the dc gain of the low-pass filter. The maximum frequency deviation of the VCO is then

$$(\Delta\omega)_{max} = K_o K V \text{ rad/s}$$

This, of course, assumes that the VCO is designed to be linear over this frequency range. If the phase-detector output can deviate between $\pm V$ volts, the tracking range will be

$$TR = 2(\Delta\omega)_{max} = 2K_o K V$$

This tracking range must be greater than the frequency deviation of the input signal. FM demodulation can then be obtained by setting the free-running frequency of the VCO equal to the center frequency of the input signal. This detection method assumes that the envelope of the input waveform has a constant amplitude. In many applications an amplifier and amplitude limiter are added before the phase-locked loop to ensure that this is the case.

A particular application of frequency-shift keying (FSK) demodulation is the detection of one of two transmitted frequencies. This detector is frequently referred to as a *touch-tone decoder*. There now exist multitone encoder and decoder integrated circuits. The decoders are able to detect a coded sequence of tones, allowing, for example, three different tones to represent eight transmitters.

Digital data are often transmitted using FSK, which consists of shifting the carrier frequency between two predetermined frequencies. The phase-locked loop can be used to demodulate the FSK signal. The voltage at the output of the loop filter shifts between two discrete voltage levels, reproducing the digital waveform. A realization of FSK demodulation is described in the following section on digital PLLs.

Amplitude Demodulation

The amplitude-modulated signal

$$S(t) = V[1 + m(t)] \sin \omega_o t$$

can be demodulated by multiplying the signal by a local oscillator signal of the same carrier frequency. The method is illustrated in Fig. 8.25. The multiplier output is

$$V(t) = V[1 + m(t)] \sin \omega_o t A \sin (\omega_o t + \theta)$$

$$= V[1 + m(t)] \frac{\cos \theta - \cos (2\omega_o t + \theta)}{2}$$

The multiplier output consists of the low-frequency modulating signal

$$V[1 + m(t)] \cos \theta$$

and the modulating signal centered about twice the carrier frequency

$$V[1 + m(t)] \cos (2\omega_o t + \theta)$$

The high-frequency term can be removed by the low-pass filter, so the filter output will be

$$V_o(t) = V[1 + m(t)] \cos \theta$$

This relatively simple direct-conversion receiver has several advantages over the more frequently used superheterodyne receiver. It eliminates the IF filter and the need for additional oscillators. The elimination of the IF filter also eliminates the need for image-rejection filters. It does, however, have problems, one of them being that if the input signal phase angle θ is not known the output voltage can be small. In order to ensure maximum output voltage the phase

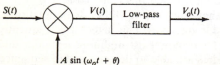

Figure 8.25 A mixer followed by a low-pass filter.

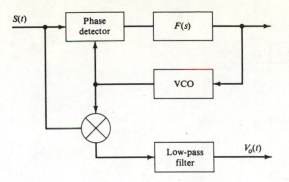

Figure 8.26 A phase-locked-loop circuit used for angle demodulation.

angle should be zero; that is, the local oscillator signal should be phase-locked to the carrier signal. In this case, the demodulation is one of coherent detection, which performs better than incoherent detection methods when the input signal-to-noise ratio is low.[8.3] The local oscillator signal, phase-locked to the input carrier, can be generated in a phase-locked loop. A complete amplitude demodulator is illustrated in Fig. 8.26. This circuit assumes that the VCO is in phase with the input. If a phase detector that causes the loop to lock with the VCO 90° out of phase with the input is used, then the VCO output must be shifted by 90° before it is mixed with incoming signals. Several phase-locked-loop integrated circuits include an additional multiplier on the chip so that amplitude demodulation can be easily realized.

Phase Shifters

Many methods of modulation and demodulation require that the local oscillator signal be phase-shifted by 90°. The circuit shown in Fig. 8.27 presents one

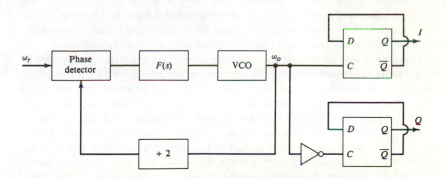

Figure 8.27 A phase-locked-loop circuit used for generating in-phase and quadrature signals.

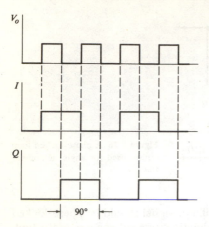

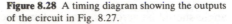

Figure 8.28 A timing diagram showing the outputs of the circuit in Fig. 8.27.

method of obtaining a signal together with the signal shifted in phase by 90°. The phase-locked loop doubles the reference frequency. The flip-flop output frequency is the same as the reference frequency, and, as can be seen from the timing diagram illustrated in Fig. 8.28, the two output signals differ by 90° in phase ($\pi/4 = 90°$).

Signal Synchronizers

The communication methods which have the lowest error rate are coherent; they require that the carrier frequency and phase be precisely known for demodulation. At the same time, it is an inefficient use of energy to transmit the carrier. Many encoding schemes transmit a low-level pilot carrier, which can be used to generate a carrier signal phase coherent with the original carrier. FM stereo and some commercial television encoding schemes use this technique. The color television signal includes a short sinusoidal burst transmitted at the rate of the horizontal synchronization pulses for synchronizing the color signals. These carrier signals can readily be recovered with a phase-locked loop.

There are other communication systems in which the carrier is suppressed, and accurate reception requires the generation of signals based on the phase information about the carrier. The receiver circuits that generate the carrier and clock signals are known as *signal synchronizers*; phase-locked loops are often used in these synchronizers. (The phase-locked loop applications that have been discussed so far all require an input signal at the frequency to be tracked.) Signal synchronizers employ a nonlinear circuit to regenerate a carrier or clock signal together with a phase-locked loop to track the signal. A frequent application is the detection of binary phase-shift keying (BPSK). If the data bit is a 1, the signal is transmitted with a phase of +90°; if the data bit is a 0, the signal phase is −90°. Each pulse is T s in duration, and the pulses are T s apart. If an equal number of

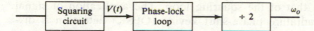

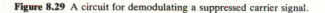

Figure 8.29 A circuit for demodulating a suppressed carrier signal.

1s and 0s are transmitted, the carrier is completely suppressed. The transmitted signal can be represented as

$$S(t) = m(t) \sin(\omega_i t + \theta_i)$$

and the average value of the modulating signal $m(t)$ is zero.

Figure 8.29 illustrates one method of recovering the carrier information. The signal at the output of the squaring circuit is

$$V(t) = [m(t) \sin(\omega_i t + \theta_i)]^2$$
$$= \frac{m^2(t)}{Z}[1 - \cos 2(\omega_i t + \theta_i)]^2$$

If the average value of $m^2(t)$ is constant, which is true in special circumstances, then $V(t)$ will contain a frequency component at twice the input frequency and a phase twice the input frequency. The phase-locked loop locks onto the signal and functions as a narrowband filter. The phase-locked loop output is divided by 2 with the frequency divider, and the output frequency is equal to the input frequency. There does exist an ambiguity in the phase of the input signal. This ambiguity can only be resolved by the special encoding of the transmitted message.

A tracking loop implementation of this recovery scheme, known as the *Costas loop*,[8.4] is shown in Fig. 8.30. The performance of the Costas loop is

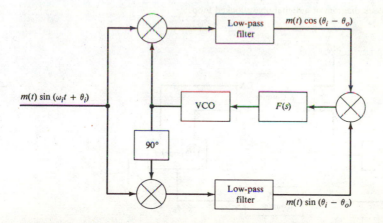

Figure 8.30 A Costas loop.

essentially the same as that of the squaring loop. The application of signal synchronization for quadrature phase-shift keying (QPSK) detection is discussed in Chap. 12.

8.6 DIGITAL PHASE-LOCKED LOOPS

In many applications, including digital communications systems, the input signal is digital and is best treated in the digital domain. Digital versions of the phase-locked loop have some advantages over their analog counterparts. A simplified block diagram of a DPLL is shown in Fig. 8.31. In addition to the phase detector, the loop consists of an accumulator and a digitally controlled oscillator. The digital phase detector is the same as those already described. The phase-detector output is a constant amplitude pulse whose width is proportional to the phase error. The digital phase-locked loop generates a digital representation of this error signal by sampling the pulse-width modulated signal at a rate that is much faster than that of the reference frequency. The digital accumulator is the digital equivalent of the analog filter. It is usually realized with an adder, multipliers, and delay units. The digitally controlled oscillator outputs a pulse at a rate that is readily controlled by an external signal.

Figure 8.32 illustrates a practical realization of a first-order DPLL. All

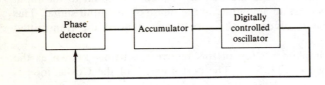

Figure 8.31 Block diagram of a digital phase-locked loop.

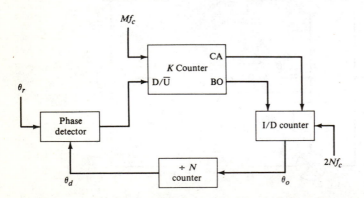

Figure 8.32 A first-order DPLL.

components except for the divide-by-N counter are realized in a single integrated circuit (74LS297). Digital phase-locked loops do not contain a voltage-controlled oscillator, which is sensitive to voltage and temperature changes. Digital integrated circuit phase-locked loops operate at higher frequencies than do their analog IC counterparts (32 MHz compared with 10 MHz), and it is far easier to generate a linear voltage-versus-frequency characteristic in a digitally controlled oscillator. Therefore, digital phase-locked loops offer easier microprocessor control.

The first function of the K counter is to convert the phase-detector output to a counter value that is proportional to the phase error. It contains an up-counter and a down-counter with carry and borrow outputs, respectively. The two input signals are the high-frequency clock and the phase-detector input that controls whether the clock is applied to the up-counter or down-counter. When the phase-detector output is low, it is applied to the up-counter, otherwise it is presented to the down-counter.

The I/D counter is clocked at the rate $2Nf_c$, where N is the modulus of the divide-by-N counter, and outputs a signal at the rate Nf_c. When a carry pulse from the K counter occurs, a half of a cycle is added to the I/D output, and a half-cycle is removed on the occurrence of a K counter borrow pulse. The frequency of the divide-by-N counter output is then f_c when there is no phase error. The output of the K counter can be interpreted as proportional to the phase error

$$K_o = \frac{K_c \phi_\epsilon M f_c}{K}$$

where Mf_c is the frequency at which the phase error is sampled, K_d is the phase-detector scale factor, and K is the modulus of the K counter. This method of implementation creates both count-up and borrow pulses, even when the phase error is zero, but the number of count-up and borrow pulses will be equal. The higher the divide-ratio K, the less often the borrow and carry pulses will occur. The length of the K counter can be controlled externally and can be varied from a divide by 2^3 to a divide by 2^{17}. A larger divide number corresponds to a narrower loop filter and thus increases the time to acquire lock, but it reduces any ripple or jitter on the output.

Applications for this DPLL include motor-speed control, noise filtering, tone recognition, frequency synthesis, frequency demodulation, and phase demodulation. A method of FSK decoding is illustrated in Fig. 8.33. Here the input f_i alternates between the two frequencies f_1 and f_2, and the loop center frequency is selected so that $f_1 < f_c < f_2$. The D flip-flop serves as a simple frequency detector. It is set if the D input is high at the moment the clock pulse goes low. The output will be low if the D input is low when the leading edge of the clock pulse occurs. If the input frequency is f_1, a negative phase error is required to reduce the output frequency from f_c to f_1. A negative phase error means that θ_r lags θ_d so that the D input of the flip-flop is not set. If f_2 occurs, the phase error will be positive (f_i leads f_o), and the D flip-flop will be set.

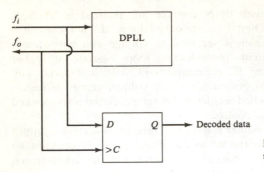

Figure 8.33 A DPLL used for frequency-shift keying (FSK) decoding.

PROBLEMS

8.1 A first-order PLL is to be used to synthesize a 1-MHz signal from a 50-kHz reference frequency. The phase-detector gain is 2 V/rad, the VCO sensitivity is 100 Hz/V, and the free-running frequency of the VCO is 1 MHz. Estimate the rise time of the system. What would the rise time be if it is desired to realize an output frequency of 1.2 MHz? How many different output frequencies can be realized if the VCO has a maximum frequency deviation of ±200 kHz?

8.2 (*a*) If the PLL of Prob. 8.1 is to synthesize an output frequency of 1 MHz, what would be the required bandwidth of a first-order low-pass filter in order for the closed-loop system transfer function to approximate that of a second-order Butterworth filter?

(*b*) What is the answer to (*a*) if the output frequency is to be 1.2 MHz?

8.3 Design the filters determined in Prob. 8.2 assuming the phase detector has a charge-pump output.

8.4 A PLL has a reference frequency of 25 MHz. Determine the harmonic with the maximum amplitude at the phase-detector output for

(*a*) An exclusive-OR phase detector. The input waveforms can be assumed to have a 50 percent duty cycle.

(*b*) A dual-D flip-flop phase detector.

8.5 Determine the condition between the phases of the two input signals for zero phase error in the output of

(*a*) A mixing-type phase detector.

(*b*) A sampling phase detector.

(*c*) An exclusive-OR phase detector.

(*d*) An *RS* flip-flop phase detector.

(*e*) A dual-D phase detector.

8.6 Describe how to construct a motor-speed controller using a phase detector and oscillator. [*Hint:* A constant-field dc motor can be described by the transfer function

$$\frac{\theta_o}{V_c} = \frac{K}{s(\tau_M s + 1)}$$

where V_c is the motor control voltage and θ_o is the motor output angle in radians.]

8.7 Use a 74LS297 integrated circuit (App. 11) to realize a digital phase-locked loop with a free-running frequency of 1 MHz. Determine the clock frequency (plus M and K) as well as the counter contents in order for the DPLL to discriminate between the two input frequencies of 0.9 and 1.1 MHz.

8.8 For the system described in Prob. 8.2 show how the system response time will vary as a function of the contents of the K register. From this result describe the number of input signals required for detection.

8.9 Calculate the maximum output phase change possible in one reference period for a second-order type I system that uses a digital phase detector with maximum output voltage V.

8.10 Show that in second-order type I systems, which use a charge-pump output phase-frequency detector, the maximum frequency change (tracking range) is given by

$$(\Delta\omega)_{max} = K_o V$$

8.11 Derive the voltage transfer characteristic for the quad-D flip-flop for the case in which $f_1 > f_2$.

8.12 Determine the loop gain $K_d K_o$ of the CD4046 phase-locked loop (see App. 7 for specifications).

8.13 Design a low-pass filter for the CD4046 (see App. 7) so that the filter bandwidth is 1 kHz using phase comparator I and no frequency offset ($R_2 = \infty$).

8.14 Design a frequency demodulator using a CD4046 integrated circuit phase-locked loop (see App. 7 for specifications). The carrier frequency is 100 kHz and the frequency deviation is 2 kHz. Use phase comparator I.

REFERENCES

8.1 de Bellescize, H.: La Reception Synchrone, *Onde Electr.*, **11**:230–240 (1932).
8.2 Barab, S., and A. McBride: Uniform Sampling Analysis of a Hybrid Phase-Locked Loop with a Sample-and-Hold Phase Detector, *IEEE Trans.*, **AES-11**:210–216 (1975).
8.3 Gardner, F. M.: *Phaselock Techniques*, 2d ed., Wiley, New York, 1979.
8.4 Viterbi, A.: *Principles of Coherent Communications*, McGraw-Hill, New York, 1966.
8.5 Costas, J. P.: Synchronous Communications, *Proc. IRE*, **44**:1713–1718 (1956).
8.6 Green, W. T. Jr., and B. Kean: Digital Phase-Locked Loops Move into Analog Territory, *Electronic Design*, March 31, 1982.
8.7 Feher, K.: *Digital Communications Satellite/Earth Station Engineering*, Prentice-Hall, Englewood Cliffs, N.J., 1983.

ADDITIONAL READING

Blanchard, A.: *Phase-Locked Loops*, Wiley, New York, 1976.
Gupta, S. C.: Phase-Locked Loops, *Proc. IEEE*, **63**:291–306 (1975).
Lindsey, W. C., and C. M. Chie: A Survey of Digital Phase-Locked Loops, *Proc. IEEE*, **69**:410–431, 1981.
——, and M. K. Simon (eds.): *Phase-Locked Loops and Their Application*, IEEE Press, New York, 1978.
Rohde, U. L.: *Digital PLL Frequency Synthesizers*, Prentice-Hall, Englewood Cliffs, N.J., 1983.

PHASE-LOCKED LOOP ANALYSIS

9.1 INTRODUCTION

In the previous chapter it was demonstrated that the phase-locked loop is an exceptionally versatile circuit with many applications. The system designer first selects the phase detector and voltage-controlled oscillator and then determines the loop gain and loop filter frequency response. These parameters determine both the loop's transient performance and the system's noise performance. As mentioned in the last chapter, there is an inverse relationship between rise time and bandwidth. The designer also faces trade-offs in establishing the system's speed of response and its noise performance. Finally, because it is a feedback loop, its loop stability must be ensured.

Loop performance is first analyzed in this chapter assuming a linear PLL model, and then the transient analysis of some nonlinear PLL models containing a digital phase detector will be determined. This latter analysis will allow an evaluation of the results obtained from the less accurate linearized model.

9.2 STEADY-STATE ERROR ANALYSIS

Steady-state error analysis determines the final error in response to inputs, which can be expressed as a time polynomial:

$$\theta_r(t) = \sum_{n=0}^{K} a_n t^n$$

For phase-locked loops the two inputs of most interest are the step input in

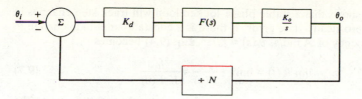

Figure 9.1 A linear PLL model.

phase

$$\theta_r(t) = \theta_o$$

and the step input in frequency

$$f_r(t) = f_o$$

or

$$\theta_r(t) = f_o t$$

It is important to know what the steady-state error will be in response to these inputs. This can be readily determined for the PLL linear model illustrated in Fig. 9.1 and derived in Chap. 8.

The error signal θ_ϵ, defined as $\theta_r - \theta_o/N$, can be expressed as

$$\theta_\epsilon(s) = \frac{\theta_r(s)}{1 + K_v F(s)/Ns} \tag{9.1}$$

where $K_v = K_o K_d$. If the systems are stable, the steady-state error for polynomial inputs $\theta_r(t) = t^n$ can be obtained from the final value theorem:

$$\lim_{t \to \infty} \theta_\epsilon(t) = \lim_{s \to 0} s\theta_\epsilon(s) \tag{9.2}$$

$$= \lim_{s \to 0} \frac{s^2 \theta_r(s) N}{K_v F(s)} \tag{9.3}$$

If $\theta_r(t)$ is a step function representing a sudden increase in phase of $\phi°$, $\theta_r(s) = \phi/s$ and

$$\lim_{t \to \infty} \theta_\epsilon(t) = \lim_{s \to 0} \frac{s\phi}{K_v F(s)} \tag{9.4}$$

$F(s)$ is either a constant or a low-pass filter that may include poles at the origin. That is,

$$\lim_{s \to 0} F(s) = \frac{K}{s^n} \neq 0 \tag{9.5}$$

Therefore, for a step increase in phase the steady-state error can be written as

$$\lim_{t \to \infty} \theta_\epsilon(t) = \lim_{s \to 0} \frac{s^{n+1} \theta_o}{K_v K} = 0 \tag{9.6}$$

This equation shows that a stable phase-locked loop will track step changes in phase with a zero steady-state error. If there is a constant amplitude change in the input frequency of A rad/s, $\theta_r(s) = f_o/s^2$, Eq. (9.4) becomes

$$\lim_{t \to \infty} \theta_e(t) = \lim_{s \to 0} \frac{f_o}{K_v F(s)} = \frac{f_o}{K_v F(0)} \tag{9.7}$$

If $F(0) = 1$, the steady-state phase error will be inversely proportional to the loop gain K_v. Recall that the larger K_v is, the larger the closed-loop bandwidth is, and thus the faster the loop response. To increase the response speed and reduce the tracking error, the loop gain should be as large as possible. If $F(0)$ is finite, there will be a finite steady-state phase error. The frequency error $f_e(t) = d/dt\, \theta_e(t)$ will be zero in the steady state. That is, the input and VCO frequencies will be proportional ($\omega_r N = \omega_o$). (These conclusions are not correct for PLLs containing digital phase detectors. The steady-state behavior of these systems will be discussed in the section on large-signal behavior.)

If it is necessary to have zero phase error in response to step changes in the input frequency, then $\lim_{s \to 0} F(s)$ must be infinite. That is, the dc gain of the low-pass filter must be infinite. This can be realized by including in $F(s)$ a pole at the origin. In this case $F(s)$ will be of the form

$$F(s) = s^{-1} \frac{s/\omega_z + 1}{s/\omega_p + 1} \tag{9.8}$$

and the system will now be type II, since the open-loop system now has two poles at the origin. However, the addition of the pole at the origin creates difficulties with the loop stability, and, in fact, the system will be unstable unless a lead network is also included in $F(s)$. Loop stability will now be examined in detail in order to determine how to design the stabilizing networks for adequate loop stability.

9.3 STABILITY ANALYSIS

Feedback systems whose open-loop transfer function has one pole at the origin are known as *type I systems*. If the open-loop system has N poles at the origin it is a type N system. Most phase-locked loops are either type I or type II systems. The PLL has an open-loop pole at the origin because of the VCO, so the system is at least type I. A second pole at the origin is often added to the filter to reduce steady-state errors and increase the noise suppression; the system is then type II. The stability characteristics of type I and type II systems will now be analyzed. The methods can be extended to higher-type systems.

A linear model block diagram for the PLL is given in Fig. 9.1. The open-loop gain (negative of the loop transmission) will be denoted as

$$G(s) = K_v \frac{F(s)}{s}$$

where
$$K_v = \frac{KK_oK_d}{N}$$

K_d is the phase detector gain, K_o is the VCO sensitivity, K is any additional loop amplification, and N is the divide ratio. K_v is known as the *velocity constant*. Stability requires that the closed-loop poles all be located in the left half of the s plane (the real part of the poles is less than zero). Stability analysis and system design are best carried out by deducing the closed-loop characteristics from the open-loop transfer function and loop gain. The application of Nyquist's criterion (see Chap. 7) to a polar plot of the open-loop frequency response is the most general method. All of the PLL systems of practical significance are minimum phase (no open-loop poles or zeros in the right-half plane) so Nyquist's stability criterion can be simplified to studying the open-loop gain and phase characteristics near the unit gain (crossover) frequency.

The easiest way to analyze the loop stability is to plot the magnitude and phase of the open-loop transfer function versus frequency. The phase margin is defined as

$$\phi_M = 180° + \arg G(j\omega_c) \qquad (9.9)$$

where ω_c, the open-loop crossover frequency, is the frequency at which the open-loop gain is unity. That is, the phase margin is equal to 180° plus the phase shift of the open-loop transfer function (a negative number) at the open-loop crossover frequency ω_c. The greater the phase margin, the more stable the system and the more phase lag from parasitic effects can be tolerated. Additional phase lag invariably arises from neglected poles or from the time delay arising in the phase detector.

Example 9.1 Consider a phase-locked loop which has $K_v = 50$ rad/s and which contains a low-pass filter with a corner frequency of 100 rad/s. What is the phase margin?

SOLUTION The magnitude and phase of the open-loop transfer function are plotted in Fig. 9.2. The system crossover frequency is approximately 50 rad/s. At this frequency the phase shift of the open-loop transfer function is $-112.5°$, so the phase margin is 67.5°.

In this example the complete phase plot was presented, but once one is familiar with phase plots, they no longer need to be included. Since these are minimum phase systems, the phase is uniquely determined by the magnitude characteristics. One can simply calculate the phase shift after determining the open-loop crossover frequency from the magnitude plot.

Example 9.2 In the previous example, if the filter corner frequency had been 10 rad/s rather than 100 rad/s, what would have been the system phase margin?

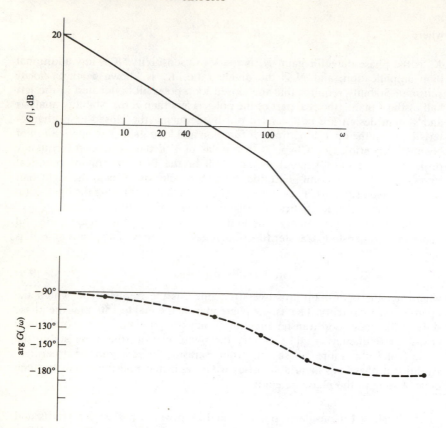

Figure 9.2 Magnitude and phase response of the open-loop system discussed in Example 9.1.

SOLUTION To determine the phase margin, first plot the magnitude of the open-loop gain and determine the crossover frequency. The straight-line approximation of the magnitude is plotted in Fig. 9.3. ω_c is found to be approximately 22 rad/s. Thus, the system phase margin is $180° - (90° + \tan^{-1} 2.2) = 23.40°$, which is too small for satisfactory loop stability. This is in agreement with Bode's rule, which states that if the magnitude of the open-loop response crosses the 0-dB line with a slope of -12 dB per octave the system is unstable.[9.1] In this example, the straight-line approximation for the gain decreases at -12 dB per octave, but the actual response crosses the 0-dB line with a slope slightly more positive than -12 dB per octave, hence the small phase margin.

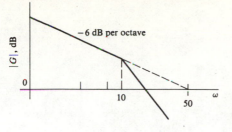

Figure 9.3 Magnitude response of the system discussed in Example 9.2.

Although the most important design parameters concerning system frequency response are the closed-loop bandwidth ω_h and the peak value M_P of the closed-loop frequency response, no design techniques exist that allow one to easily specify ω_h and M_P for higher-order systems. However, it is relatively easy to design for specified open-loop parameters ω_c and ϕ_M. There are approximations which relate ω_c and ϕ_M to ω_n, M_P, ζ, and thus, to the system rise time and overshoot. Fortunately, the conditions under which these approximations are valid are satisfied by most PLLs.

Since the interpretation is different for type I and type II systems, they will be discussed individually.

Type I Systems

Consider first the case in which $F(s)$ is a first-order low-pass filter. For this case the open-loop transfer function is

$$G(j\omega) = \frac{K_v}{j\omega(j\omega/\omega_L + 1)} \tag{9.10}$$

where ω_L is the filter bandwidth and the loop velocity constant K_v is the product of the phase detector gain K_d, the VCO gain K_o, any additional gain K, and $1/N$; that is,

$$K_v = \frac{K_d K_o K}{N}$$

If N is unity, the closed-loop transfer function is

$$\frac{\theta_o}{\theta_i} = \left(\frac{s^2}{\omega_n^2} + \frac{2\zeta}{\omega_n} s + 1 \right)^{-1} \tag{9.11}$$

where

$$\omega_n^2 = K_v \omega_L \tag{9.12}$$

and

$$\zeta = \frac{1}{2} \left(\frac{\omega_L}{K_v} \right)^{1/2} \tag{9.13}$$

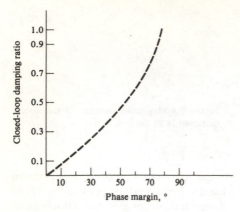

Figure 9.4 Phase margin as a function of damping ratio for a type I system.

The open-loop unity gain frequency is easily shown to be

$$\omega_c = \omega_L \left\{ \frac{[1 + 4(K_v/\omega_L)^2]^{1/2} - 1}{2} \right\}^{1/2} \tag{9.14}$$

Once ω_c is known, the phase margin

$$\phi_M = 90° - \tan^{-1} \frac{\omega_c}{\omega_L} = 90° - \tan^{-1} \left(\frac{\{1 + [(2\zeta^2)^2]^{-1}\}^{1/2} - 1}{2} \right)^{1/2} \tag{9.15}$$

can be calculated. This equation is plotted in Fig. 9.4. The parameters of the closed-loop system that are most important are adequate stability (which is related to phase margin), system bandwidth (which determines the speed of the transient response), and system transient response (which is described by rise time and overshoot). The system bandwidth for a low-pass transfer function is defined as the frequency at which the gain is equal to 0.707 times its dc value. The bandwidth of the system represented by Eq. (9.11) is

$$\omega_h = \omega_n[1 - 2\zeta^2 + (2 - 4\zeta^2 + 4\zeta^4)^{1/2}]^{1/2} \tag{9.16}$$

For the underdamped second-order system given by Eq. (9.10) ($\zeta < 1$), the peak value of the time response to a unit step input can be shown to be

$$P_o = 1 + \exp \frac{-\pi\zeta}{(1 - \zeta^2)^{1/2}} \tag{9.17}$$

The overshoot is determined solely by ζ. P_o as a function of ζ is plotted in Fig. 9.5. For zero damping the overshoot is 100 percent, and it decreases to zero for a unity-damping ratio.

For high-order systems the overshoot and bandwidth are not readily related to the open-loop system parameters, but a good first approximation is that Eqs. (9.16) and (9.17) hold true for higher-order type I systems. The damping ratio is defined in terms of the phase margin by Eq. (9.15). It is relatively easy to design

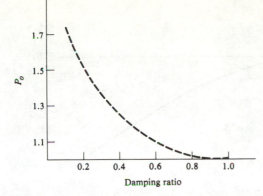

Figure 9.5 Step response overshoot as a function of damping ratio for a type I second-order system.

a system to have a given phase margin, and the design can then be evaluated by using computer simulation. If the simulation indicates that the overshoot is too high (or too low) then the phase margin can be increased (or reduced), but the relations between phase margin, damping, and overshoot are amazingly accurate for higher-order type I systems. This implies that the response of many feedback systems can be described by a second-order model. Also, the closed-loop bandwidth can be related to the open-loop crossover frequency ω_c. (It is usually about 50 percent greater.)

If a design is desired for a specified peak transient overshoot, Eq. (9.17) can be used to determine the required damping, and then Eq. (9.15) can be used to determine the required phase margin.

Example 9.3 Consider a type I PLL with the open-loop transfer function

$$G(s) = \frac{100}{s(s/\omega_L + 1)^2}$$

which corresponds to a PLL with a second-order filter for increased suppression of the high-frequency components. What should ω_L be so that the system has approximately 20 percent overshoot to a step input?

SOLUTION Although the relationships between overshoot, damping ratio, and phase margin have been derived for a second-order system, we will assume that they apply to the third-order system. From Fig. 9.5 it is seen that 20 percent overshoot to step input corresponds to the damping ratio $\zeta = 0.5$, and Fig. 9.4 shows that the damping ratio of 0.5 corresponds to a phase margin of 50°. For a phase margin of 50° the low-pass filter must have a phase shift of $-40°$ at the crossover frequency ω_c. Therefore, each filter pole can contribute $-20°$ at the crossover frequency. That is, $\tan^{-1} \omega_c/\omega_L = 20°$. Therefore, ω_L must be significantly larger than ω_c. So the crossover frequency is approximately $K_v = 100$ rad/s and $\omega_L = 298$ rad/s. The open-

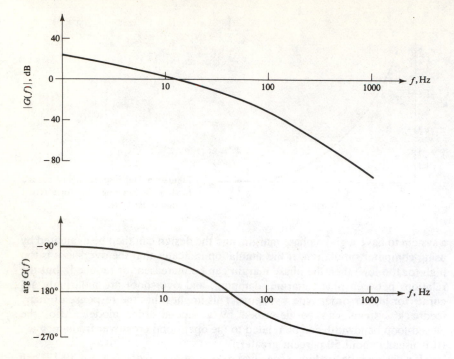

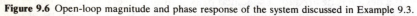

Figure 9.6 Open-loop magnitude and phase response of the system discussed in Example 9.3.

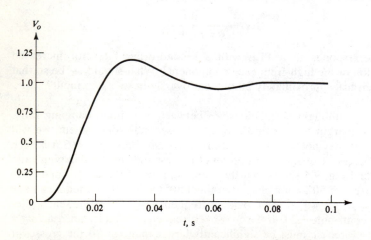

Figure 9.7 Step response of the system discussed in Example 9.3.

loop gain and phase plots are given in Fig. 9.6. It is seen that the crossover frequency is 100 rad/s and the phase margin is 50°. When the closed-loop step response of the system is simulated, the peak overshoot is found to be 14 percent. The use of the second-order approximations gives a fairly good estimate of the third-order characteristics. Also, we know that the over-shoot can be increased by decreasing the phase margin. In fact, in this case selecting $\omega_L = 233$ rad/s corresponding to phase margin of 43.5° gives an overshoot of 20 percent. The step response is plotted in Fig. 9.7. The rise time (10 to 90 percent) is approximately 13 ms.

The Pole-Zero Filter

A zero can be added to the loop filter, resulting in the transfer function

$$F(s) = \frac{1 + s/\omega_z}{1 + s/\omega_p} \tag{9.18}$$

The addition of the zero increases the ability to shape the loop's frequency response, but it decreases the amount of reference-frequency filtering. If $\omega_z > \omega_p$, then the open-loop frequency is as shown in Fig. 9.8. In this case $F(s)$ is used to reduce ω_c and the closed-loop bandwidth. For good loop stability, the frequency of the zero filter ω_z should be less than the open-loop crossover frequency.

If $\omega_p > \omega_z$, then the open-loop frequency response is as shown in Fig. 9.9. In this case the addition of the filter zero is used to increase the loop bandwidth. It also increases any high-frequency noise which may be present. However, since $F(s)$ is assumed to have a dc gain of unity, it can only be realized using active components.

In the preceding example, the loop gain K_v was fixed, and it was necessary to select the loop filter bandwidth to achieve the desired transient performance. It is often the case that the filter bandwidth selection is based on the required

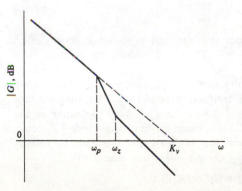

Figure 9.8 Magnitude response of a type I system with a pole-zero filter with $\omega_z > \omega_p$.

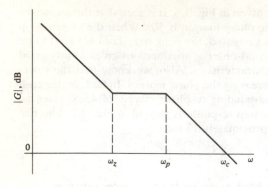

Figure 9.9 The pole of the pole-zero filter is larger than the zero frequency.

suppression of the phase detector *spurious* (the reference frequency and its harmonics). The design procedure is the same in this case, only now the loop gain is chosen to give the desired phase margin.

Control of Loop Bandwidth

In some instances it is also necessary to specify the loop bandwidth. In order to control both the loop damping and bandwidth, an amplifier can be added in series with the low-pass filter. If the filter is implemented using active components, the additional gain can be obtained without any additional components.

Example 9.4 Consider again Example 9.3, with the additional specification that the rise time in response to a unit step input be less than 1.3 ms.

SOLUTION Since the overshoot is to be 20 percent, the phase margin must still be approximately 45°. To design for the rise time specification, it often suffices to use the approximation

$$t_r = \frac{2.2}{\omega_h} \qquad (9.19)$$

which is exact only for first-order systems, but provides a good design guideline for higher-order systems. Thus, ω_h should be greater than $2.2/t_r = 1.69 \times 10^3$ rad/s.

There are several approximations for estimating the closed-loop bandwidth from the open-loop frequency response. The simplest approximation is to assume the closed-loop bandwidth is the same as the open-loop crossover frequency. This is exact for first-order type I systems. A more accurate approximation for higher-order systems is usually obtained using Eq. (9.16). In this example, assuming that $\zeta = 0.45$,

$$\omega_n = 1.27 \times 10^3 = (K_v \omega_L)^{1/2}$$

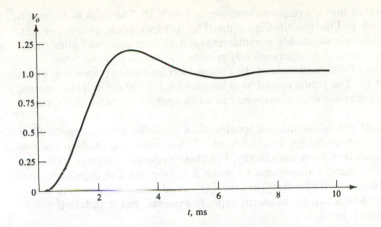

Figure 9.10 Step response of the system discussed in Example 9.4.

Since the phase margin is to be the same as in the previous example, it is reasonable to assume that the ratio K_v/ω_L remains constant. The filter frequency must also be increased to maintain the same phase margin. In this case, $K_v = 0.83 \times 10^3$ rad/s.

For this value of K_v, the closed-loop bandwidth is 1250 rad/s and the rise time is 1.6 ms. For this example, increasing K_v to 1000 and ω_L to 2330 rad/s meets the specifications.

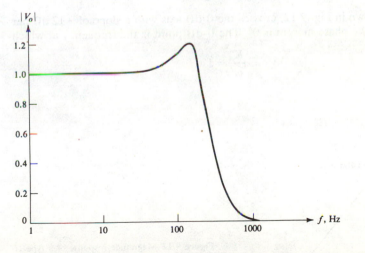

Figure 9.11 Closed-loop frequency response of the system discussed in Example 9.4.

A plot of the step response is shown in Fig. 9.10. The peak overshoot is 20 percent, and the rise time is 1.3 ms. The two specifications are now met. In general, two adjustable parameters, such as loop gain and filter bandwidth, are needed to independently specify overshoot and rise time.

A plot of the magnitude of the closed-loop frequency response is given in Fig. 9.11. The actual closed-loop bandwidth is 1500 rad/s. This example illustrates that the approximations can work well for higher-order systems.

The use of the phase margin specification provides an easy method for meeting the system design specifications. Third-order and higher systems normally cannot be solved analytically. For these systems, designing to meet a specified phase margin allows one to make a very good first approximation. Closer specifications, if necessary, are then achieved with computer simulation. A similar approach can be used for type II systems, but a different set of approximations is needed.

Type II Systems

The previous steady-state error analysis showed that for zero steady-state phase error in response to step changes in the input frequency, the low-pass filter $F(s)$ must contain a pole at the origin. The open-loop system will then have two poles at the origin. This is referred to as a *type II feedback system*. Type II systems are inherently unstable unless a phase lead network is added inside the loop. If $F(s)$ is simply $1/s$, then the open-loop transfer function is

$$G(s) = \frac{K_a}{s^2} \qquad (9.20)$$

which, as shown in Fig. 9.12, crosses the 0-dB axis with a slope of -12 dB per octave, and the phase margin is 0°. The 0-dB point is the frequency at which

$$\frac{K_a}{\omega^2} = 1$$

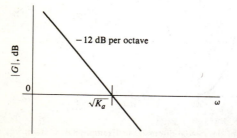

Figure 9.12 Magnitude response of a type II second-order system.

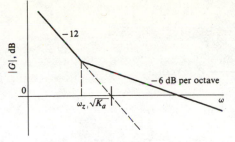

Figure 9.13 Magnitude response of a type II second-order system including a lead network.

or

$$\omega_c = K_a^{1/2} \qquad\qquad (9.21)$$

The loop gain K_a has the dimensions of (radians per second) squared.

Bode developed a rule-of-thumb for interpreting the system stability from a plot of the magnitude of the open-loop frequency response.[9.1] Specifically, if the open-loop frequency response crosses the 0-dB line with a slope of -6 dB per octave, the system is stable. If the slope is -12 dB per octave or even more negative, the system is unstable.

In order to stabilize this system, a network must be added that alters the gain so that it crosses the 0-dB line with a slope of -6 dB per octave. Such a network is referred to as a *lead network*, since it has a positive or leading phase shift. Figure 9.13 illustrates the gain modified with the simplest lead compensation that can be used. Here a zero at ω_z has been added to the open-loop transfer function. The composite low-pass filter transfer function is

$$F(s) = \frac{s/\omega_z + 1}{s} \qquad\qquad (9.22)$$

The zero location ω_z is selected to give the desired phase margin. The smaller ω_z is, the greater will be the phase margin and the greater will be the crossover frequency ω_c. Note that this system is also a second-order control system.

Figure 9.14 illustrates an easy method for realizing the filter pole and zero for the case in which the phase detector output is a charge pump. The VCO

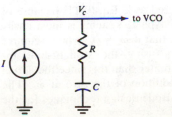

Figure 9.14 A lead network used with a charge pump.

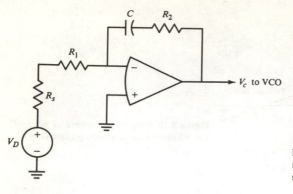

Figure 9.15 An active lead network to be used with a voltage source.

control voltage is

$$V_c(s) = I(s) \frac{RsC+1}{sC} \qquad (9.23)$$

The capacitor realizes the additional pole at the origin, and the addition of the resistor R realizes the zero at

$$\omega_z = (RC)^{-1} \qquad (9.24)$$

Note that without the resistor the system would be unstable.

If the phase detector output can be modeled as a voltage source (low output impedance), then the circuit illustrated in Fig. 9.15 can be used to realize the desired transfer function. In this case,

$$V_c(s) = V_D(s) \frac{R_2 sC + 1}{(R_1 + R_s)sC} \qquad (9.25)$$

The open-loop transfer function is

$$G = \frac{K_a(s/\omega_z + 1)}{s^2}$$

and the closed-loop transfer function is

$$\frac{G(s)}{1 + G(s)} = \frac{s/\omega_z + 1}{s^2/K_a + s/\omega_z + 1} \qquad (9.26)$$

The new transfer function is not as easily analyzed as is Eq. (9.11) because of the presence of the zero at ω_z. The design of type II systems is most easily carried out using Bode design techniques; note that if $\omega_z = \omega_c$, which requires that $\omega_c = K_a^{1/2}$, the calculated phase margin (using the straight-line approximation) is $45°$. The phase margin will be greater than this since the actual crossover frequency will be increased by the addition of the zero at ω_z. The following example illustrates the application of Bode design techniques for the design of type II systems.

Example 9.5 For a PLL with a K_a of 1000 $(\text{rad/s})^2$, design a low-pass filter such that the system will have zero steady-state phase error in response to constant frequency inputs, less than the 20 percent overshoot to step changes in the input phase and a rise time of less than 1 ms.

SOLUTION The solution to this problem calls for a type II system; the simplest open-loop transfer will be of the form

$$G(s) = \frac{K_a(s/\omega_z + 1)}{s^2}$$

Since all stable type II systems meet the steady-state error specification, the only problem is to select ω_z to meet the overshoot and rise time specifications. If ω_z is made much less than K_a, the phase-margin will approach 90°, and the open-loop crossover frequency will be that of a first-order system. The open-loop crossover frequency will then be approximately K_a/ω_z, and the system will behave as a first-order system meeting the overshoot specification, since there will be no overshoot.

If the characteristics of a second-order system are to be maintained, it will be necessary that $\omega_c \cong (K_d K_o)^{1/2} = K_a^{1/2}$. Note, however, that $\omega_c \cong K_a^{1/2} = 1000^{1/2}$ in this case, so the crossover frequency is too low to meet the rise time specification ($t_r \approx 2.2/1000^{1/2}$ s). Therefore, additional gain will be required in the loop. For the first estimate the open-loop gain K_a will be selected so that the open-loop crossover frequency $\omega_c = 2.2 \times 10^3$. This crossover frequency should result in the system meeting the rise time specification. K_a must equal $(2.2 \times 10^3)^2$. The filter zero ω_z will also be set equal to ω_c, since this will give a phase margin greater than 45° and will

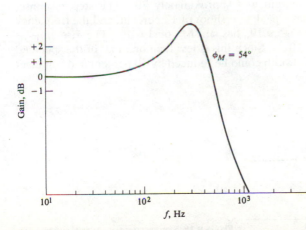

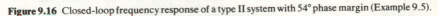

Figure 9.16 Closed-loop frequency response of a type II system with 54° phase margin (Example 9.5).

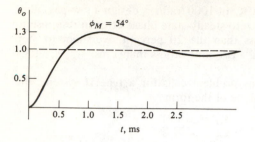

Figure 9.17 Step response of a type II system with 54° phase margin (Example 9.5).

actually increase the crossover frequency above that estimated using the straight-line approximation. The actual crossover frequency turns out to be close to 3×10^3 rad/s, and the phase margin is 54°. The magnitude of the closed-loop frequency response, plotted in Fig. 9.16, has a peak value of $M_p = 3.5$ dB. The system step response, plotted in Fig. 9.17, has a rise time of approximately 0.5×10^{-3} s and a 30 percent overshoot.

Because the overshoot is far greater than would have been predicted from the phase margin, this indicates that the required phase margin predicted from the analysis of type I systems is not always a good predictor of step response in type II systems. Nevertheless, system performance can readily be improved by increasing the phase margin. In this system the phase margin can be increased by increasing K_a or decreasing ω_z, both of which will increase the crossover frequency ω_c. For this example ω_z was first reduced to 1.665×10^3 rad/s, and the overshoot was found to be (using computer simulation) 24 percent. For the next trial ω_z was reduced to 1.25×10^3 rad/s (the open-loop crossover frequency increased to 3.8×10^3 rad/s; the phase margin was approximately 70°. The step response, plotted in Fig. 9.18, has a peak overshoot of 17 percent, and the frequency response, plotted in Fig. 9.19, has an M_p of 1.6 dB. The rise time is approximately 0.45×10^{-3} s. Since this is less than one-half of the specified rise time, the loop bandwidth could be reduced by 50 percent and still meet

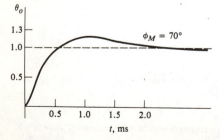

Figure 9.18 Step response of a type II system with 70° phase margin.

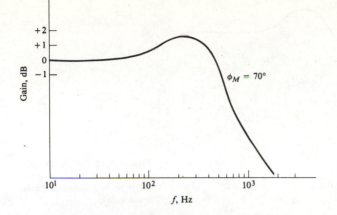

Figure 9.19 Closed-loop frequency response of a type II system with 70° phase margin.

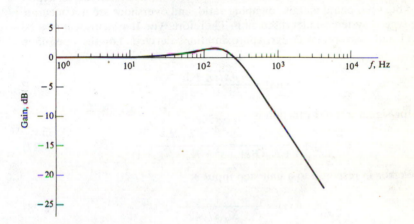

Figure 9.20 Closed-loop frequency response of a type II system that has a smaller bandwidth than the system of Fig. 9.19.

the transient response specifications. This can easily be accomplished by reducing both ω_z and the square root of K_a by 50 percent. The closed-loop frequency response and the step response are plotted in Figs. 9.20 and 9.21 for the case in which $\omega_z = 0.625 \times 10^3$ and $K_a = 1.21 \times 10^6$. The closed-loop bandwidth has been reduced to 350 Hz, the step response rise time is 0.85×10^{-3} s, and the overshoot is 17 percent.

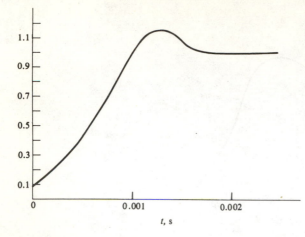

Figure 9.21 Frequency response of system discussed in Example 9.5 (smallest bandwidth case).

The preceding example illustrates that the relationships derived for type I systems between phase margin, damping ratio, and overshoot are not accurate for the type II systems under discussion. Therefore type II systems will now be analyzed and appropriate design approximations derived. For the open-loop system

$$G(s) = \frac{K_a(s/\omega_z + 1)}{s^2}$$

The error-signal transfer function is

$$\theta_\epsilon(s) = \frac{\theta_r(s)}{1 + K_a G(s)} = \frac{\theta_r(s)s^2}{s^2 + K_a(s/\omega_z + 1)} \tag{9.27}$$

and the error in response to a unit step input is

$$\theta_\epsilon(s) = \frac{s}{s^2 + K_a(s/\omega_z + 1)} \tag{9.28}$$

The corresponding time domain response is

$$\theta_\epsilon(t) = e^{-\zeta\omega_n t}\left[\cos \omega_n(1 - \zeta^2)^{1/2}t - \frac{\zeta}{1 - \zeta^2}\sin \omega_n(1 - \zeta^2)^{1/2}t\right] \tag{9.29}$$

where

$$\omega_n^2 = K_a \tag{9.30}$$

and

$$2\zeta\omega_n = \frac{K_a}{\omega_z}$$

or

$$\zeta = \frac{K_a^{1/2}}{2\omega_z} \tag{9.31}$$

The peak overshoot can be determined by calculating the minimum error signal since for a step input

$$\theta_o(t) = 1 - \theta_\epsilon(t)$$

The peak output signal is found by first finding the time at which the error signal is at a minimum; this is determined by setting the first derivative equal to zero. That is,

$$\frac{d\theta_\epsilon}{dt} = -\zeta\omega_n e^{-\zeta\omega_n t}\left[\cos\omega_n(1-\zeta^2)^{1/2}t + \frac{-\zeta}{1-\zeta^2}\sin\omega_n(1-\zeta^2)^{1/2}t\right]$$

$$- \omega_n(1-\zeta^2)^{1/2}e^{-\zeta\omega_n t}\sin\omega_n(1-\zeta^2)^{1/2}t - \zeta\omega_n e^{-\zeta\omega_n t}\cos\omega_n(1-\zeta^2)^{1/2}t = 0$$

$$(9.32)$$

The time t_p at which the minimum error occurs is found to be

$$t_p = \omega_n^{-1}[(1-\zeta^2)^{1/2}]^{-1}\tan^{-1}\frac{2\zeta(1-\zeta^2)^{1/2}}{2\zeta^2 - 1} \tag{9.33}$$

If we define

$$t'_p = \omega_n(1-\zeta^2)^{1/2}t_p = \tan^{-1}\frac{2\zeta(1-\zeta^2)^{1/2}}{2\zeta^2 - 1}$$

then the corresponding peak value of the output is

$$P_o = 1 + \exp\frac{-\zeta}{(1-\zeta^2)^{1/2}}t'_p\left[\cos t'_p - \frac{\zeta}{(1-\zeta^2)^{1/2}}\sin t'_p\right] \tag{9.34}$$

The peak overshoot P_o as a function of ζ $[(K_a^{1/2}/\omega_z = 2\zeta)]$ is plotted in Fig. 9.22. The open-loop crossover frequency of this type II system can be found by

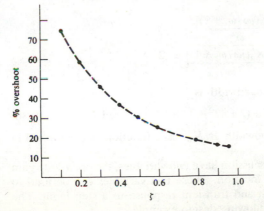

Figure 9.22 Percent overshoot of the step response of a type II second-order system.

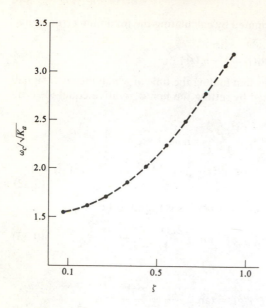

Figure 9.23 Normalized crossover frequency $\omega_c / K_a^{1/2}$ of a type II second-order system.

solving

$$\left| \frac{K_a(j\omega/\omega_z + 1)}{\omega^2} \right| = 1$$

It is readily shown that

$$\omega_c = K_a^{1/2}[2\zeta^2 + (4\zeta^4 + 1)^{1/2}]^{1/2} \qquad (9.35)$$

The normalized open-loop crossover frequency $(\omega_c / K_a^{1/2})$ as a function of ζ is plotted in Fig. 9.23. Figure 9.23 shows that increasing ζ (by either increasing K_a or decreasing ω_z) increases the open-loop crossover frequency, and hence the closed-loop bandwidth B. The closed-loop bandwidth can be determined analytically by solving

$$\frac{\left| \dfrac{K_a(j\omega/\omega_z + 1)}{\omega^2} \right|^2}{\left| 1 + \dfrac{K_a(j\omega/\omega_z + 1)}{(j\omega)^2} \right|^2} = \frac{1}{2}$$

It is found that the closed-loop bandwidth is

$$B = K_a^{1/2}[1 + 2\zeta^2 + (2 + 4\zeta^2 + 4\zeta^4)^{1/2}]^{1/2} \qquad \text{rad/s} \qquad (9.36)$$

The normalized closed-loop bandwidth $B/K_a^{1/2}$ as a function of ζ is plotted in Fig. 9.24.

The type II systems under discussion have two variables, the open-loop gain constant and ω_z, the filter zero location. These two variables can be used to determine the system bandwidth and transient response to a step input. The procedure is illustrated by the following design example.

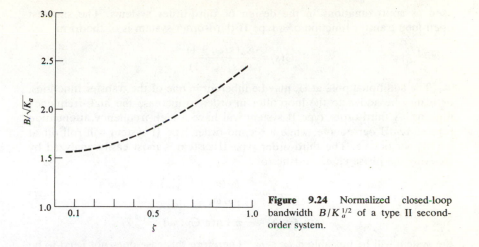

Figure 9.24 Normalized closed-loop bandwidth $B/K_a^{1/2}$ of a type II second-order system.

Example 9.6 A type II system is to be designed to have 20 percent overshoot to a unit step input of phase and a closed-loop bandwidth of 10^3 rad/s.

SOLUTION From Fig. 9.22 we see that 20 percent overshoot ($P_o = 1.2$) corresponds to $\zeta = 0.8$. For this value of ζ Fig. 9.24 indicates that the closed-loop bandwidth is

$$B = K_a^{1/2}2.18$$

so

$$K_a = \left(\frac{10^3}{2.18}\right)^2 = 0.21 \times 10^6 \text{ (rad/s)}^2$$

Therefore,

$$\omega_n = 459 \text{ rad/s}$$

and

$$\omega_z = \frac{\omega_n}{2\zeta} = 287 \text{ rad/s}$$

If this system is to be realized using the phase-locked loop filter illustrated in Fig. 9.14, the loop gain $I/CK_o = K_a$ must be 0.21×10^6, or if the gain is less, additional amplification will be required. The resistor R is selected so that

$$(RC)^{-1} = 287 \text{ rad/s}$$

Type II Third-Order Systems

Type II third-order systems cannot be solved analytically as easily as the type II second-order systems, but the results obtained for second-order systems can be used as approximations in the design of third-order systems. The simplest open-loop transfer function of a type II third-order system is of the form

$$G(s) = \frac{K_a(s/\omega_z + 1)}{s^2(s/\omega_p + 1)}$$

The additional pole at ω_p may be inherent in one of the transfer functions, or it may be added to the loop filter in order to increase the high-frequency filtering. A third-order type II system will have a high-frequency attenuation rate of 12 dB per octave, while a second-order type II system will roll off at -6 dB per octave. The third-order type II system is most easily analyzed by studying the phase characteristics of

$$G_F(j\omega) = \frac{j\omega/\omega_z + 1}{j\omega/\omega_p + 1} \tag{9.37}$$

Since
$$\arg G(j\omega) = -\pi + \arg G_F(j\omega)$$

the system will be unstable if $\omega_p \leq \omega_z$. Therefore this case does not need to be considered. If $\omega_z < \omega_p$ then the phase margin is

$$\phi_M = \arg G_F(j\omega_c) = \tan^{-1}\frac{\omega_z}{\omega_c} - \tan^{-1}\frac{\omega_p}{\omega_c} \tag{9.38}$$

A plot of $\arg G_F(j\omega)$ for various values of α equals ω_p/ω_z is contained in Fig. 9.25. The frequency at which the peak value of $\arg G_F(j\omega)$ occurs can be found

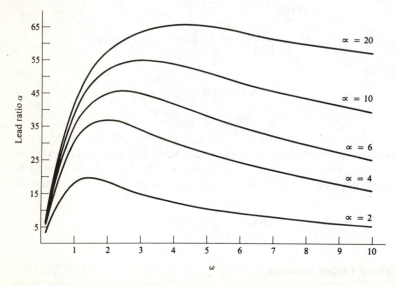

Figure 9.25 Lead network phase shift as a function of frequency for selected lead ratios α.

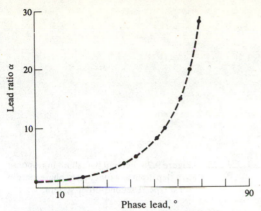

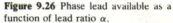

Figure 9.26 Phase lead available as a function of lead ratio α.

by setting the derivative $d/d\omega$ arg $G_F(j\omega)$ equal to 0. It is readily shown that the frequency ω at which the arg $G_F(j\omega)$ is a maximum is

$$\omega_m = (\omega_z\omega_p)^{1/2} \tag{9.39}$$

If this form of transfer function is to be used as the stabilizing filter, then selecting the crossover frequency equal to

$$\omega_c = (\omega_z\omega_p)^{1/2} \tag{9.40}$$

will result in the maximum phase margin. If $\omega_c \neq (\omega_z\omega_p)^{1/2}$, then a larger lead ratio α will be required to obtain the same phase margin. A plot of the maximum phase lead available from $G_F(j\omega)$ as a function of α is plotted in Fig. 9.26. Once

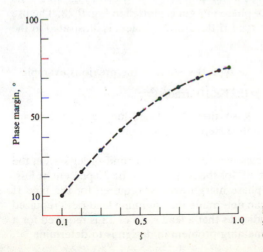

Figure 9.27 Phase margin of a type II second-order system as a function of damping ratio.

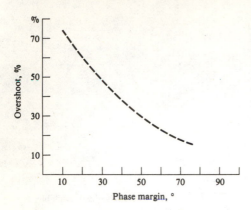

Figure 9.28 Percent overshoot in response to a step input as a function of phase margin for a type II second-order system.

the designer knows the necessary phase margin, the necessary lead ratio can be obtained from that figure. The required phase margin can be estimated by again considering the characteristics of the type II second-order system. Equation (9.35) gives the open-loop crossover frequency of this system as a function of K_a and ζ (or ω_z). For the second-order type II system the phase margin is

$$\phi_M = \tan^{-1} \frac{\omega_c}{\omega_z}$$

Substituting ω_c from Eq. (9.35) we obtain

$$\phi_M = \tan^{-1} \frac{K_a^{1/2}}{\omega_z} [2\zeta^2 + (4\zeta^4 + 1)^{1/2}]^{1/2}$$
$$= \tan^{-1} 2\zeta [2\zeta^2 + (4\zeta^4 + 1)^{1/2}]^{1/2} \tag{9.41}$$

The corresponding phase margin as a function of ζ is plotted in Fig. 9.27. The percent overshoot as a function of phase margin is plotted in Fig. 9.28. The use of these curves in the design of a type II third-order system is illustrated in the following example.

Example 9.7 Consider again the type II system of the previous example.

$$K_a = 0.21 \times 10^6 \text{ (rad/s)}^2$$

Determine a lead-lag network so that the closed-loop system has approximately the same overshoot to step changes in phase.

SOLUTION Figure 9.28 indicates that for a type II second-order system the phase margin must be at least 70° for the overshoot to be 20 percent or less. If we assume that the same phase margin will be required for the type II third-order system, then we can determine the minimum lead ratio required from Fig. 9.26. This figure indicates that a lead ratio of 28 is required for a phase margin of 70°. The remaining problem in design is to determine ω_z, which is to be placed so that Eq. (9.39) is satisfied. At ω_c the magnitude of the lead network response is (in decibels)

$$20 \log G_F(j\omega_c) = 20 \log \frac{|j\omega_c/\omega_z + 1|}{|j\omega_c/\alpha\omega_z + 1|} = 10 \log \frac{\omega_c^2/\omega_z^2 + 1}{\omega_c^2/\alpha^2\omega_z^2 + 1} \quad (9.42)$$

and since $\omega_c^2 = \alpha\omega_z^2$,

$$20 \log G_F(j\omega) = 10 \log \frac{\alpha + 1}{\alpha^{-1} + 1} \approx 20 \log \alpha^{1/2} \quad (9.43)$$

That is, the magnitude of the gain of the lead-lag network is equal to $\alpha^{1/2}$ at the crossover frequency, provided Eq. (9.40) is satisfied. In this example, with α equal to 28, $G_F(j\omega_c)$ increases the gain by

$$20 \log 28^{1/2} = 14.5 \text{ dB}$$

at ω_c, so we must determine where the uncompensated frequency response is -14.5 dB. That is,

$$20 \log \frac{K_a}{\omega^2} = -14.5 \text{ dB}$$

which occurs at

$$\omega_c = 28^{1/4} K_a^{1/2} = 1.06 \times 10^3 \text{ rad/s}$$

The 28:1 lead ratio will increase the crossover frequency by a factor of 2.3.

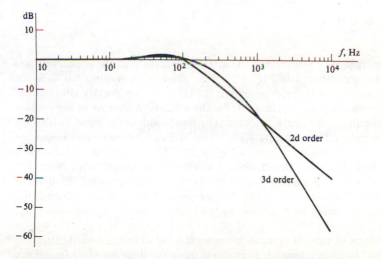

Figure 9.29 Frequency response of second- and third-order type II systems.

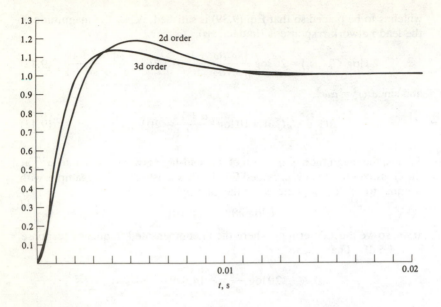

Figure 9.30 Step response of second- and third-order type II systems.

The zero is placed at

$$\omega_z = \frac{\omega_c}{28^{1/2}} = 200 \text{ rad/s}$$

and the pole at

$$\omega_p = 28\omega_z = 5600 \text{ rad/s}$$

The frequency response of the systems described in the preceding two examples is compared in Fig. 9.29. The third-order system has a wider bandwidth, and for frequencies above 10^3 Hz it also has greater attenuation. This attenuation could be desirable for the additional filtering of input noise that is created in the phase detector. The bandwidth of the third-order loop could be reduced by reducing K_a (plus ω_z and ω_p). The transient responses of the two systems are shown in Fig. 9.30. The third-order system has a shorter rise time, as expected, since it has the greater bandwidth. Note that its overshoot is 13 percent, significantly below the specified 20 percent maximum. The overshoot could be increased, if desired, by reducing the phase margin either by reducing the lead ratio or simply by increasing ω_p.

The analysis of type II systems has selected the filter bandwidth after the loop gain has been determined. It is often the case that the filter frequency response is first determined based on other considerations, particularly suppression of the high-frequency components of the phase detector output. If the filter is specified, then the loop gain is selected so that the desired phase margin is obtained.

Loops Including a Time Delay

Many phase-locked loops include a time delay, which can seriously degrade loop stability. Time delays, for example, can occur in sampling phase detectors and digital dividers. Figure 9.31 illustrates in block-diagram form a time-delay network. For an input $f(t)$ to a time-delay network, the output is

$$e_o(t) = f(t-T) \tag{9.44}$$

Therefore,

$$E_o(s) = \int_0^\infty f(t-T)e^{-sT}\,dt$$

$$= F(s)e^{-sT}$$

That is, the transfer function of a time-delay network is

$$\frac{E_o(s)}{F(s)} = e^{-sT} \tag{9.45}$$

The delay network is shown in block-diagram form in Fig. 9.32.

The magnitude of the frequency response of a time-delay network is $|e^{-j\omega T}| = 1$, so the time delay does not affect the magnitude of the frequency response. However, the phase shift

$$\arg e^{-j\omega T} = -\omega T$$

is a linearly decreasing function of frequency. The effect that a time delay can have on stability is illustrated by the following examples.

Example 9.8 A simple linear model of a PLL consisting of a VCO and phase detector (modeled as a time delay) is illustrated in Fig. 9.33. The problem is to determine for what value of delay T the loop will become unstable.

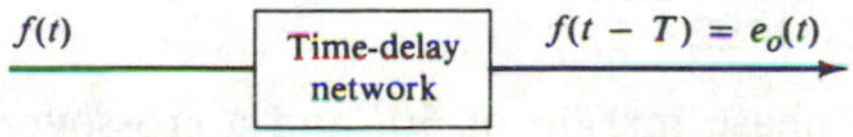

$f(t)$ → [Time-delay network] → $f(t-T) = e_o(t)$

Figure 9.31 A time-delay network.

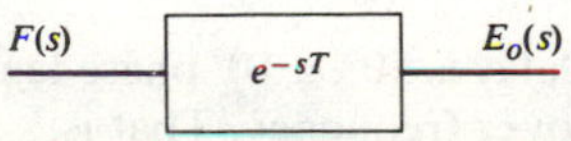

$F(s)$ → [e^{-sT}] → $E_o(s)$

Figure 9.32 Block diagram of a time-delay network.

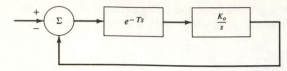

Figure 9.33 Linear model of a type I PLL including a time delay.

SOLUTION Since the open-loop transfer function is

$$G(s) = \frac{e^{-Ts}K_o}{s}$$

the crossover frequency is found from

$$\left| \frac{e^{-j\omega T}K_o}{j\omega} \right| = 1$$

or
$$K_o = \omega_c$$

The phase shift at the crossover frequency is

$$\phi = -\frac{\pi}{2} - \omega_c T$$

and the system will be marginally stable if

$$\omega_c T = \frac{\pi}{2} \qquad \text{or} \qquad T = \frac{\pi}{2K_o}$$

For smaller values for T the system will be stable, and for larger values of T it will be unstable.

In the preceding example the system would have been stable for all values of gain if it had not been for the phase lag due to the time delay. In more complex systems it is usually not possible to arrive analytically at a relationship between loop gain and time delay T, but the effect of the time delay on phase margin is readily determined. This is illustrated by the following example.

Example 9.9 Consider the phase-locked loop system with the open-loop transfer function

$$G(s) = \frac{1000}{s(s/1192 + 1)}$$

This system is found to have a phase margin of 50° and a crossover frequency of approximately 100 rad/s. What time delay can the phase detector introduce and still have a phase margin of 40°?

SOLUTION Since the phase margin without time delay is 50°, a 10° phase lag can be introduced by the time delay at the crossover frequency. That is,

$$\phi_T = -\omega_c T \le -0.174 \text{ rad } (10°)$$

So
$$T \le \frac{0.174}{\omega_c} = 0.174 \times 10^{-3}$$

In a well-designed system with adequate phase margin, an additional phase lag of 0.1 rad can usually be tolerated. In such a system any time delay T introduced will not be a significant factor, provided $T \le 0.1/\omega_c$.

Loops Containing a Sample-and-Hold Phase Detector

Consider the phase-locked loop shown in Fig. 9.34, which contains a sampling phase detector and zero-order data hold. It has been shown in the previous chapter that this PLL can be replaced by the linear model shown in Fig. 8.22.

The open-loop transfer function is

$$G(j\omega) = \frac{K_v e^{-j\omega T/2}}{j\omega} \frac{\sin \omega T/2}{\omega T/2} \tag{9.46}$$

This system appears to be a type II system since there are two poles at the origin, but since

$$\lim_{\omega \to 0} \frac{\sin \omega T/2}{\omega T/2} \to 1$$

there is also a zero at the origin. Hence the system is type I. At the crossover frequency ω_c, the open-loop phase shift is

$$\phi = -\frac{\pi}{2} - \frac{\omega_c T}{2}$$

and the phase margin is

$$\phi_M = \pi + \phi = \frac{\pi}{2} - \frac{\omega_c T}{2}$$

Since the magnitude of the open-loop gain at the crossover frequency is unity,

$$\frac{T}{2} K_v \frac{\sin \omega_c T/2}{(\omega_c T/2)^2} = 1$$

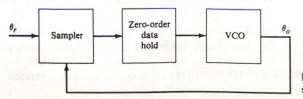

Figure 9.34 A PLL including a sampling phase detector.

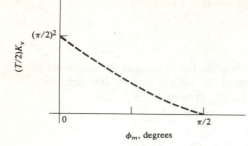

Figure 9.35 Phase margin as a function of sampling rate for a second-order PLL using a sampling phase detector.

or

$$\frac{T}{2} K_v = (\omega_c T/2)^2 / \sin \omega_c T/2$$

$$= \frac{(\pi/2 - \phi_M)^2}{\sin (\pi/2 - \phi_M)} \qquad (9.47)$$

This equation describes the relation between phase margin ϕ_M, sampling period T, and loop gain K_v for a PLL composed of an ideal sampling phase detector, zero-order data hold, and VCO. The plot of $T/2 K_v$ as a function of ϕ_M given in Fig. 9.35 shows that for each value of ϕ_M there is a single value of $T/2 K_v$. For a $T/2 K_v$ of $(\pi/2)^2$ the phase margin is 0°. As K_v is decreased the phase margin increases and reaches 90° for $K_v = 0$.

The effect on loop performance of changes in the other system variable (the sampling rate T) can be determined in the same manner. Since at the crossover frequency ω_c the magnitude of the open-loop gain is unity,

$$\frac{T}{2} K_v \frac{\sin \omega_c T/2}{(\omega_c T/2)^2} = 1$$

If $K_v T$ remains constant and the sampling rate T is changed, ω_c must change such that $\omega_c T$ is constant. That is, if the sampling rate is decreased (T increases), and K_v decreases so that $T K_v$ remains constant, the crossover frequency must decrease so that $\omega_c T$ remains constant.

Example 9.10 Calculate the value of K_v required for a 45° phase margin in a PLL whose open-loop transfer function is given by Eq. (9.46).

SOLUTION In order to have a 45° phase margin, the phase lag of the sample and hold must be 45° at the crossover frequency. Therefore, $\omega_c T/2 = \pi/4$ and the crossover frequency must be $\omega_c = \pi/2T$. Since the magnitude of the open-loop gain is unity at the crossover frequency, K_v is determined from

$$\frac{T}{2} K_v \frac{\sin \pi/4}{(\pi/4)^2} = 1$$

or

$$K_v = \frac{(\pi/4)^2 2^{3/2}}{T}$$

An important characteristic of the sample-and-hold phase detector is the time delay, which is equal to one-half of the reference period. The effect on the magnitude of the frequency response can usually be ignored, but the time delay can significantly affect the loop stability.

9.4 PLL TRANSIENT PERFORMANCE

Since the PLL is a nonlinear device over much of its operating range, the complete performance of the loop depends especially on the type of phase comparator used in the system. Nevertheless, an approximate analysis of the system performance provides insight for formulating generalized design guidelines. Before considering the dynamic performance of the loop, it is instructive to consider the range of input frequencies over which the loop can remain locked. First consider a type I loop with a sinusoidal phase detector. If the PLL is in lock, small changes in the input frequency will cause a change in the phase-detector output voltage in the direction that drives the error signal back toward zero. The change in frequency will be the forward loop gain times the error voltage, or

$$\Delta f = K_d K_o \cos \theta_\epsilon$$

where θ_ϵ is the phase error. This equation assumes that the signals are changing so slowly that there is no attenuation in the low-pass filter. Since $\cos \theta_\epsilon \leq 1$

$$\Delta f \leq K_d K_o = K_v$$

That is, the maximum change in frequency for which a type 1 loop can remain locked, referred to as the *lock range* or *tracking range*, is less than, or equal to, the forward loop gain. Lock range refers to how far the input frequency can change (slowly) without the loop losing frequency lock. The same analysis holds for digital phase detectors, except that K_d is replaced by the detector output voltage in the expression for K_v. *Capture range* refers to how close the input frequency must come to the free-running frequency of the VCO before lockup can occur. When the frequency error is other than zero, the PD output voltage is attenuated by the low-pass filter before it reaches the VCO. The capture range is thus less than the lock range. The actual value depends on the type of low-pass filter used. A general expression for loop capture range is not available as the system is highly nonlinear.

For type II loops, K_v is theoretically equal to infinity. The dc value of the phase detector output is integrated over time and, if saturation does not occur, can become arbitrarily large. The tracking range of type II loops is limited by the dynamic range of the voltage-controlled oscillator. The loop capture range cannot exceed the frequency range of the VCO.

Transient Analysis of the Linearized PLL

In studying the transient response of phase-locked loops one must consider the linear (small-signal) and nonlinear (large-signal) operation of the loop. We will consider both the linear and nonlinear responses of type I and type II PLLs. For the linear model, the error signal $\theta_e(s)$ is obtained from Eq. (9.1), and the output signal is

$$\theta_o(s) = \frac{\theta_r(s) K_o K_d F(s)/s}{1 + [K_o K_d F(s)/s]N} \tag{9.48}$$

In the following analysis it is assumed that the initial phase error is zero and that it remains sufficiently small so that the loop operation remains linear. N will be assumed equal to 1 in order to simplify the notation.

Type I Systems: Phase Step Response

If at $t = 0$ the input phase is suddenly changed by an amount $\phi[\theta_r(s) = \phi/s]$, the error signal is

$$\theta_e(s) = \frac{\phi}{s[1 + K_v/sF(s)]}$$

If no low-pass filter is used in the loop $[F(s) = 1]$, then

$$\theta_e(s) = \frac{\phi}{s + K_v}$$

and

$$\theta_e(t) = \phi e^{-K_v t} \tag{9.49}$$

The error voltage decays exponentially toward zero with increasing time. The larger the loop gain, the faster the loop responds to the change in phase. If $F(s)$ is a simple low-pass filter, then $F(s) = (\tau s + 1)^{-1}$, and the error voltage will be

$$\theta_e(s) = \frac{\phi(\tau s + 1)}{s(\tau s + 1) + K_v} = \frac{\phi(s + 2\zeta\omega_n)}{s^2 + 2\zeta\omega_n s + \omega_n^2}$$

where

$$2\zeta\omega_n = \tau^{-1}$$

and

$$\omega_n^2 = \frac{K_v}{\tau}$$

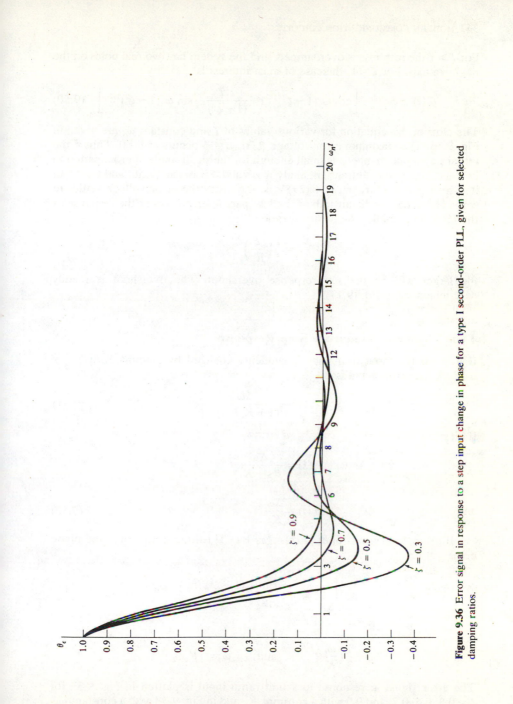

Figure 9.36 Error signal in response to a step input change in phase for a type I second-order PLL, given for selected damping ratios.

For $\zeta > 1$ the response is overdamped, and the system has two real poles on the negative axis. For $\zeta < 1$, the case of most interest is .

$$\theta_\epsilon(t) = \phi e^{-\zeta\omega_n t}\left[\cos \omega_n(1-\zeta^2)^{1/2}t + \frac{\zeta}{(1-\zeta^2)^{1/2}}\sin \omega_n(1-\zeta^2)^{1/2}t\right] \quad (9.50)$$

The plots of this equation for various values of ζ and constant ω_n are given in Fig. 9.36. The maximum error voltage $\theta_\epsilon(t)_{max} = \phi$ occurs at $t = 0$. Thus if the initial step input change ϕ is small enough for the small-angle approximation to be valid, the complete transient analysis is valid. From Eq. (9.50) and Fig. 9.36 it is seen that the larger $\zeta\omega_n = (2\tau)^{-1}$ is, the faster the error voltage settles to zero. The wider the bandwidth of the low-pass filter, the faster the response of the loop. The smaller the damping ratio

$$\zeta = \frac{1}{2}\left(\frac{2}{K_v}\right)^{1/2}$$

the larger will be the step-response overshoot. The overshoot is readily determined using Eq. (9.17).

Type I Systems: Frequency Step Response

If at $t = 0$ the input frequency is suddenly changed by amount $\Delta\omega$ $[\theta_r(s) = \Delta\omega/s^2]$, the error signal is

$$\theta_\epsilon(s) = \frac{\Delta\omega}{s^2[1 + K_vF(s)/s]} \quad (9.51)$$

If $F(s) = 1$ (no filter),

$$\theta_\epsilon(s) = \frac{\Delta\omega}{s(s + K_v)}$$

and

$$\theta_\epsilon(t) = \frac{\Delta\omega}{K_v}(1 - e^{-K_v t}) \quad (9.52)$$

If a simple low-pass filter $[F(s) = (\tau s + 1)^{-1}]$ is used in the loop, the phase error will be

$$\theta_\epsilon(t) = 2\zeta\frac{\Delta\omega}{\omega_n} + \frac{\Delta\omega}{\omega_n}e^{-\zeta\omega_n t}\left[\frac{1-2\zeta^2}{(1-\zeta^2)^{1/2}}\sin \omega_n(1-\zeta^2)^{1/2}t - 2\zeta\cos\omega(1-\zeta^2)^{1/2}t\right] \quad (9.53)$$

where $\quad \zeta = \frac{1}{2}\left(\frac{\omega_L}{K_v}\right)^{1/2} \quad$ and $\quad \omega_n = (K_v\omega_L)^{1/2}$

The error signal in response to a unit ramp input is plotted in Fig. 9.37 for $\zeta = 0.3, 0.5, 0.7,$ and 0.9 with a constant K_v and in Fig. 9.38 with a constant ω_n.

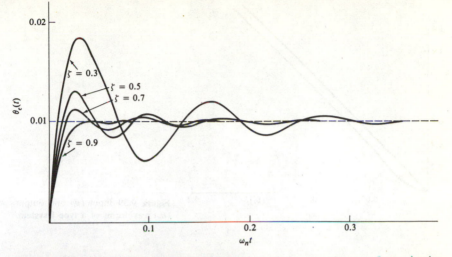

Figure 9.37 Error signal in response to a step input change in frequency for a type I second-order PLL, given for selected damping ratios; K_v is constant.

In Fig. 9.37 the damping ratio is varied by changing the filter bandwidth, and in Fig. 9.38 the product $2K_v$ is kept constant. An example of the input and output waveforms is given in Figure 9.39. For the error expression given by Eq. (9.51) the steady-state error is

$$\epsilon_{ss}(t) = \lim_{t \to 0} \theta_\epsilon(t) = \lim_{s \to 0} s\theta_\epsilon(s) = \frac{\Delta \omega}{K_v} = \frac{2\zeta}{\omega_n} \Delta \omega \qquad (9.54)$$

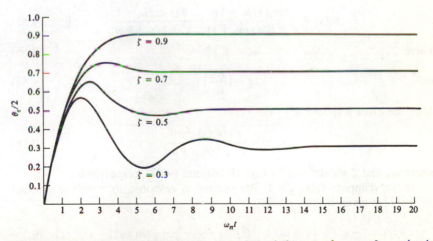

Figure 9.38 Error signal in response to a step input change in frequency for a type I second-order PLL, given for selected damping ratios; ω_n is constant.

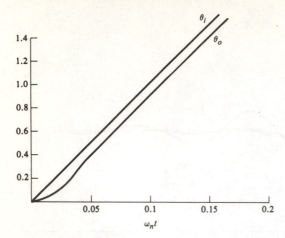

Figure 9.39 Input (θ_i) and output (θ_o) waveforms of a type I system.

provided the limit exists. Therefore, the larger the loop gain K_v, the smaller the steady-state error, but this implies a smaller damping ζ and hence more overshoot. If the steady-state error and transient performance are both important parameters, it may be necessary to use a type II loop.

Type II Loop

For the type II second-order loop,

$$\theta_o(s) = \frac{\theta_r(s)(s/\omega_z + 1)}{s^2 + K_a(s/\omega_z + 1)} = \frac{\theta_r(s)(s/\omega_z + 1)}{s^2 + 2\zeta\omega_n s + \omega_n^2} \qquad (9.55)$$

where

$$\omega_n = K_a^{1/2} \qquad (9.56)$$

and

$$2\zeta = \left(\frac{K_a}{\omega_z}\right)^{1/2} \qquad (9.57)$$

Also, the error response is

$$\theta_\epsilon(s) = \frac{\theta_r(s)s^2}{s^2 + 2\zeta\omega_n s + \omega_n^2} \qquad (9.58)$$

where ω_n and ζ are defined by Eqs. (9.56) and (9.57), respectively.

If the damping ratio $\zeta < 1$, the output in response to a unit step input $[\theta_r(t) = 1]$ is given by

$$\theta_o(t) = 1 - e^{-\zeta\omega_n t}\left[\cos \omega_n (1 - \zeta^2)^{1/2} t - \frac{\zeta}{(1 - \zeta^2)^{1/2}} \sin \omega_n (1 - \zeta^2)^{1/2} t\right] \qquad (9.59)$$

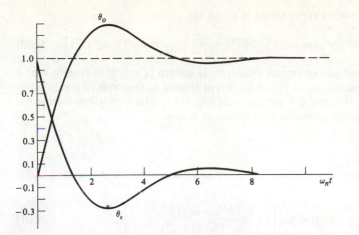

Figure 9.40 Error and output response of a type II PLL ($\zeta = 0.5$) to a step change in phase.

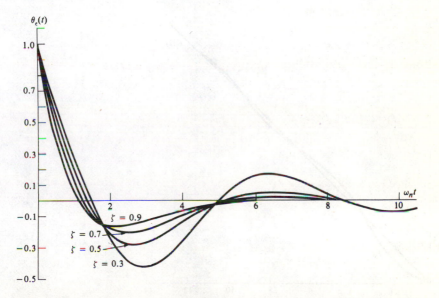

Figure 9.41 Error response of a type II PLL to a step change in phase, given for selected damping ratios.

and the corresponding error signal is given by

$$\theta_\epsilon(t) = e^{-\zeta\omega_n t}\left[\cos \omega_n(1 - \zeta^2)^{1/2}t - \frac{\zeta}{(1 - \zeta^2)^{1/2}}\sin \omega_n(1 - \zeta^2)^{1/2}t\right] \quad (9.60)$$

The error and output signals of a type II system ($\zeta = 0.5$) in response to a unit step input are shown in Fig. 9.40. Error signals in response to a step input for $\zeta = 0.3$, 0.5, 0.7, and 0.9 are given in Fig. 9.41. The steady-state error for the type II system in response to step inputs is zero.

Ramp Inputs

For a step change in frequency $[\theta_r(t) = \Delta\omega t]$, the output of the second-order type II system is

$$\theta_o(t) = \Delta\omega\left[t - \frac{e^{-\zeta\omega_n t}}{\omega_n}\frac{\sin \omega_n(1 - \zeta^2)^{1/2}t}{(1 - \zeta^2)^{1/2}}\right] \quad (9.61)$$

The input and output waveforms of an underdamped system are illustrated in Fig. 9.42, and the error signals of a type II second-order system in response to a ramp input are illustrated in Fig. 9.43. It is seen that the steady-state error of type II systems in response to step changes in frequency decays to zero.

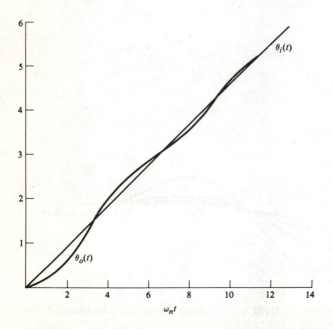

Figure 9.42 Input (θ_i) and output (θ_o) waveforms of a type II PLL.

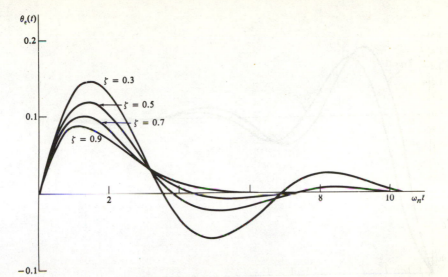

Figure 9.43 Error response of a type II PLL to a step change in frequency, given for selected damping ratios.

Comparison of Type I and Type II Loops

When it is necessary to have zero steady-state error to ramp inputs, a type I system cannot be used, but either a type I or type II system can realize zero steady-state error to step inputs. For step inputs a type II system will have a shorter rise time but also more overshoot than a type I system. Figure 9.44 compares the step response of type I and II systems with the same damping ratio and ω_n. The type II systems has a shorter rise time and more overshoot, as was predicted from the stability analysis of the system. In many applications it is the settling time which is of the most importance and not the rise time. Settling time in a phase-locked loop is usually defined as the time required for the phase error to settle below a specified time for which the frequency output is unusable while switching from one frequency to another. It is difficult to measure the instantaneous frequency, but relatively easy to measure the instantaneous phase, so phase settling time is usually used for frequency synthesizer specifications. The phase error of a type II system in response to a step change frequency of $\Delta\omega$ rad/s is obtained from Eq. (9.61). At any time the phase error satisfies the inequality

$$|\theta_\epsilon(t)| \le \frac{\Delta\omega}{\omega_n} \frac{e^{-\zeta\omega_n t}}{(1 - \zeta^2)^{1/2}}$$

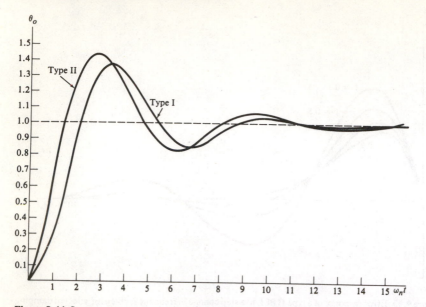

Figure 9.44 Step responses of second-order type I and type II PLL systems with identical ζ and ω_n.

So, for a specified maximum steady-state error θ_M,

$$t_s \le (\zeta\omega_n)^{-1} \ln\left\{ \frac{\Delta\omega}{\omega_n\theta_M} [(1-\zeta^2)^{1/2}]^{-1} \right\} \tag{9.62}$$

The normalized settling time $\omega_n t_s$ is plotted in Fig. 9.45 for a change in frequency ($\Delta\omega/\omega_n = 0.5$) and for three different values of steady-state error θ_M:

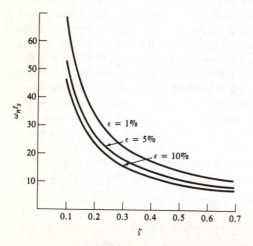

Figure 9.45 Normalized settling time ($\omega_n t_s$) for $\Delta\omega/\omega_n = 0.5$.

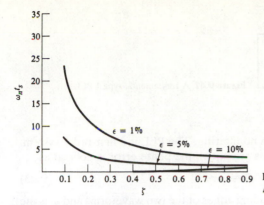

Figure 9.46 Normalized settling time for $\Delta\omega/\omega = 0.1$.

0.01, 0.05, and 0.1. The settling time to a step change in frequency of $\Delta\omega/\omega_n = 0.1$ is plotted in Fig. 9.46.

Large-Signal Behavior

When the error signal is large, the linear PLL model is no longer valid. The main source of the nonlinearity is the phase detector (the nonlinear characteristics being different for each type of phase detector), and the system's performance must be analyzed in the time domain. The following sections illustrate the analysis of the large-signal performance of phase-locked loops.

Digital Phase Detectors

The output of a digital phase or phase-frequency detector is either a two- or three-amplitude level signal. As the phase error increases, the amplitude does not change; however, the duration of the output pulse does. The analysis and performance of such systems will be illustrated by the analysis of type I phase-locked loops. The analysis provides insight into the design of modern phase-locked loops.

First-Order Systems

It is assumed that the phase-frequency detector outputs a pulse of $+V$ volts when the phase error is positive and $-V$ volts when the phase error is negative. The width of the phase detector output pulse τ is equal to the time between the leading edges of the two waveforms if the output leads in phase. However, if the phase error is positive (reference leads in phase) a positive voltage will be applied to the VCO, the VCO will speed up, and the output phase will reset the phase detector in a shorter time than would elapse if the phase error were negative. To illustrate this operation we will first consider a type I loop,

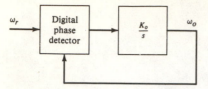

Figure 9.47 A fundamental type I PLL.

illustrated in Fig. 9.47, which uses no filtering. The PLL is a first-order system. If the input and output frequencies are equal, the phase error is defined as

$$\theta_\epsilon = \omega \tau \qquad (9.63)$$

where τ is the time between the leading edges of the two waveforms and ω is the waveform frequency. The output frequency is

$$\omega_o = \omega_c + K_o V_c \qquad (9.64)$$

where ω_c is the free-running frequency of the VCO. Since the control voltage V_c is either $\pm V$ or zero, the VCO frequency can only take on three distinct values. The output phase is

$$\theta_o(t) = \omega_c t + \int_0^t K_o V_c(t)\, dt + \phi_o \qquad (9.65)$$

where ϕ_o is the phase at $t = 0$, or $\theta_o(0) = \phi_o$.

We will first consider the case in which the input leads in phase. It is assumed that the input frequency ω_r is equal to ω_c. (If the input frequency is not equal to ω_c, the output frequency cannot adjust to ω_r, since this loop does not have the capability of continuously changing the frequency.) At $t = 0$, the phase of the reference is taken as zero. That is, $\theta_r(0) = 0$. Then for positive phase error the output phase is negative, or $\phi_o(0) = -\phi_o$, and the leading edge of the reference frequency sets the phase detector. The leading edge of the output lags by τ s, so the phase error is

$$\theta_\epsilon(0) = \phi_o$$

where $\tau = \phi_o / \omega_c$.

The phase detector outputs a pulse of $+V$ volts until the leading edge of the output signal resets the phase detector at time t_o. At $t = 0$ the VCO frequency instantaneously increases to

$$\omega_o = \omega_c + K_o V \qquad \text{rad/s}$$

and the output phase is

$$\theta_o = \omega_c t + K_o \int_0^t V_c(t)\, dt - \phi_o$$

$$= \omega_c t + K_o V_c t - \phi_o \qquad 0 \le t \le t_o$$

The output phase becomes zero at the time

$$t_o = \frac{\phi_o}{\omega_c + K_o V} \qquad (9.66)$$

at which time the phase detector resets and the VCO control voltage V_c becomes zero. The phase error at t_o is $\theta_\epsilon(t_o) = \omega_c t_o$, and since the VCO and the reference signal are again operating at the same frequency this will also be the phase error at the time T when the reference frequency again sets the phase detector and the VCO again increases in frequency. At the end of the nth reference period the phase error will be

$$\theta_\epsilon(nT) = \frac{\phi_o}{(1 + K_o V/\omega_c)^n} \qquad (9.67)$$

(Note that the larger $K_o V$ is, the smaller will be the phase error, but once the phase error is positive it remains positive.)

If the initial phase error is negative (the VCO output leads the reference waveform by τ s), the phase detector output will be set to $-V$ volts and will remain at this level until the leading edge of the reference waveform resets the phase detector after τ s. At this time the output phase will be

$$\theta_\epsilon(\tau) = -(\omega_c \tau - K_o V \tau) = -(\phi_o - K_o V \tau) \qquad (9.68)$$

and if the loop gain $K_o V = \omega_c$, the phase error will be zero when the reference waveform resets the phase detector. The phase error will remain negative, provided $K_o V < \omega_c$. This will always be the case, since the maximum frequency deviation of a VCO below its free-running frequency ω_c is equal to ω_c. The frequency deviation is also $K_o V_c$, so $K_o V/\omega_c$ cannot be larger than 1 for frequency deviations below the free-running frequency of the VCO. The phase error after n output cycles will be

$$\theta_\epsilon(nT) = -\phi_o \left(1 - \frac{K_o V}{\omega_c}\right)^n \qquad (9.69)$$

The analysis of this simple system illustrates that the large-signal response of PLL systems that utilize a digital phase detector is not symmetrical, since it differs depending upon whether the initial phase error is positive or negative.

If $K_o V/\omega_c$ is close to 1, negative phase errors will be reduced faster than positive phase errors. This suggests that to reduce the settling time the phase detector logic could be modified to provide a larger gain for positive phase errors. Figure 9.48 illustrates the phase-error transient response of a type I first-order loop to a step input of phase. The normalized loop gain $K_o V/\omega_c = 0.5$ in this example. The negative phase error diminishes faster than the positive phase error, but it must be kept in mind that the phase detector cycle time is longer for negative error signals.

The analysis of this simple system also illustrates some of the limitations of linearized analysis. For example, for this type I system the final value theorem

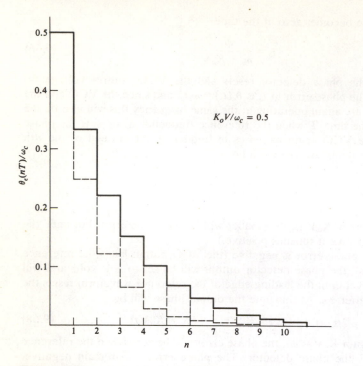

Figure 9.48 Step response of a fundamental type I PLL.

predicts that the final frequency error in response to step changes $\Delta\omega$ in frequency will be

$$\lim_{t\to\infty}[f_r(t) - f_o(t)] = \lim_{s\to\infty}\frac{s(\Delta\omega/s^2)}{1 + K_v/s} = \frac{\Delta\omega}{K_v}$$

However, it is obvious from the large-signal analysis that VCO cannot track arbitrary changes in the input frequency since the VCO can only take on three different values of frequency.

Another feature of loop performance brought out by this analysis is that the time for the phase detector to reset is different for positive and negative phase errors. If the output leads in phase and sets the phase detector, the phase detector is set for a time equal to the initial phase error, but if the input leads in phase and sets the phase detector, it is set for a time given by Eq. (9.66). In this case, the time is a function of the loop parameters.

Type I Loop Including a Low-Pass Filter

The preceding analysis has shown that when the loop does not contain a low-pass filter the VCO frequency jumps discontinuously from ω_c to $\omega_c \pm K_o V$. A

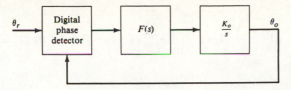

Figure 9.49 A type I PLL including a filter $F(s)$.

low-pass filter can be added to the loop to smooth out the frequency transitions and improve the spectral purity of the output signal. It will be shown that the filter can also reduce the system settling time. It is assumed that the low-pass filter is first-order, with a transfer function of

$$F(s) = \left(\frac{s}{a} + 1\right)^{-1}$$

If the charge-pump phase detector and filter illustrated in Fig. 9.14 are used, the filter is realized by selecting the capacitor C such that $RC = a^{-1}$.

The system illustrated in Fig. 9.49 will now be analyzed. The case in which the reference leads in phase by ϕ_o rad will be considered first. The phase detector output voltage will consist of a pulse of width t_o and amplitude V so the voltage applied to the VCO is

$$V_c(s) = \frac{V(1 - e^{-t_o s})}{s(s/a + 1)} \tag{9.70}$$

or

$$V_c(t) = V(1 - e^{-at}) \qquad 0 \le t \le t_o \tag{9.71}$$

At time t_o the phase detector resets to zero output, and

$$V_c(t) = Ve^{-at}(e^{at_o} - 1) \qquad t_o \le t \le T \tag{9.72}$$

The output phase during the time the phase detector output is $+V$ is

$$\theta_o(t) = \int_0^t (\omega_c + KV(1 - e^{-at})) \, dt - \phi_o$$

$$= \omega_c t + K_o V \left(t + \frac{e^{-at} - 1}{a}\right) - \phi_o \qquad 0 \le t \le t_o \tag{9.73}$$

The output phase will be zero at t_o, so t_o can be found by solving

$$0 = \omega_c t_o + K_o V \left(t_o + \frac{e^{-at_o} - 1}{a}\right) - \phi_o \tag{9.74}$$

The value of t_o can be found using numerical approximation techniques to solve Eq. (9.74). For $t_o \le t \le T$ the VCO control voltage is

$$V_c(t) = V(1 - e^{-at_o})e^{-a(t - t_o)} \qquad t_o \le t \le T \tag{9.75}$$

The output phase at time T is

$$\theta_o(T) = \theta_o(t_o) + \int_{t_o}^T (\omega_c + K_o V(1 - e^{-at_o})e^{-a(t - t_o)}) \, dt \tag{9.76}$$

and since $\theta_o(t_o) = 0$,

$$\theta_o(T) = \omega_c(T - t_o) + \frac{K_o V}{a}[1 - e^{-at_o} + e^{-aT}(1 - e^{-at_o})] \qquad (9.77)$$

The phase error at time T is

$$\theta_\epsilon(T) = \theta_\epsilon(t_o) - K_o V(1 - e^{-at_o}) \int_{t_o}^{T} e^{-a(t-t_o)} \, dt$$

$$= \omega_c t_o - \frac{K_o V}{a}[1 - e^{-at_o} + e^{-aT}(1 - e^{at_o})] \qquad (9.78)$$

This derivation assumes that the phase error $\theta_\epsilon(T)$ remains positive. To see that it does, the expression $\theta_\epsilon(T)$ will be rewritten, substituting Eq. (9.74) for $\omega_c t_o$ in Eq. (9.78). That is,

$$\theta_\epsilon(T) = \phi_o - K_o V \left(t_o + \frac{e^{-at_o} - 1}{a} \right) - \frac{K_o V}{a}(1 - e^{-at_o} + e^{-aT})(1 - e^{at_o})$$

$$= \phi_o - K_o V t_o - \frac{K_o V}{a} e^{-aT}(1 - e^{at_o})$$

$$= \phi_o - K_o V t_o + \alpha$$

where $$\alpha = \frac{K_o V}{a} e^{-aT}(e^{at_o} - 1) > 0$$

Also $\phi_o = \omega_c \tau$ and $K_o V \le \omega_c$, so

$$\phi_o - K_o V t_o \ge \omega_c(\tau - t_o)$$

Since $\tau > t_o$, we have that $\theta_\epsilon(T) > 0$, provided $\theta_\epsilon(0) > 0$. Therefore, as was the case when no filter was used, if in a type I loop containing a simple low-pass filter the phase error becomes positive, then it will remain positive in the absence of further external input. Note that this is a marked difference from the linear model, which predicts that the response is underdamped if the loop gain is sufficiently large.

Figure 9.50 illustrates how the system settling time varies as a function of the filter bandwidth for positive phase errors. For this example the loop gain $K_o V / \omega_c = 1$, and the initial phase error is 180°. The vertical axis represents how many reference cycles it takes for the time difference between the two signals to be less than 10^{-6} s ($\phi_\epsilon < 10^{-6} \omega_r$). If the filter bandwidth is too narrow, a large number of reference cycles is needed before the phase error settles below the prescribed limit. As the filter bandwidth increases, the settling time reaches a minimum at $B = 0.5\omega_r$. As the bandwidth is further increased, the settling time again increases, approaching the limit of 12 reference cycles for the case of infinite filter bandwidth. In the case where the reference leads in phase, there is an optimum value of filter bandwidth for minimizing the settling time. The minimum in Fig. 9.50 is relatively broad. For positive phase errors the

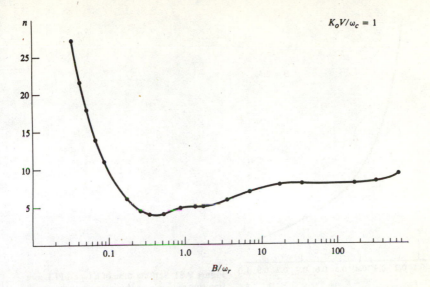

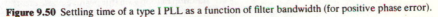

Figure 9.50 Settling time of a type I PLL as a function of filter bandwidth (for positive phase error).

bandwidth of this type I system should be between the limits

$$0.25\omega_r \le B \le 2.5\omega_r \qquad \text{rad/s}$$

for the fastest response.

Figure 9.51 illustrates how the settling time varies as a function of the normalized loop gain (K_v/ω_r). The filter bandwidth is $2.5\omega_r$, and curves are plotted for two different values of initial error. The larger the loop gain, the shorter the settling time. Larger loop gain implies a larger frequency deviation of the VCO. The most important factor in reducing the loop settling time is the frequency deviation of the VCO.

If the output leads the reference in phase by τ s, the phase error is negative; for this case the width of the phase-detector pulse is τ s because the phase detector resets on the leading edge of the reference waveform, which is independent of the VCO waveform. Therefore,

$$\theta_o(\tau) = \int_o^\tau \left[\omega_c + K_o V_c(t) \right] dt$$

In this case the output phase is taken to be zero at $t = 0$,

$$\theta_o(0) = 0$$

and the phase error is

$$\theta_e(0) = -\omega_c \tau$$

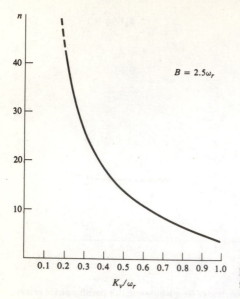

Figure 9.51 Settling time of a type I PLL as a function of loop gain (for positive phase error).

The phase detector is set by the leading edge of the output waveform, and the control voltage is

$$V_c(t) = -V(1 - e^{-at}) \qquad 0 \le t \le \tau \qquad (9.79)$$

Therefore, the output phase at time τ

$$\theta_o(\tau) = \omega_o\tau - K_o V\left(\tau + \frac{e^{-a\tau} - 1}{a}\right) \qquad (9.80)$$

After the phase detector is reset, the VCO control voltage decays toward zero:

$$V_c(t) = -K_o Ve^{-at}(e^{-a\tau} - 1) \qquad \tau \le t \le t_o \qquad (9.81)$$

The output phase is

$$\theta_o(t) = \theta_o(\tau) - \frac{K_o V(e^{a\tau} - 1)(e^{-a\tau} - e^{-at})}{a} + \omega_c(t - \tau) \qquad t \le t \le t_o \quad (9.82)$$

If Eq. (9.80) is substituted for $\theta_o(\tau)$,

$$\theta_o(t) = \omega_c t - K_o V\left[\tau + \frac{e^{-a\tau} - 1}{a} + \frac{(e^{a\tau} - 1)(e^{-a\tau} - e^{-at})}{a}\right] \qquad (9.83)$$

The phase error at time t_o

$$\theta_\epsilon(t_o) = \theta_o(t_o) - \theta_i(t_o)$$

$$= \theta_o(t_o) - \omega_c(t_o - \tau) \qquad (9.84)$$

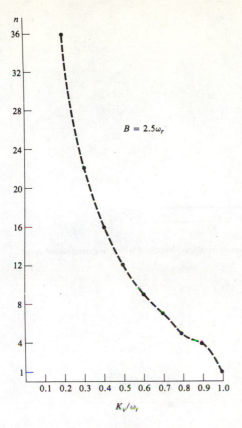

$B = 2.5\omega_r$

Figure 9.52 Settling time of a type I PLL as a function of loop gain (for negative phase error).

The problem now is to find the time t_o at which the output phase is again 0 or 2π. That is, $\theta_o(t_o) = 2\pi$. The time t_o is readily found by numerical methods. (The following results were obtained using the Newton-Raphson root-finding algorithm.)

How the settling time of a type I system varies as a function of normalized loop gain (K_v/ω_r) when the initial phase error is negative is illustrated in Fig. 9.52 for the case in which the low-pass filter bandwidth $B = 2.5\omega_r$. The curve is similar to the case in which the initial phase error is positive (Fig. 9.51). In order to minimize the loop settling time, the normalized loop gain needs to be close to unity. How the settling time varies as a function of low-pass filter bandwidth is illustrated in Fig. 9.53 for a normalized loop gain of unity. The settling time reaches a minimum value of 1 when the filter bandwidth is approximately $2\frac{1}{2}$ times the reference frequency. The settling time remains at this value as the filter bandwidth is further increased. Figure 9.54 is similar to Fig. 9.53 except that the normalized loop gain $K_v/\omega_r = 0.5$. There is little interaction between the minimum filter bandwidth and the normalized loop gain. A comparison of Figs.

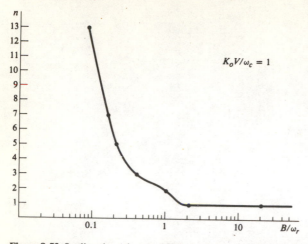

Figure 9.53 Settling time of a type I PLL as a function of loop bandwidth (for negative phase error).

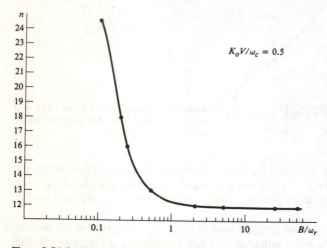

Figure 9.54 Settling time of a type I PLL as a function of loop bandwidth (for negative phase error and reduced loop bandwidth).

9.50 and 53 shows that if only the loop settling time is considered in selecting the filter bandwidth, the bandwidth should be approximately $2\frac{1}{2}$ times the reference frequency.

Sinusoidal Phase Detection[9.2]

If a sinusoidal phase detector is used without a low-pass filter, and the upper frequency term $(\omega_r + \omega_c)$ is ignored, the phase detector output voltage will be equal to the voltage applied to the VCO:

$$V_c = K_d \sin \theta_\epsilon \tag{9.85}$$

The phase error is

$$\theta_\epsilon = \theta_r - \theta_o$$

so the frequency error is

$$\frac{d\theta_\epsilon}{dt} = \omega_r - \frac{d\theta_o}{dt}$$

If ω_c is the free-running frequency of the VCO, then

$$\theta_o(t) = \omega_c + K_o V_c(t)$$

so

$$\frac{d\theta_\epsilon}{dt} = \omega_r - \omega_c - K_o V_c(t)$$

$$= \Delta\omega - K_o V_c(t) \tag{9.86}$$

where $\Delta\omega = \omega_r - \omega_c$ is the difference between the reference frequency and the free-running frequency of the VCO. If Eq. (9.85) is substituted in Eq. (9.86),

$$\frac{d\theta_\epsilon}{dt} = \Delta\omega - K_o K_d \sin \theta_\epsilon \tag{9.87}$$

This is the nonlinear equation describing the phase error in a phase-locked loop containing a sinusoidal phase detector and no filter. Also, the high-frequency term has been ignored. For the sake of generality, assume the loop also contains an amplifier; then Eq. (9.87) can be rewritten as

$$\frac{d\theta_\epsilon}{dt} = \Delta\omega - K_v \sin \theta_\epsilon(t) \tag{9.88}$$

where

$$K_v = K_o K_d K$$

If $\Delta\omega > K_v$, the loop cannot acquire phase lock since the maximum change of VCO frequency $(\Delta\omega)_{max} = K_v$ and it is assumed that $K_v \geq \Delta\omega$. Equation (9.88) can be rewritten as

$$\frac{d\theta_\epsilon}{\Delta\omega - K_v \sin \theta_\epsilon(t)} = dt$$

which has the solution

$$t - t_o = \frac{2}{[(\Delta\omega)^2 - K_v^2]^{1/2}} \tanh^{-1} \frac{\Delta\omega + K_v}{\Delta\omega - K_v} \tan\left(\frac{\pi}{4} - \frac{\theta_\epsilon}{2}\right) \tag{9.89}$$

where t_o, the initial time, will be assumed to be zero. This equation can be rearranged to express the phase error as a function of time.[9.2]

$$\theta_\epsilon(t) = 2 \tan\left(\frac{K_v - \Delta\omega}{K_v + \Delta\omega}\right)^{1/2} \frac{1 - \exp\{-[K_v^2 - (\Delta\omega)^2]^{1/2} t\}}{1 + \exp\{-[K_v^2 - (\Delta\omega)^2]^{1/2} t\}} + \frac{\pi}{2} \tag{9.90}$$

Note that

$$\lim_{t \to \infty} \theta_\epsilon(t) = 2 \tan^{-1} \left(\frac{K_v - \Delta\omega}{K_v + \Delta\omega} \right)^{1/2} + \frac{\pi}{2} \tag{9.91}$$

Numerical Example

Consider the simplest case where $\Delta\omega = 0$. Then Eq. (9.88) becomes

$$\frac{d\theta_\epsilon}{-K_v \sin \theta_\epsilon(t)} = dt$$

which is readily solved to give

$$\frac{1}{2} \ln \frac{1 - \cos \theta_\epsilon}{1 + \cos \theta_\epsilon} = -K_v(t - t_o) + C$$

or

$$\frac{1 - \cos \theta_\epsilon}{1 + \cos \theta_\epsilon} = A e^{-2K_v(t - t_o)}$$

Without loss of generality, let the initial error be θ_o at $t = t_o = 0$. Then

$$\frac{1 - \cos \theta_o}{1 + \cos \theta_o} = A$$

and

$$\cos \theta_\epsilon = \frac{1 - A e^{-2K_v(t)}}{1 + A e^{-2K_v t}} = \frac{1 + \cos \theta_o - (1 - \cos \theta_o) e^{-2K_v t}}{1 + \cos \theta_o + (1 - \cos \theta_o) e^{-2K_v t}}$$

Several significant results for the large-signal behavior of PLLs containing a sinusoidal phase detector have also been obtained using phase-plane techniques.[9.3] Since the emphasis here is on the response of loops containing digital phase detectors, those results will not be discussed, except to emphasize that the large-signal behavior of PLLs is markedly different for different types of phase detectors.

PROBLEMS

9.1 A simple phase-locked loop has $K_d K_o = 1000$ rad/s, $F(s) = 1$, and $N = 20$. What is the closed-loop bandwidth? What is the closed-loop bandwidth if the filter described by the transfer function

$$F(s) = \left(\frac{s}{50} + 1 \right)^{-1}$$

is added inside the loop?

9.2 The system of Prob. 9.1 (including the filter) is to be designed so that it will be able to track step changes in frequency ($\Delta\omega = 1$ rad/s) with less than 0.01 rad steady-state error. Is the loop gain adequate? If not, how much additional gain must be added in order to meet the steady-state error specification?

9.3 Graph the open-loop frequency response, determine the crossover frequency, and calculate the phase margin of the PLL described in Prob. 9.1, both with and without the low-pass filter in the loop.

9.4 Derive the formula [Eq. (9.16)] for the closed-loop bandwidth of a type I system.

9.5 Calculate the overshoot of the system described in Prob. 9.1 in response to a 2° step change in the input phase.

9.6 What is the maximum bandwidth that the filter of Prob. 9.1 can have if the system peak overshoot to step inputs in phase is to be less than 20 percent? Solve the problem using both straight-line approximations and the analytic method.

9.7 Modify the system of Prob. 9.1 so that it has less than 20 percent overshoot to step inputs and also has a rise time of less than 10 ms. Design the circuit to be added to the loop.

9.8 The following open-loop system $G(s)$ has zero steady-state phase error to constant frequency inputs. Determine the filter frequency ω_z so that the closed-loop system has approximately 20 percent overshoot to step inputs. What will the closed-loop bandwidth be?

$$G(s) = \frac{10^6(s/\omega_z + 1)}{s^2}$$

9.9 Design a type II system that has a closed-loop bandwidth of 10^3 rad/s and a 30 percent overshoot to a step input.

9.10 For a type II system with $K_a = 10^6$ (rad/s)^2 determine a lead-lag network so that the closed-loop system has less than 30 percent overshoot to step changes in phase. The network pole location should be minimized in order to maximize the high-frequency filtering.

9.11 Determine the transfer function of the amplifier shown in Fig. P9.11. The transconductance $g_m = 10^{-4}$ S. If this amplifier is connected as a unity-gain voltage follower, what is the phase margin? How large a capacitor must be connected between the compensation terminals to obtain a 45° phase margin? The transfer function of the dependent source is

$$V_o = \frac{(-2 \times 10^3) V_a}{2 \times 10^{-5}s + 1}$$

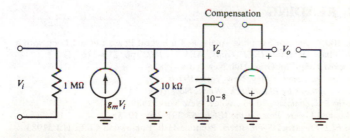

Figure P9.11 A two-stage amplifier.

9.12 An open-loop transfer function is given by

$$G(s) = e^{-sT} \frac{10^4(s/100 + 1)}{s^2}$$

For what value of time delay T will the system become unstable?

9.13 A type I system has the open-loop transfer function of

$$G(s) = \frac{K_v}{s(s/10^5 + 1)^2}$$

Select K_v so that the closed-loop system has approximately 25 percent overshoot to a step input. Estimate the system's closed-loop bandwidth.

9.14 A type II system has the open-loop transfer function of

$$G(s) = \frac{K_a(s/200 + 1)}{s^2(s/2000 + 1)}$$

Determine the value of K_a that will minimize the overshoot to a step input. What will the minimum overshoot be?

9.15 For the simple phase-locked loop consisting of a phase-frequency detector and a VCO, sketch the phase error as a function of time for the case where

$$\frac{K_o V}{\omega_c} = \frac{3}{4}$$

For what value of loop gain will the system become unstable?

9.16 Design an FM demodulator using the CD4046 integrated-circuit PLL (see App. 7 for specifications). The carrier frequency is to be 100 kHz and the frequency deviation is to be 4 kHz. Select the loop filter so that the closed-loop bandwidth is approximately 8 kHz.

REFERENCES

9.1 Bode, H.: *Network Analysis and Feedback Amplifier Design*, Van Nostrand, New York, 1945.
9.2 Kroupa, V. F.: *Frequency Synthesis*, Wiley, New York, 1973.
9.3 Viterbi, A.: *Principles of Coherent Communications*, McGraw-Hill, New York, 1966.

ADDITIONAL READING

Barab, S., and A. McBride: Uniform Sampling Analysis of a Hybrid Phase-Locked Loop with a Sample-and-Hold Phase Detector, *IEEE Trans. AES*-11, 1975, pp. 210–216.
Best, R. E.: *Phase-Locked Loops*, McGraw-Hill, New York, 1984.
Blanchard, A.: *Phase-Locked Loops*, Wiley, New York, 1976.
Egan, W. F.: *Frequency Synthesis by Phase Lock*, Wiley, New York.
Gardner, F. M.: *Phaselock Techniques*, 2d ed., Wiley, New York, 1979.
Gupta, S. C.: Phase-Locked Loops, *Proc. of the IEEE*, **63**:291–306 (1975).
Rohde, U. L.: *Digital PLL Frequency Synthesizers*, Prentice-Hall, Englewood Cliffs, N.J., 1983.

FREQUENCY SYNTHESIZERS

10.1 INTRODUCTION

A frequency synthesizer is a device that generates a large number of precise frequencies from a single reference frequency. The term "frequency synthesis" was first used by Finden[10.1] in 1943 for the generation of frequencies that were a harmonic of a submultiple of a reference frequency. Recent advances in integrated-circuit design include the development of inexpensive frequency synthesizers and their subsequent application in most communication receivers.

A frequency synthesizer can replace the expensive array of crystal resonators in a multichannel radio receiver. A single crystal oscillator provides a reference frequency, and the frequency synthesizer generates the other frequencies. Because they are relatively inexpensive and because they can be easily controlled by digital circuitry, frequency synthesizers are being included in many new communication system designs.

The oldest synthesis method, first described by Finden, is referred to as *direct frequency synthesis*; it utilizes mixers, frequency multipliers, dividers, and bandpass filters. Direct synthesis has been superseded in almost all applications by indirect (coherent) synthesis, which utilizes a phase-locked loop that may be analog or digital. The newest method, direct digital frequency synthesis (DDFS), uses a digital computer and digital-to-analog converter to generate the signals. Each of these methods has advantages, as well as disadvantages, and if the specifications are sufficiently stringent, it may be necessary to incorporate all three methods into the synthesizer design. In this chapter, the three methods of frequency synthesis are described, and a design example that combines the different synthesis methods to meet the overall specifications is presented. Because one of the most demanding synthesizer specifications is output noise, the oscillator being a primary source of the random noise associated with frequency synthesis, a brief description of the random noise occurring in quality oscillators and phase-locked loops is included in this chapter so that the reader can better assess the noise performance of synthesizers. First we will examine the various methods of frequency synthesis.

10.2 DIRECT FREQUENCY SYNTHESIS

Direct frequency synthesis is the oldest of the frequency synthesis methods. It synthesizes a specified frequency from one or more reference frequencies from a combination of harmonic generators, filters, multipliers, dividers, and frequency mixers. Bipolar transistors, because of the exponential base-to-emitter voltage characteristics, are well suited for use as harmonic generators.

One method of using a harmonic generator is shown in Fig. 10.1. The desired frequency is obtained with a filter tuned to the desired output frequency. Highly selective filters are required with this method. The multiple-oscillator approach is an alternative method. The oscillators are usually easier to realize than the bandpass filters. Figure 10.2 illustrates a method of generating 99 discrete frequencies from 18 crystal oscillators. One switch selects one of the nine oscillators that cover the frequency range 1 to 9 kHz in 1-kHz steps, and the other switch covers the frequency range 10 to 90 kHz in 10-kHz steps. The

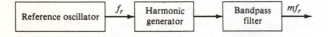

Figure 10.1 A direct frequency synthesizer.

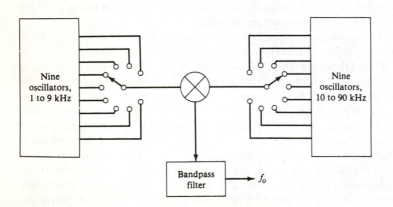

Figure 10.2 A two-decade direct frequency synthesizer.

two signals are then combined in a frequency mixer, and the bandpass filter selects the higher of the two mixer output frequencies.

Direct frequency synthesis refers to the generation of new frequencies from one or more reference frequencies using a combination of multipliers, dividers, bandpass filters, and mixers. A simple example of direct synthesis is shown in Fig. 10.1. The new frequency $\frac{2}{3}f_o$ is realized from f_o by using a divide-by-3 circuit, a mixer, and bandpass filter. In this example $\frac{2}{3}f_o$ has been synthesized by operating directly on f_o.

One of the foremost considerations in the design of direct-frequency synthesizers is the mixing ratio

$$r = \frac{f_1}{f_2} \tag{10.1}$$

where f_1 and f_2 are the two input frequencies to the mixer. If the mixing ratio is too large or too small, the two output frequencies will be too close together, and it will be difficult to remove one of the signals with filtering.

Example 10.1 If the two mixer-input frequencies are 100 and 1 MHz ($r = 100$), the mixer-output frequencies will be 99 and 101 MHz. The removal of one of these frequencies would require an extremely complex filter.

The filter requirements can be reduced by using an offset frequency. This approach is utilized in the next direct synthesis method to be described.

Figure 10.3 illustrates a type of direct synthesis module frequently used in direct-frequency synthesizers. The method is referred to as *double-mix-divide*. An input frequency f_i is combined with a frequency f_1, and the upper frequency $f_1 + f_i$ is selected by the bandpass filter. This frequency is then mixed with a switch-selectable frequency $f_2 + f^*$. (In the following f^* refers to any one of 10 switch-selectable frequencies). $f_2 + f^*$ can be realized with one of the methods illustrated in Figs. 10.1 and 10.2. The output of the second mixer consists of the two frequencies $f_i + f_1 + f_2 + f^*$ and $f_i + f_1 - f_2 - f^*$; only the higher frequency appears at the output of the bandpass filter. If the frequencies f_i, f_1, and f_2 are selected so that $10f_i = f_i + f_1 + f_2$, then the frequency at the output of the

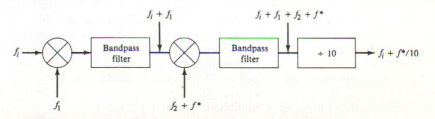

Figure 10.3 A double-mix-divide module.

divide-by-10 will be

$$f_o = f_i + \frac{f^*}{10} \tag{10.2}$$

The double-mix-divide module has increased the input frequency by the switch-selectable frequency increment $f^*/10$. Double-mix-divide modules can be cascaded to form a frequency synthesizer with any degree of resolution. The double-mix-divide modular approach has the additional advantage that the frequencies f_1, f_2, and f_i can be the same in each module so that all modules can contain identical components.

Considered solely from a theoretical viewpoint the double-mix-divide module appears unnecessarily complicated, since the output frequency $f_i + f^*/10$ could be realized using one mixer and bandpass filter. The advantages of the approach shown in Fig. 10.3 are practical; it allows better mixing ratios (with relaxed filtering criteria) and allows for the same bandpass filters in each stage. The effect of deleting f_2 will be illustrated after first discussing a three-digit synthesizer.

Example 10.2 A direct-frequency synthesizer with three digits of resolution can be realized using three double-mix-divide modules. Each decade switch selects one of 10 frequencies $f_2 + f^*$. In this example the output of the third module is taken before the decade divider. For example, it is possible to generate the frequencies between 10 and 19.99 MHz (in 10-kHz increments), using the three-module synthesizer, by selecting

$$f_i = 1 \text{ MHz}$$

$$f_1 = 3 \text{ MHz}$$

$$f_2 = 6 \text{ MHz}$$

Since

$$f_i + f_1 + f_2 = 10f_i \tag{10.3}$$

the output frequency before the last division by 10 will be

$$f_o = 10f_i + f_3^* + \frac{f_2^*}{10} + \frac{f_1^*}{100} \tag{10.4}$$

Since f^* occurs in 1-MHz increments, $f_1^*/100$ will provide the desired 10-kHz frequency increments. The output is taken before the last decade divider, as this provides a sine-wave output. The divider output has a square-wave waveform. If, for example, the frequency 14.86 MHz is required, f_1^* will be 6 MHz, f_2^* will be 8 MHz, and f_3^* will be 4 MHz.

Theoretically, either f_1 or f_2 could be eliminated, provided

$$f_i + f_1 \text{ (or } f_2\text{)} = 10f_i \tag{10.5}$$

but the additional frequency is used in practice to provide additional frequency separation at the mixer output. This frequency separation eases the bandpass filter requirements. For example, if f_2 is eliminated, $f_1 + f_i$ must equal $10f_i$, or 10 MHz in the preceding example. If f_1^* of 1 MHz is selected, the output of the first mixer will consist of the two frequencies, 9 and 11 MHz. The lower of these closely spaced frequencies must be removed by the filter. The filter required would be extremely complex in order to achieve such selectivity. If instead a 5-MHz signal f_2 is also used so that $f_i + f_1 + f_2 = 10$ MHz, the two frequencies at the first mixer input will be $f_i + f_1 = 5$ MHz and $f_2 + f_1^* = 6$ MHz. Therefore, the frequencies present at the mixer output (for an f_1^* of 1 MHz) will be 1 and 11 MHz. In this case the two frequencies will be much easier to separate with a bandpass filter. The ancillary frequencies f_1 and f_2 can only be selected in each design after considering all possible frequency ratios at the mixer output.

Direct synthesis can produce fast frequency switching, almost arbitrarily fine frequency resolution, low phase noise, and the highest frequency of operation of any of the methods. Direct frequency synthesis requires considerably more hardware (oscillators, mixers, and bandpass filters) than the two other synthesis techniques to be described. The hardware requirements result in direct synthesizers being larger and more expensive to construct. Another disadvantage of the direct synthesis technique is that unwanted (spurious) frequencies can appear at the output. The wider the frequency range the more likely that spurious components will appear in the output. These disadvantages must be weighed against the versatility, speed, and flexibility of direct synthesis.

10.3 FREQUENCY SYNTHESIS BY PHASE LOCK

The disadvantages associated with direct synthesis are greatly diminished with the frequency synthesis technique (often referred to as *indirect synthesis*) that employs a phase-locked loop (PLL). A simple PLL is illustrated in Fig. 10.4. A detailed analysis of PLL characteristics is given in Chaps. 8 and 9, but for the present discussion it is sufficient to state that when the PLL is functioning properly the two phase-detector input frequencies are equal. That is,

$$f_r = f_d \qquad (10.6)$$

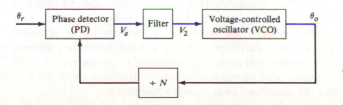

Figure 10.4 An indirect frequency synthesizer.

The frequency f_d is obtained by dividing the VCO output frequency f_o by N:

$$f_d = \frac{f_o}{N} \qquad (10.7)$$

Therefore, the output frequency f_o is an integral multiple of the reference frequency, or

$$f_o = Nf_r \qquad (10.8)$$

The PLL with a frequency divider in the loop thus provides a method for obtaining a large number of frequencies from a single reference frequency. If the divide ratio N is realized using a programmable divider, it is possible to easily change the output frequency in increments of f_r. The PLL with a programmable divider provides an easy method for synthesizing a large number of frequencies, all of which are an integral multiple of the reference frequency. There are, however, problems associated with this method. The major difficulties will first be discussed, and then some of the methods presently used to circumvent these problems will be described.

From Eq. (10.8) we note that the frequency resolution is equal to f_r. That is, the output frequency can be changed in increments as small as f_r; however, this is in conflict with the requirement of a short time interval for changing frequencies. Although an exact expression for the switching time has yet to be derived, a frequently used rule of thumb is that the switching time

$$t_s = \frac{25}{f_r} \qquad (10.9)$$

It takes approximately 25 reference periods to switch frequencies. The frequency resolution is therefore inversely proportional to the switching speed. A contemporary specification for satellite communication systems, which use frequency hopping, is that the frequency resolution be equal to 10 Hz and the switching time be less than 10 μs! Since the above rule of thumb predicts a switching time of 2.5 s, it is clear that the simple PLL frequency synthesizer cannot meet both specifications. The choice of reference frequency dominates loop performance.

Effects of Reference Frequency on Loop Performance

The expression for the output frequency [Eq. (10.8)] shows that in order to obtain fine frequency resolution the reference frequency must be small. This creates conflicting requirements. One problem is that to cover a broad frequency range requires a large variation in N. Even if the hardware problems can be overcome, some method will normally be needed to compensate for the variations in loop dynamics that occur for widely varying values of N. It is shown in Chap. 9 that the linearized loop transfer function is

$$\frac{f_o(s)}{f_i(s)} = \frac{\theta_o(s)}{\theta_i(s)} = \frac{K_v F(s)/s}{1 + K_v F(s)/Ns} \qquad (10.10)$$

where $F(s)$ is the transfer function of the low-pass filter. If N is to assume a large number of values, say from 1 to 1000, then there will be a 60-dB variation in the open-loop gain and a correspondingly wide variation in the loop dynamics, unless some method (such as the use of a programmable amplifier) is employed to alter the loop gain for different N values.

A second problem encountered with a low reference frequency is that the loop bandwidth must be less than or equal to the reference frequency, because the low-pass filter must filter out the reference frequency and its harmonics present at the phase detector output. Thus, the filter bandwidth must be less than the reference frequency. It is explained in Sec. 9.3 (PLL stability analysis) that loop bandwidth is normally less than the filter bandwidth for adequate stability. Therefore, a low reference frequency results in a frequency synthesizer that will be slow to change frequency.

Another problem introduced by a low reference frequency is its effect on noise introduced in the VCO. Figure 10.5 shows a linearized model of a PLL with the three main sources of noise. ϕ_{N_r} is the noise on the reference signal. ϕ_{N_d} is the noise created in the phase detector. The largest phase-detector noise components are at the reference frequency and the harmonics of this frequency. ϕ_{N_o} is the noise introduced by the VCO. Figure 10.6 illustrates a frequency spectrum typical of VCO noise. Most of the energy content of VCO noise is near the oscillator frequency; in the PLL model it can be interpreted as a low-frequency noise. The total noise of the closed-loop system at the VCO output ϕ_N is given by

$$\phi_N = \frac{(\phi_{N_r} + \phi_{N_d})K_v F(s)/s}{1 + K_v F(s)/Ns} + \frac{\phi_{N_o}}{1 + K_v F(s)Ns} \tag{10.11}$$

$$= G(s)(\phi_{N_r} + \phi_{N_d}) + G_r(s)\phi_{N_o}$$

Since $F(s)$ is either unity or a low-pass transfer function, $G(s)$ is a low-pass transfer function and $G_r(s)$ is a high-pass transfer function. The PLL functions as a low-pass filter for phase noise arising in the reference signal and phase detector, and it functions as a high-pass filter for phase noise originating in the VCO. Since the VCO noise is a low-frequency noise, the output noise due to

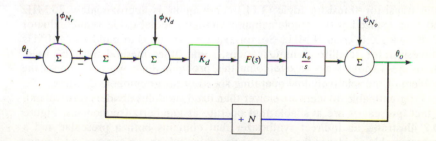

Figure 10.5 A PLL synthesizer including three noise sources.

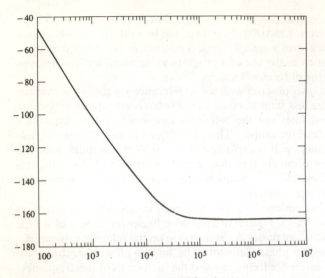

Figure 10.6 Frequency spectrum of VCO noise.

ϕ_{N_o} is minimized by having the loop bandwidth as wide as possible. At the same time, the loop bandwidth should be less than the reference frequency in order to minimize the effect of ϕ_{N_d}, which is dominated by spurious frequency components at the reference frequency and its harmonics.

Therefore, the desire to have a low reference frequency f_r in order to obtain fine frequency resolution is offset by the need to have f_r large in order to reduce the loop settling time and also to minimize the amount of noise contributed by the VCO.

Variable-Modulus Dividers

Another difficulty with the system illustrated in Fig. 10.4 is that the maximum operating speed of programmable dividers is slower than that required in many communication systems. The upper limit of a programmable divider realized from transistor–transistor logic (TTL) components is approximately 25 MHz and that realized with complementary-symmetry metal-oxide semiconductor (CMOS) logic is about 4 MHz. So, for example, if one is to build a 2×10^9 Hz synthesizer for satellite communications, some other method must be used. There are various ways to overcome this problem. First we will discuss the problem of the relatively low operating speed of programmable dividers.

Programmable dividers are slower than fixed-modulus dividers (prescalars). In fact, prescalars are available that operate at gigahertz frequencies. Figure 10.7 illustrates an indirect synthesizer that contains both a prescalar and a programmable divider in the loop. The prescalar, which can operate at frequen-

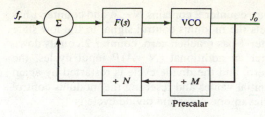

Figure 10.7 A PLL including a pre-scalar.

cies into the gigahertz region, first reduces the output frequency by the factor M before it is applied to the programmable divider. When the loop is in lock,

$$f_r = \frac{f_o}{MN} \qquad \text{or} \qquad f_o = N(Mf_r) \qquad (10.12)$$

Although the use of the prescalar allows the loop to operate with higher output frequencies, the output frequency can only be changed in increments of Mf_r. In order to obtain the same resolution, the reference frequency must be decreased by the prescalar factor M. A method available for obtaining good frequency resolution while operating at high output frequencies uses variable-modulus dividers, which, although not as fast as prescalars, are much faster than programmable dividers.

The two-modulus divider system illustrated in Fig. 10.8 contains a divider that divides by the modulus $P + Q$ when the modulus control is high and divides by P when the modulus control is low. In this particular scheme the output of the variable-modulus divider simultaneously drives the two programmable dividers 1 and 2. The programmable dividers operate at the input clock rate f_i divided by P or $P + Q$. The divide cycle begins with counter 1 preset to A, counter 2 preset to N, and the modulus control high so that the two-modulus divider output frequency is equal to the input frequency divided by $P + Q$.

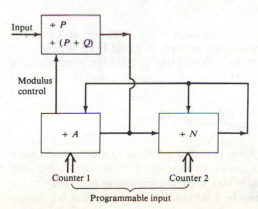

Figure 10.8 A programmable divider realized with a variable-modulus divider.

After $A(P+Q)$ cycles of the input, counter 2 contains $N-A$ and counter 1 has reached zero, at which time it sets the modulus control signal to the low state and ceases to count. After counter 1 has reached zero, counter 2 counts down each P cycle of the input. After an additional $(N-A)P$ input cycles, the contents of counter 2 are also zero, and the divide cycle is restarted by again loading the counters with their initial values and resetting the modulus control signal. The number of input cycles in one complete divide cycle is

$$D = (P+Q)A + P(N-A) = AQ + PN \qquad (10.13)$$

(Note that N must be greater than A for the method to work.) If $Q=1$, then the divide ratio, although it has a minimum value $D_{\min} = PN$, can be incremented in unit steps. A frequently used divide ratio is $P=10$ and $P+Q=11$. Then Eq. (10.13) becomes

$$D = 10N + A \qquad (10.14)$$

which shows that the 10/11 divider can be used to obtain division ratios with increments of 1, provided $N > A$. Since $A_{\max} = 9$, N must be at least 10 and $D_{\min}$ is 100. The minimum divide ratio is not usually a problem in frequency-synthesizer design.

Example 10.3 If it is desired to design a frequency synthesizer to cover the frequency range from 100 to 109 MHz in 1-MHz increments, a reference frequency of 1 MHz is suitable (a higher reference frequency would not be). Since 100 MHz is too fast for a programmable divider, a 10/11 variable-modulus divider will be considered. A will vary from 0 to 9, so N must be at least 10. The minimum value of D will be, using the 10/11 divider,

$$D_{\min} = \frac{100 \times 10^6}{10^6} = 100$$

which will provide $(f_o)_{\min} = 100$ MHz. Thus the desired division ratio can be obtained using a 10/11 variable-modulus divider together with the programmable dividers.

Other variable-modulus division ratios such as 5/6, 6/7, 8/9, and 100/101 are also frequently used. In the preceding example if it had been necessary to cover the frequency interval 100 to 100.99 MHz in 10-kHz increments, a maximum reference frequency of 10 kHz would be needed, and the minimum divide ratio required would be

$$D_{\min} = \frac{100 \times 10^6}{10^4} = 10^4 = \frac{(f_o)_{\min}}{f_r}$$

since $A_{\max} = 99$ and $N_{\min}$ is 100. Also, $D = AQ + PN$, so Q must equal 1 for a frequency resolution of 10 kHz. It is possible to select $P=10$ and $N=10^3$, but it is better to select $P=100$ and $N=100$, since the maximum frequency to the two programmable dividers will then be 1.0099 MHz. This will allow for the use of low-noise CMOS logic for the programmable dividers. Therefore, a 100/101 variable-modulus divider is suited for this design. If $P = 10$ is selected, the maximum frequency to the programmable dividers would be 10.0099 MHz.

Down Conversion

Another approach to circumventing the high-frequency limitation of the programmable dividers is to shift the output frequency down by mixing the output frequency with a local oscillator frequency. Figure 10.9 illustrates a single down-conversion synthesizer. The low-pass filter following the mixer is used to filter out the higher mixer-output frequency ($f_o + f_L$). The divider output frequency is

$$f_d = f_r = \frac{f_o - f_L}{N}$$

so

$$f_o = f_L + Nf_r$$

The main disadvantages of this method are that the complexity and size are increased, the possibility of spurious components being introduced by the mixer is increased, and the phase lag of the filter used in the feedback path can degrade the loop performance.

Methods for Reducing Switching Time

There are methods available for circumventing the conflict between the need for fine frequency resolution and the need to quickly change frequencies. A method of reducing response time is to include a coarse steering signal. When the frequency is changed by altering the divide ratio N, a steering signal can be generated and applied immediately to direct the VCO to the new frequency (see Fig. 10.10). The steering signal can be obtained from a lookup table stored in memory with the D/A convertor used to generate the analog steering signal. Another frequently used method is to incorporate multiple phase-locked loops in the synthesizer.

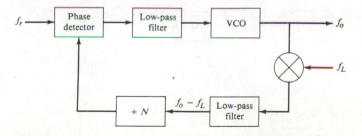

Figure 10.9 A PLL frequency synthesizer with down conversion.

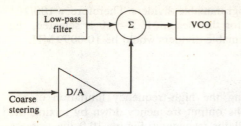

Figure 10.10 Coarse steering can be used to reduce PLL switching time.

Multiple-loop frequency synthesizers. One possible method of obtaining fine frequency resolution with a high reference frequency is illustrated in Fig. 10.11. The output frequency is obtained by dividing the VCO output frequency by M. That is,

$$f_o = \frac{Nf_r}{M}$$

and the frequency resolution is

$$f = \frac{f_r}{M}$$

Hence fine resolution is obtained by making sure that M is sufficiently large. A problem inherent to this technique is that the loop frequency may become too large. The difficulty is illustrated by the following numerical example.

Example 10.4 Consider the design of a frequency synthesizer to cover the frequency range from 10 to 10.1 MHz with 1-kHz resolution. The reference frequency is to be 100 kHz. To obtain the 1-kHz frequency resolution from the 100-kHz reference frequency requires that the VCO output frequency be divided by 100. An output frequency of 10 MHz will require that the VCO be operating at 1 GHz!

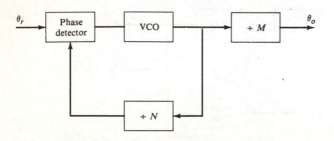

Figure 10.11 A PLL frequency synthesizer with a post-divider for increased frequency resolution.

While the idea of adding a postdivider is not a good solution in general, the concept does find practical application in multiple-loop synthesizers. A multiple-loop synthesizer uses one or more loops to obtain the fine frequency resolution and combines the outputs of these loops with that of another loop, which generates the high-frequency components of the desired output frequency. The principles involved can be easily understood by examining one such synthesizer.

Example 10.5 Consider the design of a frequency synthesizer to cover the frequency range from 35.40 to 40.00 MHz in 1-kHz increments. If this is to be accomplished in a single-loop synthesizer, a reference frequency of 1 kHz would be required (with a response time of approximately 25 ms), together with the divide ratio N;

$$35.40 \times 10^6 \le N \le 40.00 \times 10^6$$

An alternate design is shown in Fig. 10.12.[10.2] The synthesizer consists of three PLLs. PLLs A and B both use the 100-kHz reference frequency. Loop C locks the divided output of loop A (f_A) to the difference between the output frequency f_o and the output of loop B (f_B). That is,

$$f_A = f_o - f_B \tag{10.15}$$

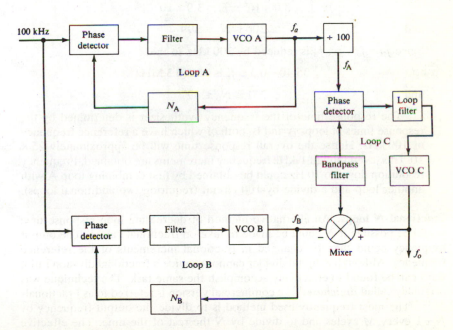

Figure 10.12 The three-loop frequency synthesizer discussed in Example 10.5.

or $$f_o = f_B + f_A \qquad\qquad (10.16)$$

Phase-locked loop C serves as a mixer and filter for f_A and f_B. If f_A and f_B are directly combined in a mixer, the sum and difference frequencies will be too close together to be adequately separated with a bandpass filter. The present technique of using a phase-locked loop for frequency mixing does accomplish good separation.

Since the reference frequency of loop A is 100 kHz, its output frequency f_a can be varied in 100-kHz increments, and

$$f_A = \frac{f_a}{100} = N_a \times 10^3$$

varies in 1-kHz increments. Loop A is used to generate the 1- and 10-kHz increments of output frequency and loop B the 0.1- and 1-MHz changes in output frequency. f_A could be selected to vary between 0 and 99 kHz in 1-kHz increments, but f_A also serves as the reference frequency for loop C. If, for example, $f_A = 1$ kHz this would require that loop C be a relatively slow loop and would determine the overall response time of the synthesizer. To reduce the response time of loop C, f_A is increased by 300 kHz so that

$$300 \text{ kHz} \leq f_A \leq 399 \text{ kHz}$$

Therefore,

$$3.0 \times 10^6 \leq f_a \leq 3.9 \times 10^6$$

and

$$300 \leq N_a \leq 399$$

Since $f_B = f_o - f_A$, f_B is reduced by 300 kHz so that

$$35.40 - 0.3 \leq f_B \leq 40 - 0.3 \text{ MHz}$$

and

$$351 \leq N_B \leq 397$$

The response time of the frequency synthesizer is determined by the response times of loops A and B, both of which have a reference frequency of 100 kHz. Hence the overall response time will be approximately 25×10^{-2} ms, even though 1-kHz frequency increments are obtained. Frequency resolution down to 10 Hz could be obtained by first combining loop A with another loop and a divide-by-100 circuit (requiring two additional loops).

Fractional-N loops. An alternative method of decreasing loop response time would be possible if N could be made to take on fractional values. The output frequency could then be changed in fractional increments of the reference frequency. Although a digital divider cannot provide a fractional division ratio, ways can be found to effectively accomplish the same task. The technique was originally called *digiphase*;[10.3] a commercial version is referred to as Fractional-N.[10.4] The most frequently used method is to divide the output frequency by $N+1$ every M cycles and to divide by N the rest of the time. The effective division ratio is then $N + M^{-1}$, and the average output frequency is given by

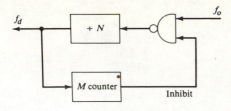

Figure 10.13 A simplified method of implementing fractional division.

$$f_o = (N + M^{-1})f_r \qquad (10.17)$$

This expression shows that f_o can be varied in fractional increments of the reference frequency by varying M. A simplified method for generating the fractional division is shown in Fig. 10.13. The divider divides the input frequency by N, and the counter counts the number of cycles of waveform output. Each time the counter reaches a count of $M-1$ the counter output goes low for one input cycle, and one input cycle does not reach the divider. Therefore, the divider requires $(N+1)$ input cycles to change state.

The number of output cycles during one complete cycle of the M counter is

$$f_o = f_d N(M-1) + f_d(N+1) = f_d(NM+1)$$

and the average output frequency per cycle of the M counter is

$$(f_o)_{av} = f_d(N + M^{-1})$$

The average frequency at the divider output is the output frequency divided by $N + M^{-1}$, so a form of fractional division has been realized. The method of implementing fractional N division shown in Fig. 10.13 will work as long as M is an integer, but normally it will not be. A more general method of implementing fractional division can be obtained using a phase accumulator. The phase-accumulator approach is illustrated by the following example.

Example 10.6 Consider the problem of generating 455 kHz using a fractional N loop with a 100-kHz reference frequency. The integral part of the division is $N = 4$ and the fractional part is $M^{-1} = 0.55$, or $M = 1.8$. M is not an integer; the VCO output is to be divided by 5 $(N+1)$ every 1.8 cycles, or 55 times every 100 cycles. Although M is not an integer, the fractional division can be easily implemented by adding the number 0.55 (M^{-1}) to the contents of an accumulator every output cycle. Each time the accumulator overflows (the contents exceed 1) the divider divides by 5 rather than 4. Only the fractional value of the addition is retained in the phase accumulator.

The phase-accumulator realization of fractional division is illustrated in Fig. 10.14. Fine frequency resolution can be arbitrarily obtained by increasing the length of the phase accumulator. The previous example used a 100-kHz reference frequency. A resolution of $10^5/10^5 = 1$ Hz could be obtained by using

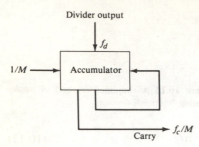

Divider output

Figure 10.14 A phase accumulator used for fractional division.

a 5-binary-coded decimal (BCD) accumulator. The performance of a fractional N synthesizer will be further illustrated with another numerical example.

Example 10.7 Consider the problem of incrementing the output frequency of a 1-MHz synthesizer by 1000 Hz, the reference frequency being 10 kHz. Since

$$f_o = (N + M^{-1})f_r$$

$N = 100$; if f_o is to be increased to 1.001×10^6 Hz, then $M = 10$. That is, at every 10 reference cycles (10^3 output cycles), the output frequency is divided by 101. The average output frequency is then 100.1×10^4 Hz, which is the desired frequency. While the reference signal goes through one period, the VCO signal goes through 100.1 cycles and the output of the divider ($\div 100$) goes through 1.001 cycles; its phase relative to the reference frequency advances by $0.001 \times 2\pi$ rad each reference cycle. After the 10 reference cycles the divider reference output leads the reference signal by $0.01 \times 2\pi$ rad. At this time one VCO cycle is skipped; the skipping of one VCO cycle delays the divider output by $0.01 \times 2\pi$ rad, which is exactly how much the divider output had increased in phase.

Although the average output frequency was 1.001 MHz in the preceding example, the instantaneous output frequency changes with each reference cycle because of the increasing phase difference between the divider output and the reference signal. The timing diagram of Fig. 10.15 illustrates this point. If the

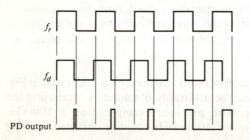

Figure 10.15 Timing signals in a PLL frequency synthesizer using fractional division.

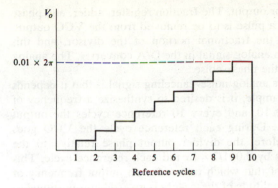

Figure 10.16 Typical waveform of the average value of the phase detector output of a fractional-N synthesizer.

divider output frequency is slightly faster than the reference frequency, the phase detector output will consist of pulses of increasing width, and the dc value of these pulses will appear as shown in Fig. 10.16. This voltage will create fluctuations in the output frequency if it is not eliminated before it reaches the VCO.

Figure 10.17 contains a simplified diagram of a fraction-N synthesizer that eliminates the deterministic noise occurring at the phase detector output by adding a signal equal in magnitude and opposite in sign to the deterministic

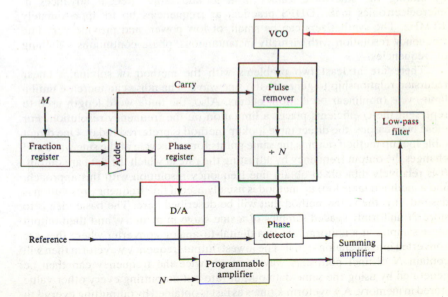

Figure 10.17 A complete fractional-N frequency synthesizer.

voltage present at the detector output. The fraction register, adder, and phase register determine how often a pulse is to be removed from the VCO output. The phase register contains the fractional portion of the divisor, and this information is converted to an analog signal in the D/A converter. The analog signal is then used to reduce the phase noise.

One further feature of this analog noise-canceling signal is that it depends on both M and N. If, for example, it is desired to synthesize a frequency of 2.001 MHz, then M is again 10, and every 10 reference cycles the output frequency is divided by 201. During each reference cycle the VCO goes through 200.1 cycles. Therefore, the divider output phase relative to the reference frequency advances by $0.001/2 \times 2\pi$ rad each reference cycle. This phase increment is one-half of that which occurs for an output frequency of 1.001 MHz. In general, the amplitude of these steps is inversely proportional to frequency. Therefore, the D/A output amplitude must be adjusted by a programmable gain amplifier, with the gain inversely proportional to N. The analog signal is subtracted from the phase detector output in order to provide a low-noise VCO control signal.

10.4 DIRECT DIGITAL SYNTHESIS[10.5]

Direct digital frequency synthesis (DDFS) is achieved either by solving a digital recursion relationship using a general-purpose computer or microcomputer or by storing the sine-wave values in a lookup table. Recent advances in microelectronics make DDFS practical at frequencies up to approximately 10 MHz. The synthesizers can be small, of low power, and provide very fine frequency resolution with virtually instantaneous, phase-continuous switching of frequencies.

There are at least two problems with the method of solving a linear recursion relationship to generate the sine wave. The noise can increase until a limit cycle (nonlinear oscillation) occurs. Also, the finite word length used to represent the coefficients places a limitation on the frequency resolution. For these two reasons, the direct table lookup method is preferred today. One direct table lookup method outputs the same points for each cycle of the sine wave and changes the output frequency by adjusting the rate at which the data are output. It is relatively difficult to obtain fine frequency resolution with this approach, and a modified table lookup method is usually used if fine frequency resolution is desired. It is the latter method that will be described here. The basic idea is to store N uniformly spaced samples of a sine wave in memory, and then output these samples at a uniform rate to a digital-to-analog converter where they are converted into an analog signal. The lowest output frequency waveform then will contain N distinct points. A waveform of twice the frequency can then be generated by using the same data output rate but outputting every other value stored in memory. A waveform k times as fast is obtained by outputting every kth point at the same rate. The frequency resolution is the same as the lowest frequency f_L. There is an upper frequency limit that is determined by the number of points stored in memory. Theoretically it is only necessary to output two samples of the sine wave and recover the fundamental frequency with analog filtering on the output of the D/A converter. Normally four or more points are used in the highest frequency signal, as this somewhat eases the requirements of the analog filter at the output. The architecture of a complete DDFS is shown in Fig. 10.18. The system consists of a phase accumulator, which is simply a digital accumulator, a read-only memory, a reference oscillator, a D/A converter, and a low-pass filter. To generate the lowest frequency the value 1 is added to the phase accumulator each reference cycle, and the next value from the lookup table is outputted. To output the frequency which is k times as fast as the lowest frequency, the value k is added to the phase accumulator each time and the corresponding value from the lookup table is outputted.

If P points are used in the waveform at the highest output frequency f_u, then $N = f_u/f_1 \times P$ points are used in the lowest frequency waveform. The number N is limited by the amount of available memory, and P, which must be greater than 2, is determined by the output low-pass filtering requirements. For the period of the highest output frequency,

$$T_u = f_u^{-1} = PT \quad \text{or} \quad f_u = (PT)^{-1} \quad (10.18)$$

Therefore, the highest possible obtainable output frequency is determined by the fastest sampling rate possible. The single most important factor limiting the high-frequency performance of direct frequency synthesizers is the speed of the D/A converter. It not only limits the maximum output frequency, but it introduces noise and harmonic distortion. For frequency synthesizers realized with a microprocessor the upper frequency limit will be determined by the number of computer clock cycles required to do the phase accumulation and memory lookup transfer. For the new high-speed digital signal-processing integrated circuits this time can be reduced to less than 20 ns. There is no lower limit on the lowest output frequency with this method. It will subsequently be shown that the lower frequency limit can be extended simply by extending the size of the phase accumulator.

To complete the DDFS, memory size and the length (number of bits) of each word must be determined. Word length is determined by the system noise

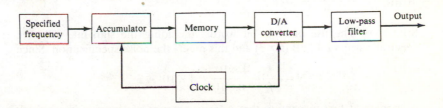

Figure 10.18 A direct digital frequency synthesizer (DDFS).

requirements. The D/A output samples are those of an exact sinusoid corrupted with deterministic noise due to the truncation caused by the finite length of the digital words. It can be shown that if an $(n + 1)$-bit word length is used (including one bit as the sign bit) the worst-case noise power (relative to the signal) due to the truncation will be approximately

$$\sigma^2 = (2^n)^{-1} \quad \text{or} \quad \sigma^2 = -6n \text{ dB} \qquad (10.19)$$

For each bit added to the word length the spectral purity improves by 6 dB.

Example 10.8 What word length will be required in a DDFS if the output spectral purity is to be at least 80 dB?

SOLUTION Since the noise power is $-6n$ dB, n must be at least 14. One additional bit is needed for the sign; therefore, the minimum word length needed is 15 bits for 80-dB signal-to-noise ratio.

For four output samples at the highest frequency, the memory size is determined from Eq. (10.18)

$$N = 4 \frac{f_u}{f_L}$$

where N is the number of points in the lowest frequency sinusoid. Clearly N words of memory would be sufficient for storing the data. However, the amount of memory required can usually be markedly reduced. First of all, it is only necessary to store the values for the first quadrant (0 to 90°) of the sine wave, since the values for the other three quadrants can be computed directly from these values; so a maximum of $N/4 = f_u/f_L$ memory points are required. The amount of memory can also be reduced by including one or more multipliers, but since multiplication is relatively slow, particularly with microprocessors, and memory is small and inexpensive, multiplication is rarely used to reduce the memory requirements. The amount of memory may still be reduced from that specified by Eq. (10.19) when the spectral purity requirements are not too severe. This point is illustrated in the following example.

Example 10.9 Design a DDFS to cover the frequency range 0 to 10 kHz with a frequency resolution of 0.001 Hz. The spectral purity is to be at least 40 dB.

SOLUTION The use of eight-bit words, including the sign bit, will give spectral purity of 42 dB (6×7) and this meets the noise specification. Since

$$N = 4 \frac{f_u}{f_L} = \frac{4 \times 10^4}{0.001} = 4 \times 10^7 < 2^{26}$$

it appears at first inspection that a large amount of memory is required. However, only $2^8 = 256$ different words can be realized using eight-bit

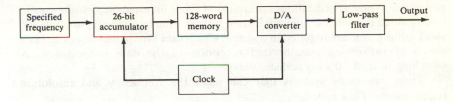

Figure 10.19 The direct digital frequency synthesizer discussed in Example 10.9.

words, so 256 memory locations should suffice. The explanation of this apparent contradiction is that although 4×10^7 different points are specified, the phase increments $\Delta\theta = \omega T$ are so small that approximately $2^{26} \div 2^8 = 2^{18}$ increments are needed before a change is registered in the eight-bit word. (A 26-bit word would be required to represent all 2^{26} words.) The complete design is illustrated in Fig. 10.19. A 26-bit accumulator is required, but only seven bits are used to address the 128-word memory. The sign bit directly controls the polarity of the D/A output. The least significant bits of the accumulator are not used in the addressing. Greater frequency resolution could be obtained simply by increasing the length of the accumulator.

At the upper frequency limit f_u the output waveform will consist of only four samples, and so it will not look much like a sine wave unless the harmonics of the fundamental frequency are removed by a low-pass filter. This filter should have a bandwidth slightly greater than f_u and a steep attenuation rate outside the passband. Although the harmonic filtering will not be as great for lower-frequency waveforms, these waveforms contain more sample points, and hence the harmonic content will be less.

The main drawback of DDFS is that it is limited to relatively low frequencies. The upper frequency is directly related to the maximum possible clock frequency. An upper frequency limit of approximately 10 kHz can be realized with microprocessors and one of approximately 10 MHz with special-purpose logic. DDFS is also noisier than the other methods, but adequate spectral purity can be obtained if sufficient low-pass filtering is used at the output. DDFS systems are easily constructed using readily available micro-processors. The main advantages of the method are its flexibility, easy realization of very low frequencies, and virtually instantaneous switching time. Combining DDFS for fine frequency resolution with other synthesis methods to obtain high-frequency performance is discussed in the next section.

10.5 SYNTHESIZER DESIGN EXAMPLE

Consider the design of a frequency synthesizer to cover the range 198 to 200 MHz in 10-Hz increments. The frequency switching time should be as short as possible. These specifications are typical of those imposed on a synthesizer to be contained in a satellite communication system. Fine frequency resolution and short settling time are required in a system which uses "frequency hopping" as a means of preventing unauthorized reception of the data transmission. A switching time of 10 μs is realistic today.

There are many systems that can meet the frequency and resolution requirements. They include

1. A single-loop indirect synthesizer
2. A multiple-loop indirect synthesizer
3. A fractional-N synthesizer combined with one or more local oscillators for up conversion
4. A direct frequency synthesizer
5. A PLL-DDFS combination
6. A combination of DDFS, direct, and indirect synthesizers

The following discussion considers some of these possibilities.

The single-loop indirect (PPL) synthesizer is not a good choice for several reasons. First of all, N must vary from 19.8×10^7 to 20×10^7, and programmable dividers that operate at 198 MHz are not yet available. Also, the 10-Hz frequency resolution specification requires that the reference frequency be 10 Hz. For a 10-Hz reference frequency the loop settling time would be on the order of 2.5 s, which is much too slow.

A two-loop synthesizer that can cover the specified frequency spectrum is shown in Fig. 10.20. The output frequency is the sum of the local oscillator frequency f_L and the frequency of VCO 1 (f_1). That is,

$$f_o = f_L + f_1 \tag{10.20}$$

The output f_1 is found from

$$f_r = \frac{f_1 - f_2/1000}{M} \tag{10.21}$$

and

$$f_r = \frac{f_2 - f_L/2}{N} \tag{10.22}$$

Therefore,

$$f_o = f_L + M_r + \frac{f_2}{1000}$$

$$= f_L + Mf_r + N\frac{f_r}{1000} + \frac{f_L}{2000} \tag{10.23}$$

for a reference frequency, $f_r = 10$ kHz, and

$$f_o = 1.0005f_L + 10^4 M + 10N$$

N could be a three-digit decimal number (1 to 999) to select the three least

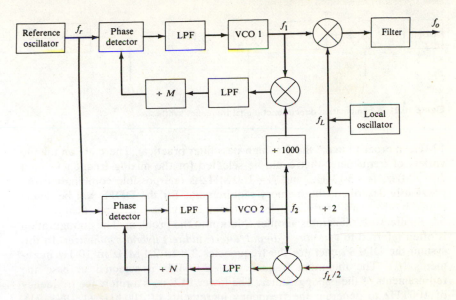

Figure 10.20 A hybrid frequency synthesizer.

significant digits of the frequency, and M could vary between 0 and 200 to select the three most significant digits. But then the output bandpass filter requirements would be too stringent. Therefore it is better to place a minimum value on M and reduce f_L such that Eq. (10.23) is satisfied. For example, M could vary from 800 to 1000; then

$$1.0005 f_L = 198 \times 10^6 - 800 \times 10^4$$

or

$$f_L = 189.90504 \times 10^6 \text{ Hz}$$

For this synthesizer the reference frequency for each loop is 10 kHz, so the settling time would be approximately $t_s = 25/10^4 = 2.5$ ms, which is a marked improvement over the single-loop system. If a shorter settling time is required, other alternatives need be considered. Another possibility would be a three-loop synthesizer. The design of such a system is left as an exercise.

A direct frequency synthesizer could be designed to meet the specifications, but the hardware would be complex. A direct synthesizer using double-mix-and-divide modules would require seven such modules. A DDFS cannot be used because of the high output frequency required, but a direct digital frequency synthesizer could be used to obtain the fine frequency resolution with very short settling time. The DDFS could then be combined with a direct or indirect synthesizer to obtain the high output frequency.

Figure 10.21 illustrates one possible solution to the design problem. The DDFS realizes 1 to 3 MHz in 10-Hz increments. The lower frequency is offset to

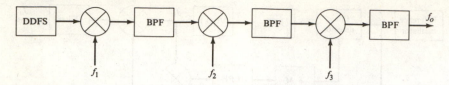

Figure 10.21 A combination direct/direct digital frequency synthesizer.

1 MHz in order to make the first high-pass filter practical. There are an infinite variety of frequencies that could be selected for the mixing frequencies, but $f_1 = 7$ MHz, $f_2 = 30$ MHz, and $f_3 = 160$ MHz is one possible combination. As previously described, the memory requirements for the DDFS will be determined by the word length, which is determined by the noise specification.

Figure 10.22 illustrates another solution to the problem. The configuration is often referred to as a *direct digital/direct/indirect hybrid synthesizer*. In this system the DDFS generates the frequencies 2.0 to 2.1 MHz in 10-Hz increments (f_d). The minimum frequency of 2.0 MHz is selected to ease the requirements of the bandpass filter. The lower PLL uses a reference frequency of 100 kHz to generate the frequency increments of 100 kHz ($1 \leq N \leq 21$). Since the DDFS responds almost instantaneously, the overall settling time is determined primarily by the loop with the 100-kHz reference frequency; the settling time is estimated to be 250 μs.

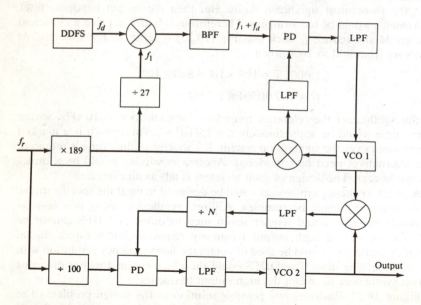

Figure 10.22 A hybrid frequency synthesizer.

10.6 PHASE NOISE

The preceding discussion of frequency synthesizers emphasized that the output noise is an important design consideration. The main sources of output noise in PLL synthesizers are spurious components at the reference frequency, and its harmonics (due to the phase detector), and the noise originating in the VCO. This noise creates a theoretical noise floor, which is a minimum against which actual systems can be compared.

A Model for Oscillator Phase Noise

If the power spectral density (power as a function of frequency) is measured at the output of an oscillator, a curve such as that of Fig. 10.6 is observed. Rather than all of the power being concentrated at the oscillator frequency, some of the power is distributed in frequency bands on both sides of the oscillator frequency. These unwanted frequency components are referred to as *oscillator noise*.

Oscillator noise will have a different impact on system performance, depending upon the application. The noise of a synthesizer used in a transmitter is transmitted on frequencies above and below the desired frequency of transmission. A similar process occurs in a receiver. The local oscillator phase noise can mix with an unwanted signal to create an unwanted signal in the IF passband. This process is referred to as *reciprocal mixing*. The phase noise is one of the limiting factors on determining how closely spaced (in the frequency domain) two communication channels can be.

With a spectrum analyzer, it will not be possible to measure the noise characteristics of a signal unless the spectrum-analyzer oscillators have substantially less noise than that of the signal to be measured. Leeson[10.6] developed a model that describes the origins of phase noise in oscillators, and since it closely fits experimental data, the model is widely used and will be described later in this chapter; first a relation between the observed power spectral density function and $\theta(t)$ will be developed.

The oscillator output $S(t)$ can be expressed by

$$S(t) = V(t) \cos[\omega_o t + \theta(t)] \tag{10.24}$$

where $V(t)$ describes the amplitude variation as a function of time and $\theta(t)$ the phase variation. $\theta(t)$ is referred to as *phase noise*. A well-designed, high-quality oscillator is very amplitude-stable, and $V(t)$ can be considered constant. For a constant amplitude signal all oscillator noise is due to $\theta(t)$.

A carrier signal of amplitude V and frequency f_o, which is frequency-modulated by a sine wave of frequency f_m, can be represented by

$$S(t) = V \cos\left(\omega_o t + \frac{\Delta f}{f_m} \sin \omega_m t\right) \tag{10.25}$$

where Δf is the peak frequency deviation and $\theta_p = \Delta f/f_m$ is the peak phase deviation—often referred to as the *modulation index β*. Equation (10.25) can be expanded as

$$S(t) = V[\cos(\omega_o t) \cos(\theta_p \sin \omega_m t) - \sin \omega_o t \sin(\theta_p \sin \omega_m t)] \quad (10.26)$$

If the peak phase deviation is much less than 1 ($\theta_p \ll 1$), then

$$\cos(\theta_p \sin \omega_m t) \approx 1$$

and

$$\sin(\theta_p \sin \omega_m t) \approx \theta_p \sin \omega_m t$$

That is, for $\theta_p \ll 1$, the signal $S(t)$ is approximately equal to

$$S(t) = V[\cos(\omega_o t) - \sin \omega_o t(\theta_p \sin \omega_m t)] \quad (10.27)$$

$$= V\left\{\cos(\omega_o t) - \frac{\theta_p}{2}[\cos(\omega_o + \omega_m)t - \cos(\omega_o - \omega_m)t]\right\}$$

Equation (10.27) shows that if the peak phase deviation is small, the phase deviation results in frequency components on each side of the carrier of amplitude $V\theta_p/2$. This development has shown that a constant amplitude signal of frequency f_o phase-modulated with a signal of constant frequency f_m and peak phase deviation θ_p results in frequency sidebands at the frequencies $f_o \pm f_m$. The ratio of the peak sideband voltage V_n to the peak carrier voltage V is

$$\frac{V_n}{V} = \frac{\theta_p}{2}$$

and the power ratio is

$$\left(\frac{V_n}{V}\right)^2 = \frac{\theta_p^2}{4} = \frac{\theta_{\text{rms}}^2}{2} \quad (10.28)$$

It is customary to extend this result to the interpretation of the power spectral density of a constant-amplitude signal. Consider the normalized power spectral

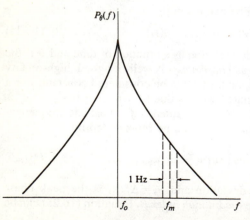

Figure 10.23 Oscillator-noise-power spectral density.

density plot shown in Fig. 10.23. If the normalized power spectral density $P_\theta(f)$ is approximately constant over a unit bandwidth, then the power in that bandwidth S'_θ is

$$S'_\theta(f_m) = \int_{f_m-1/2}^{f_m+1/2} P_\theta(f)\, df = P_\theta(f_m) \tag{10.29}$$

Since $P_\theta(f)$ is symmetrical about the carrier frequency f_o, the power in both sidebands is

$$S_\theta(f_m) \simeq 2P_\theta(f_m) \tag{10.30}$$

This noise power is interpreted as due to a phase-modulating noise at the frequency f_m:

$$S_\theta(f_m) = \theta_{\text{rms}}^2(f_m) = \frac{\theta_p^2(f_m)}{2} \tag{10.31}$$

where θ_p is the peak value of the phase modulation. $S_\theta(f)$ then is the ratio of power in the unit bandwidth centered at f_m to the carrier power. With this interpretation of the noise-power spectral density, the noise can now be described in terms of its origins.

We will assume that the oscillator is composed of an amplifier with gain A and a high-Q resonant circuit, as illustrated in block diagram form in Fig. 10.24. Since the gain of the resonant circuit has been normalized to unity at the resonant frequency f_o, the amplifier gain A must also be unity in order for the circuit to oscillate. Let S_θ represent the amplifier-noise-power spectral density referenced to the amplifier input. The available white noise N power per unit bandwidth at the amplifier input is given by

$$N = N_i + N_a = FkT \tag{10.32}$$

where F is the amplifier noise figure. Therefore, the ratio of white noise power per unit bandwidth to signal power P_s is FkT/P_s, which is a component of S_θ. In addition, amplifiers generate an additional "flicker," or $1/f$ phase noise, about the carrier frequency due to carrier density fluctuations in the transistor.

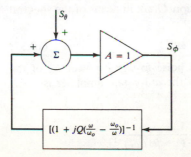

Figure 10.24 A model used to characterize oscillator noise.

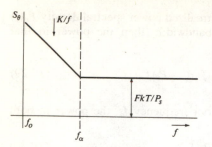

Figure 10.25 Amplifier-noise-power spectral density.

A plot of a typical S_θ spectrum is contained in Fig. 10.25. For frequencies below f_α, S_θ has a $1/f$ spectrum. At higher frequencies the spectrum is flat and equal to FkT/P_s. The frequency f_α below which the spectrum has a $1/f$ shape depends upon the characteristic of the individual amplifier. For the circuit of Fig. 10.25 with positive feedback and an A of 1, the closed-loop steady-state transfer function between the amplifier input and output is given by

$$\frac{C(j\omega)}{S(j\omega)} = [1 - H(j\omega)]^{-1} \qquad (10.33)$$

where

$$H(j\omega) = \left[1 + jQ\left(\frac{\omega}{\omega_o} - \frac{\omega_o}{\omega}\right)\right]^{-1} \qquad (10.34)$$

Since $H(j\omega)$ is a high-Q filter and we are interested in describing the noise-power distribution about the center frequency ω_o, $H(j\omega)$ can be replaced by its low-pass equivalent[10.7]

$$H_L(j\omega) = \left(1 + \frac{j\omega}{\omega_L}\right)^{-1} \qquad (10.35)$$

where

$$\omega_L = \frac{\omega_o}{2Q} \qquad (10.36)$$

is the equivalent bandwidth. The noise-power spectral density $S_o(\omega)$ at the output of a filter with a voltage transfer function $G(\omega)$, in terms of the spectral density $S_i(\omega)$ of the input noise, is given by

$$S_o(\omega) = S_i(\omega)|G(\omega)|^2 \qquad (10.37)$$

Therefore, the ratio of the equivalent phase noise-to-signal power S_ϕ of the closed-loop system measured at the output of the unity gain amplifier is

$$S_\phi = S_\theta[|1 - H(\omega)|^2]^{-1}$$

$$= \frac{S_\theta}{[1 - (1 + j\omega/\omega_L)^{-1}][1 - (1 - j\omega/\omega_L)^{-1}]}$$

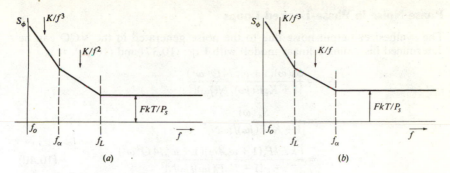

Figure 10.26 (a) Output noise of an oscillator with a low-Q resonator; (b) output noise of an oscillator with a high-Q resonator.

$$= \frac{S_\theta(1 + \omega^2/\omega_L^2)}{\omega^2/\omega_L^2}$$

$$= S_\theta \left(1 + \frac{\omega_L^2}{\omega^2}\right) \tag{10.38}$$

which can be written [using Eqs. (10.28), (10.32), and (10.36)] as

$$S_\phi(\omega) = S_\theta(\omega)\left(1 + \frac{\omega_o^2}{4Q^2\omega^2}\right) \tag{10.39}$$

This is the expression proposed by Leeson[10.6] for describing the noise at the output of an oscillator.

For S_θ as depicted in Fig. 10.25, the output phase noise spectrum S_ϕ is as shown in Fig. 10.26a, provided the filter bandwidth is greater than the f_α of the amplifier. At frequencies close to the carrier frequency f_o, the noise power decreases with a $1/f^3$ (-18 dB per octave) slope; between f_α and f_L the power spectral density (PSD) decreases as $1/f^2$ (-12 dB per octave), and for frequencies above the filter bandwidth the output phase noise is white. If the filter bandwidth f_L is less than f_α, the noise-power spectral density is depicted in Fig. 10.26b. In this case the PSD decreases as $1/f$ (-6 dB per octave) for frequencies between f_L and f_α and is independent of frequency for frequencies greater than f_α. Equation (10.39) provides a quantitative measure for comparing an oscillator's noise performance to a theoretical minimum based on the amplifier's noise figure and f_α.

At high frequencies the oscillator noise floor is proportional to the noise figure of the amplifier used in realizing the oscillator. Since the minimum noise figure is 1, the minimum noise floor is kT, or -174 dB/Hz. At lower frequencies close to the oscillating frequency, the noise increases, but it is seen from Eq. (10.39) that the actual amplitude is inversely proportional to Q^2 of the resonator. So the higher the Q, the smaller will be the phase noise near the oscillating frequency.

Phase Noise in Phase-Locked Loops

The synthesizer output noise due to the noise generated in the VCO can be determined (assuming a linear model) with Eqs. (10.37) and (10.39). It is

$$
\begin{aligned}
S_{\phi_o} &= \frac{S_\theta(\omega)(1 + \omega_o^2/4Q^2\omega^2)}{|1 + K_v F(j\omega)/N(j\omega)|^2} \\
&= \frac{S_\phi(\omega)}{|1 + K_v F(j\omega)/j\omega N|^2} \\
&= \frac{FkT/P_s(1 + \omega_\alpha/\omega)(1 + \omega_o^2/4Q^2\omega^2)}{|1 + K_v F(j\omega)/j\omega N|^2}
\end{aligned}
\tag{10.40}
$$

The denominator can be approximated by

$$
[|1 + G(j\omega)|^2]^{-1} \approx [|G(j\omega)|^2]^{-1} \qquad \text{for } |G(j\omega)| \gg 1
$$

(that is, inside the loop bandwidth).

For frequencies above the loop bandwidth,

$$
[|1 + G(j\omega)|^2]^{-1} \approx 1 \qquad \text{for } |G(j\omega)| \ll 1
$$

The frequency at which the two approximations coincide is the open-loop crossover frequency, which is approximately the closed-loop bandwidth. Therefore, for frequencies higher than the open-loop crossover frequency, the closed-loop noise due to the VCO noise S_{ϕ_o} is approximately the same as the VCO noise, or

$$
S_{\phi_o} \approx S_\phi(\omega) \qquad \omega > \omega_c
\tag{10.41}
$$

and for lower frequencies

$$
S_{\phi_o} \approx S_\phi(\omega) \frac{\omega^2 N^2}{K_v^2 |F(j\omega)|^2}
\tag{10.42}
$$

If a type I loop is employed with a filter bandwidth greater than ω_c and the VCO output phase noise spectrum is as shown in Fig. 10.26a, then the synthesizer output noise spectrum due to the VCO noise will be as shown in Fig. 10.27. The open-loop noise power spectrum decreases at -18 dB per octave, so the closed-loop noise power spectrum decreases at a rate of -6 dB per octave until the frequency ω_α; then the noise power spectrum does not change with frequency until the PLL filter frequency is reached. The noise power density then increases at $+12$ dB per octave until the open-loop crossover frequency ω_c is reached. At higher frequencies the power spectral density plot assumes the same shape as S_ϕ.

For a type II loop the open-loop transfer function is

$$
G(j\omega) = \frac{K_v(j\omega/\omega_z + 1)}{N(j\omega)^2}
\tag{10.43}
$$

so the closed-loop noise spectrum (due to the VCO noise) will increase with a

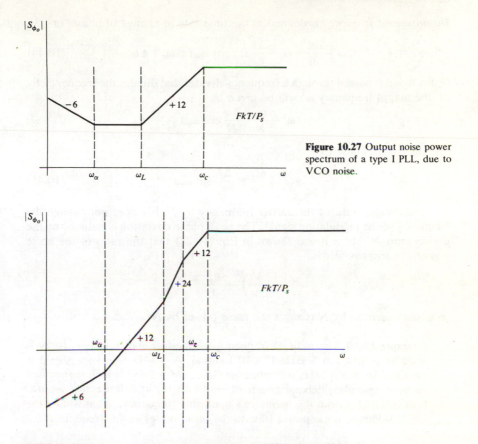

Figure 10.27 Output noise power spectrum of a type I PLL, due to VCO noise.

Figure 10.28 Output noise power spectrum of a type II PLL, due to VCO noise.

rate of +6 dB per octave at low frequencies. The shape of the normalized closed-loop noise power spectrum (due to the VCO noise only) is shown in Fig. 10.28. The slope increases from +6 to +12 dB per octave at ω_α and to +24 dB per octave at ω_L, then decreases to +12 dB per octave at ω_z and assumes the same shape at S_ϕ for frequencies above ω_c. The type II system provides substantially more filtering of the VCO noise within the loop bandwidth.

Effect of Frequency Division and Multiplication on Phase Noise

Equation (10.25) states that the instantaneous phase $\theta_i(t)$ of a carrier frequency modulated by a sine wave of frequency f_m is given by

$$\theta_i(t) = \omega_o t + \frac{\Delta f}{f_m} \sin \omega_m t$$

Instantaneous frequency is defined as the time rate of change of phase, or

$$\omega = \frac{d\theta_i}{dt} = \omega_o + \frac{\Delta f}{f_m} \omega_m \cos \omega_m t \leq \omega_o + \Delta\omega \tag{10.44}$$

If this signal is passed through a frequency divider that divides the frequency by N, the output frequency ω^1 will be given by

$$\omega^1 = \frac{\omega_o}{N} + \frac{\Delta\omega}{N} \cos \omega_m t$$

and the output phase by

$$\theta_i(t) = \frac{\omega_o t}{N} + \frac{\Delta f}{N f_m} \sin \omega_m t \tag{10.45}$$

The divider reduces the carrier frequency by N, but does not change the frequency of the modulating signal. The peak phase deviation is reduced by the divide ratio N. Since it was shown in Eq. (10.28) that the ratio of the noise power to carrier power is

$$\left(\frac{V_n}{V}\right)^2 = \frac{\theta_p^2}{4}$$

frequency division by N reduces the noise power by N^2.

Example 10.10 The indirect frequency synthesizer shown in Fig. 10.29 is used to generate a 5-GHz (5×10^9) signal. A 1-kHz reference signal is obtained from a 5-MHz reference oscillator ($M = 5000$) which is specified to have a single sideband power of -140 dB/Hz at a frequency separation of 0.5 kHz from the oscillator's operating frequency. What will be the single sideband noise power (due to the input noise) at this frequency?

SOLUTION If the loop bandwidth is assumed to be approximately 1 kHz, then the noise from the reference oscillator 0.5 kHz from the carrier frequency will not be reduced by the low-pass filtering of the PLL. The approximate loop transfer function is

$$\theta_o = \frac{\theta_r K_v F(s)/s}{1 + K_v F(s)/sN} \approx N\theta_r = \frac{N}{M} \theta_i \tag{10.46}$$

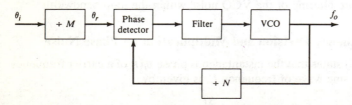

Figure 10.29 A PLL for synthesizing high-frequency signals.

for reference frequencies below the loop bandwidth of 1 kHz. Although the divider M reduces the input noise power, the net effect is that the output noise power is the reference oscillator noise power multiplied by $(N/M)^2$. N must equal 5×10^6 in order to obtain the output frequency of 5×10^9 Hz, and the output noise power due to the reference oscillator is therefore

$$N_o = -140 \text{ dB/Hz} + 10 \log \left(\frac{5 \times 10^6}{5 \times 10^3}\right)^2 = -80 \text{ dB/Hz}$$

at a frequency offset of 0.5 kHz.

Actually, the noise would be much worse than predicted in the previous example. The reference oscillator is already a low-noise device, and the noise cannot be reduced below some noise floor by additional division. This illustrates a problem inherent in PLL frequency synthesizers used to generate an output frequency much higher than the reference oscillator frequency. Although the reference oscillator noise power may be small, the same noise power appears on the output signal amplified by the factor N^2, where N is the output-frequency to reference-oscillator-frequency ratio.

PROBLEMS

10.1 A relationship was developed in Ch. 3 that expressed the filter attenuation of the nth multiple of the center frequency of a second-order tuned circuit in terms of the circuit Q. For the frequency synthesizer shown in Fig. 10.1, what must be the Q of the bandpass filter in order to obtain 80-dB suppression of the other mixer product?

10.2 Design a direct frequency synthesizer to generate 15.8×10^6 Hz from a 1×10^6 Hz reference oscillator.

10.3 Design a direct frequency synthesizer, using double-mix-and-divide modules, to cover the frequency spectrum 25 to 29.999 MHz in 1-kHz increments. Specify all frequencies used and the maximum frequency mixing ratio of all mixers.

10.4 Design a phase-locked loop synthesizer to meet the specifications of Prob. 10.3.

    (*a*) What is the reference frequency?

    (*b*) What is the range of the divide ratio N?

    (*c*) The 25-MHz output frequency is too high for a programmable divider, and a variable-modulus divider should be used. Use a 10/11 divider and determine the initial values of the counters required to synthesize 26.111 MHz.

10.5 Could a 100/101 variable-modulus divider be used in Prob. 10.4? Explain your answer.

10.6 For the three-loop frequency synthesizer illustrated in Fig. 10.12 determine N_A and N_B required to obtain an output frequency of 38.912×10^6 Hz.

10.7 Design a multiple-phase-locked-loop synthesizer to cover the frequency range 35.4 to 40.0 MHz in 10-Hz increments. The reference frequency is to be 100 kHz. No loop should operate with a reference frequency below 100 kHz.

10.8 Use a fractional N frequency synthesizer to synthesize a frequency of 2.33 kHz using a 1-kHz reference frequency.

(a) What must be the size of the phase accumulator?

(b) Sketch the output of the phase detector ($N = 2$, $M = 3$).

(c) Sketch the phase detector output if the output frequency is to be 4.33 kHz ($N = 4$, $M = 3$).

(d) What must the phase accumulator size be to realize a frequency resolution of 10 Hz?

10.9 It is desired to design a direct digital frequency synthesizer with a maximum output frequency of 10 kHz and a step size of 10 Hz.

(a) What must the clock frequency be if four samples are to be used in the highest-frequency waveform?

(b) How many bits must the accumulator contain?

(c) What number must be added to the accumulator at each cycle to generate an output frequency of 100 kHz?

(d) What must the accumulator size be if the minimum step size is to be reduced to 1 Hz?

(e) What word length is required for a 50-dB signal-to-noise ratio?

10.10 Design a DDFS to cover the frequency range 0 to 5 kHz in 0.01-Hz increments. The spectral purity is to be at least 50 dB. Specify the accumulator size, memory requirements, sampling rates, and characteristics of the output low-pass filter.

10.11 Design a frequency synthesizer to cover the frequency range 100 to 100.999 MHz in 10-Hz increments. The frequency switching time is to be less than 100 μs. Discuss different configurations which can be used to meet the specifications.

10.12 Design a three-loop synthesizer to cover the frequency range 198 to 200 MHz with a frequency resolution of 10 Hz. The loop frequency switching time should be as short as possible.

10.13 Figure P10.13 illustrates another method for covering the 198- to 200-MHz frequency range with a frequency resolution of 10 Hz. Select frequency ranges for the two direct digital frequency synthesizers and specify the frequencies of the four oscillators used to mix the frequency up to the specified operating frequency.

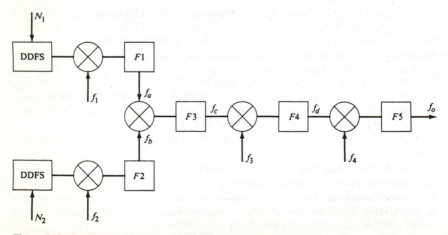

Figure P10.13 A hybrid direct/direct digital frequency synthesizer.

10.14 A 10-MHz oscillator has a noise figure of 10 dB, an f_α of 20 kHz, and a Q of 100.

(a) What is the noise power (relative to the carrier) in a 1-Hz bandwidth 1 MHz above the carrier frequency?

(b) What is the relative noise power 100 kHz above the carrier frequency?

(c) What is the relative noise power 10 kHz above the carrier frequency?

10.15 If f_α in Prob. 10.14 had been 1 MHz above the carrier frequency, what would have been the answers to parts (a), (b), and (c) of the problem?

10.16 A 10-MHz reference oscillator is to be used with a phase-locked loop to synthesize an output frequency of 10^9 Hz. If the output noise is to be -100 dB/Hz relative to the carrier, what must be the output noise level of the reference oscillator?

10.17 Discuss various methods of realizing a frequency synthesizer to cover the frequency spectrum 10 Hz to 1 kHz in 10-Hz increments. The discussion should include the reference frequency used in each case, the synthesizer switching speed, and the advantages and disadvantages of each method.

REFERENCES

10.1 Finden, H. J.: The Frequency Synthesizer, *J. IEEE*, **90**(3):165–180 (1943).

10.2 Peterson, M. E.: The Design and Performance of an Ultra Low-Noise Digital Frequency Synthesizer for Use in VLF Receivers, *Proc. 26th Annual Symposium on Frequency Control*, U.S. Army Electronics Command, Fort Monmouth, N.J., Nat. Tech. Infor. Serv. Accession Nr. AD 771043, 1972, pp. 55–70.

10.3 Gillette, G. C.: The Digiphase Synthesizer, *Proc. 23rd Annual Symposium on Frequency Control*, U.S. Army Electronics Command, Fort Monmouth, N.J., Nat. Tech. Infor. Serv. Accession Nr. AD 746209, 1969, pp. 201–210.

10.4 Gibbs, J., and R. Temple: Frequency Domain Yields Its Data to Phase-Locked Synthesizer, *Electronics*, April 27, 1978, pp. 107–113.

10.5 Tierney, J., C. M. Rader, and B. Gold: A Digital Frequency Synthesizer, *IEEE Trans. on Audio and Electroacoustics*, AU-19, 1971, pp. 48–57.

10.6 Leeson, D. B.: A Simple Model of Feedback Oscillator Noise Spectrum, *Proc. IEEE*, **54**:329–330 (1966).

10.7 Clark, E. K., and D. Hess: *Communication Circuits: Analysis and Design*, Addison-Wesley, Reading, Mass., 1971.

BIBLIOGRAPHY

Frequency Synthesis

Apetz, B., B. Scheckel, and G. Weil: A 120 MHz AM/FM PLL-IC with Dual on-Chip Programmable Charge Pump/Filter Op-Amp, *IEEE Trans. on Cons. Elect.*, **27**:234–242 (1981).

Beyers, B. E.: Frequency Synthesis Tuning Systems with Automatic Offset Tuning, *IEEE Trans. on CE*, vol. CE-24, no. 3, August 1978, pp. 419–428.

Bjerede, B. E., and G. Fisher: An Efficient Hardware Implementation for High Resolution Frequency Synthesis, *Proc. 31st Ann. Symp. on Frequency Control*, U.S. Army Electronics Command, Fort Monmouth, N.J., Natl. Tech. Infor. Serv. Accession Nr. AD 771043, 1977, pp. 318–321.

——— and ———: A New Phase Accumulator Approach for Frequency Synthesis, *Proc. IEEE NAECON '76*, May 1976, pp. 928–932.

Blachowicz, L. F.: Dial Any Channel to 500 MHz, *Electronics*, May 2, 1966, pp. 60–69.

d'Andrea, G., V. Libal, and G. Weil: Frequency Synthesis for Color TV-Receivers with a New Dedicated μ Computer, *IEEE Trans. on Cons. Elect.*, **27**:272–283 (1981).

Dayoff, I., and B. Krischner: A Bulk CMOS 40-Channel CB Frequency Synthesizer, *IEEE Trans. on CE*, vol. CE-23, no. 4, November 1977, pp. 518–521.

Egan, W. F.: *Frequency Synthesis by Phase Lock*, Wiley, New York, 1981.

Fukui, K.: A Portable All-Band Radio Receiver Using Microcomputer Controlled PLL Synthesizer, *IEEE Trans. on CE*, vol. CE-26, August 1980, pp. 299–310.

Furuno, K., S. Mitra, K. Hirano, and Y. Ito: Design of Digital Sinusoidal Oscillators with Absolute Periodicity, *IEEE Trans. AES*, 11, 1975, pp. 1286–1299.

Gorski-Popiel, J.: *Frequency Synthesis: Techniques and Applications*, IEEE, New York, 1975.

Ichinose, K.: One Chip AM/FM Digital Tuning System, *IEEE Trans. on CE*, vol. CE-26, August 1980, pp. 282–288.

Kroupa, V. F.: *Frequency Synthesis: Theory, Design & Applications*, Wiley, New York, 1976.

Manassewitsch, V.: *Frequency Synthesizers: Theory and Design*, Wiley, New York, 1981.

Mills, T. B.: An AM/FM Digital Tuning System, *IEEE Trans. on CE*, vol. CE-24, no. 4, November 1978, pp. 507–513.

Mueller, K. J., and C. P. Wu: A Monolithic ECL/I^2L Phase-Locked Loop Frequency Synthesizer for AM/FM TV, *IEEE Trans. on CE*, vol. CE-25, no. 3, August 1979, pp. 670–676.

Rhodes, R., W. Hutchinson, and B. Hutchinson: Frequency Agile Phase-Locked Loop Synthesizer for a Communications Satellite. *Proc. Natl. Telecommunications Conference*, 1980, pp. 22.3.1–22.3.6.

Rohde, U. L.: *Digital PLL Frequency Synthesizers*, Prentice-Hall, Englewood Cliffs, N.J., 1983.

Rohde, U. L.: Modern Design of Frequency Synthesizers, *Ham Radio*, July 1976, pp. 10–22.

Rzezewski, T., and T. Kawasaki: A Microcomputer Controlled Frequency Synthesizer for TV, *IEEE Trans. on CE*, vol. CE-24, no. 2, May 1978, pp. 145–153.

Sample, L.: A Linear CB Synthesizer, *IEEE Trans. on CE*, vol. CE-23, no. 3, August 1977, pp. 200–206.

Stinehelfer, J., and J. Nichols: A Digital Frequency Synthesizer for an AM and FM Receiver, *IEEE Trans.*, BTR-15, 1969, pp. 235–243.

Tanaka, K., S. Ike Guichi, Y. Nakayama, and Osamuu Ikeda: New Digital Synthesizer LSI for FM/AM Receivers, *IEEE Trans. on Cons. Elect.*, **27**:210–219 (1981).

Yamada, T.: A High Speed NMOS PLL-Synthesizer LSI with On-Chip Prescaler for AM/FM Receivers, *IEEE Trans. on CE*, vol. CE-26, August 1980, pp. 289–298.

Yuen, G. W. M.: An Analog-Tuned Digital Frequency Synthesizer Tuning System for FM/AM Tuner, *IEEE Trans. on CE*, vol. CE-23, no. 4, November 1978, pp. 507–513.

Phase Noise

Baghdady, E. J., R. N. Lincoln, and B. D. Nelin: Short-Term Frequency Stability: Characterization, Theory, and Measurements, *Proc. IEEE*, **53**:704–722 (1965).

Barnes, J. A., and R. C. Mockler: The Power Spectrum and Its Importance in Precise Frequency Measurements, *IRE Trans. on Instrumentation*, **9**:149–155 (1960).

——— et al.: Characterization of Frequency Stability, U.S. Dept. of Comm., NBS Technical Note 394, 1970.

Cutler, L. S., and C. L. Searle: Some Aspects of the Theory and Measurement of Frequency Fluctuations in Frequency Standards, *Proc. IEEE*, **54**:136–154 (1966).

Edson, W. A.: Noise in Oscillators, *Proc. IRE*, **48**:1454–1466 (1960).

Hafner, Erich: The Effects of Noise in Oscillators, *Proc. IEEE*, **54**:179–198 (1966).

Leeson, D. B.: Short-Term Stable Microwave Sources, *Microwave J.*, 59–69 (June 1970).

———: A Simple Model of Feedback Oscillator Noise Spectrum, *Proc. IEEE*, **54**:136–154 (1966).

Lindsey, W. C., and C. M. Chie: Specification and Measurement of Oscillator Phase Noise Instability, *Proc. Ann. Frequency Control Symp.*, 1981, pp. 302–310.

Reynolds, Chuck: Measure Phase Noise, *Electronic Design*, February 15, 1977, pp. 106–108.

Rutman, Jaques: Characterization of Frequency Stability: A Transfer Function Approach and Its Application to Measure via Filtering of Phase Noise, *IEEE Trans. on Instrumentation and Measurement*, **22**:40–48 (1974).

ELEVEN

POWER AMPLIFIERS

11.1 INTRODUCTION

Power amplifiers are those amplifiers whose design concerns are based on a combination of output power, drive level, power dissipation, distortion, size, weight, and efficiency (power output divided by power supplied). Simultaneously, the transistors used in power amplifiers have requirements based on breakdown voltage, current limitations, and maximum power dissipation. The output power of power amplifiers can range from the milliwatt region for small, portable transistor amplifiers to the megawatt region for large broadcast stations.

The power amplifier is invariably the last stage in the amplifier chain because the power level is highest at this point; no intermediate-stage amplifier would be operated with a power gain significantly less than 1. Because the signal level is the largest at this point, it results in the maximum amount of distortion due to the nonlinear characteristics of the device. These nonlinearities produce unwanted frequency components (harmonics) and intermodulation distortion (IMD) products. However, there are various methods of designing circuits, methods which lead to different levels of efficiency and create different amounts of distortion. Because various modulation techniques can tolerate different amounts of distortion, power-amplifier design depends on the type of signal (modulation) to be amplified. In the least efficient design, the maximum amount of power is dissipated in the transistor, requiring larger and more expensive transistors than would otherwise be required. However, good power-amplifier design techniques can result in more economical and reliable electronics.

Power amplifiers are classified according to their mode of operation; the most frequently used classes are discussed in this chapter. The class of operation is determined by how the transistor is biased and the nature of the output circuit.

The original classification of operating modes included class A, B, AB, and C amplifiers. A class A amplifier is a linear amplifier. Theoretically, it will produce a sine-wave output in response to a sine-wave input. The output frequency will be the same as the input frequency, and the output amplitude will be a linear function of the input amplitude. If the amplifier output is a linear function of the input over 50 percent (180°) of the input waveform, it is categorized as class B. If the linear conduction angle is less than 50 percent, it is class C; and if the conduction angle is greater than 180° but less than 360°, it is referred to as class AB.

Today, additional classes of amplifiers exist, most of them using the transistor as a switch. The more popular forms of switching amplifiers are considered after the class A, B, and C amplifiers have been described.

11.2 CLASS A AMPLIFIERS

Class A power amplifiers are no different in behavior from the linear amplifiers studied up to this point, except that their power and distortion levels are of primary importance. For class A operation the output will be a sine wave in response to a sine-wave input. The class of generation is determined by the input signal level and how the transistor is biased. Figure 11.1 describes an ac coupled amplifier which can be biased for class A, B, or C.

For the amplifier shown in Fig. 11.1 the transistor quiescent voltage (no ac collector current) is

$$V_{ce} = V_Q = V_{cc} - I_c R_E \tag{11.1}$$

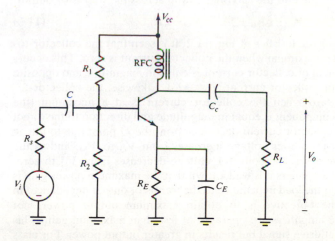

Figure 11.1 An ac coupled amplifier.

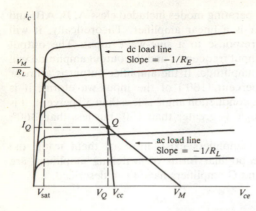

Figure 11.2 Transistor collector-emitter characteristics including the ac and dc load lines.

The slope of the dc load line dI_c/dV_{ce} is then $-1/R_E$, as illustrated in Fig. 11.2. In class A power amplifiers the dc resistance R_E will normally be much less than the ac resistance. R_E is kept small to limit the dc power dissipated in the resistor. In some designs R_E is zero, but a small R_E is often included for bias stabilization of bipolar transistor amplifiers and self-biasing of field-effect transistor amplifiers. If R_E is small, the Q (quiescent) -point voltage is

$$V_Q \approx V_{cc}$$

For ac operation the coupling and emitter bypass capacitors act as short circuits, and the ac collector current i_c is given by

$$-i_c R_L = V_o$$

The output voltage is equal to the ac voltage drop across the transistor output:

$$V_{cc} - i_c R_E + V_{ce} \approx V_Q - i_c R_L \qquad (11.2)$$

From the transistor characteristics of Fig. 11.2, it is seen that the collector-to-emitter voltage is at a maximum when the collector current is zero. This occurs when the ac component of collector current is equal in magnitude and opposite in direction to the dc collector current ($i_c = -I_C$). Likewise, the collector-to-emitter voltage is zero when the collector current is at a maximum (the alternating current component is equal in magnitude and direction to the direct current I_C). As the collector current decreases from the Q point due to the ac signal, the collector-to-emitter voltage increases from V_Q to V_M; and as the collector current increases, the collector voltage decreases from V_Q to zero.

For small-signal amplifiers it is well known that the maximum power gain is obtained by matching the load impedance to the transistor output impedance. In power amplifiers the objective is to obtain maximum output power, not maximum gain. The amplifiers are operated at less than maximum gain. This requires a large input drive signal but results in greater output power. For class A operation maximum output power is obtained by selecting the ac load impedance such that the maximum signal swing can be obtained from the device. The output must be symmetrical to avoid distortion. In an ideal transistor $V_{sat} = 0$, the collector-to-emitter voltage can decrease from V_{cc} to zero; so for symmetrical operation it can also increase to

$$V_M = 2V_Q = 2V_{cc}$$

Likewise, the output current can decrease from I_Q to zero, so the maximum output current is

$$I_M = \frac{V_M}{R_L} = \frac{2V_{cc}}{R_L}$$

and

$$I_Q = \frac{V_{cc}}{R_L}$$

The slope of the ac load line is $-1/R_L$ and

$$-R_L^{-1} = -\frac{I_M}{V_M} = \frac{-I_Q}{V_{cc}}$$

This is the value of the load resistance that results in maximum output power.

A surprising characteristic of this amplifier is that the maximum voltage dropped across the transistor is twice the supply voltage, and the peak-to-peak output voltage is also $2V_{cc}$. How this is possible can be seen from the equivalent circuit shown in Fig. 11.3. Here the RF choke has been replaced by a constant current source since no alternating current flows through the device and the direct current is I_Q. Also, no direct current flows through the capacitor; therefore the dc voltage drop across the device is V_{cc}, and there is no ac voltage drop across the capacitor. Here the capacitor can be replaced by a battery V_{cc}. At the instant at which the load current is

$$i_L = -i_C = -I_Q = \frac{V_{cc}}{R_L}$$

the output voltage is

$$V_o = I_Q R_L = V_{ce}$$

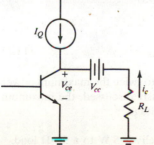

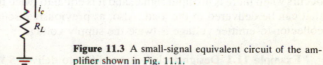

Figure 11.3 A small-signal equivalent circuit of the amplifier shown in Fig. 11.1.

If the emitter-resistor is sufficiently small,

$$V_{ce} = V_o + V_{cc}$$

so

$$(V_{ce})_{\text{max}} \approx 2 V_{cc}$$

For a sinusoidal input signal the collector current consists of a dc component I_Q, which does not flow through the load, and an ac component i_L, which does flow through the load. That is,

$$i_C = I_Q + i_L$$

where $i_L = I_p \sin \omega t$. Since the collector current cannot become negative, $I_p \le I_Q$. The average power will be

$$(P_o)_{\text{av}} = \frac{I_p^2 R_L}{2} \le \frac{I_Q^2 R_L}{2}$$

and the power supplied by the battery will be (neglecting the small amount of power dissipated in the base bias circuitry)

$$P_{cc} = V_{cc} I_Q = \frac{V_{cc}^2}{R_L}$$

so the efficiency is

$$\eta = \frac{P_o}{P_{cc}} = \frac{I_p^2 R_L^2}{2 V_{cc}^2} \tag{11.3}$$

The maximum efficiency occurs for $I_p = I_Q$ and is

$$\eta_{\text{max}} = \frac{I_Q^2 R_L^2}{2 V_{cc}^2} = 50\%$$

The maximum operating efficiency for a class A ac coupled power amplifier is 50 percent and occurs with the maximum input signal. If the output signal decreases, I_p decreases and so does the efficiency.

The power dissipated in the transistor is

$$P_T = P_{cc} - P_o = \frac{V_{cc}^2}{R_L} \left(1 - \frac{I_p^2}{2 I_Q^2} \right) \tag{11.4}$$

The maximum power dissipated in the transistor

$$(P_T)_{\text{max}} = \frac{V_{cc}^2}{R_L} = 2(P_o)_{\text{max}} \tag{11.5}$$

occurs when there is no input signal, and it is equal to twice the maximum power that can be delivered to the load. Also, as previously discussed, the maximum collector-to-emitter voltage is twice the supply voltage.

Example 11.1 Design a class A amplifier to deliver 5 W to a 50-Ω load.

SOLUTION For 5-W power in a 50-Ω resistor, the peak value of the sinusoidal voltage across the resistor is

$$V_p = (2 \times 5 \times 50)^{1/2} = 22.4 \text{ V}$$

Since the peak voltage cannot exceed V_{cc}, a standard supply $V_{cc} = 24$ V would be suitable. The corresponding peak value of the ac load current is

$$I_p = \frac{22.4}{50} = 0.448 \text{ A}$$

Therefore, the transistor must be biased so that

$$I_Q \geq 0.448 \text{ A}$$

For a 24-V supply with no emitter (or source) resistance,

$$I_Q = \frac{V_{cc}}{50} = 0.48 \text{ A}$$

The power supplied is then

$$P_{cc} = V_{cc}I_Q = 11.52 \text{ W}$$

and the efficiency is

$$\eta = \frac{5}{11.52} = 43.4\%$$

The transistor that is selected must be able to dissipate 11.52 W in case the input power drops to zero, and the transistor collector-to-emitter breakdown voltage must be at least 48 V ($2V_{cc}$).

The effect of the saturation voltage is to create signal distortion and reduce efficiency. The maximum value of collector current that can be applied without V_{ce} decreasing to V_{sat} has the peak value

$$(I_p)_{max} = \frac{V_{cc} - V_{sat}}{R_L}$$

Therefore, when the saturation voltage is considered, the maximum efficiency is

$$\eta_{max} = \frac{1}{2}\left(1 - \frac{V_{sat}}{V_{cc}}\right)^2 \qquad (11.6)$$

Transformer-Coupled Class A Amplifiers

If a load impedance is specified, a transformer can be used to improve the power gain by transforming the load impedance to that required for maximum output power. A class A transformer-coupled FET amplifier is shown in Fig. 11.4, and the load lines for the amplifier are shown in Fig. 11.5. The characteristic curves

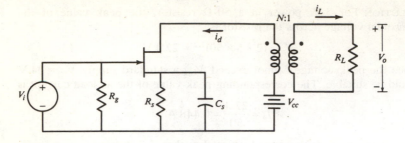

Figure 11.4 A transformer-coupled class A amplifier.

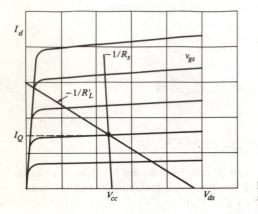

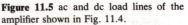

Figure 11.5 ac and dc load lines of the amplifier shown in Fig. 11.4.

are similar to those of the bipolar transistor amplifier, except that the controlling signal for the field-effect transistor is the gate-to-source voltage.

If the transformer is ideal,

$$i_L = N i_d$$

where i_d is the ac drain current. The ac voltage across the drain-to-source junction is

$$v_d = N v_o$$

Therefore, the ac load impedance seen by the transistor is

$$R'_L = N^2 R_L \qquad (11.7)$$

The slope of the dc load line is $-1/R_S$, and if $R_S \ll R'_L$, the slope of the ac load line is $-1/R'_L$. The maximum signal swing is again $2V_{cc}$ (ignoring V_{sat}) and the peak drain current is

$$I_M = (i_d)_p = \frac{2V_{cc}}{R'_L} \qquad (11.8)$$

The power calculations for the transformer-coupled load are the same as for the capacitive-coupled load. The transformer provides additional flexibility for matching the load to the source, but the power dissipated in the transformer can significantly reduce the amplifier efficiency. Transformer- or inductor-coupled bipolar transistor amplifiers are subject to a phenomenon known as *thermal runaway*.[11.1] The heating up of the transistor causes more current to flow, which causes greater self-heating, which can cause the device to self-destruct. The problem rarely occurs with resistive loads since the increased current results in a reduced collector-to-emitter voltage, and eventually the circuit will reach equilibrium. Nor does the problem arise with FET amplifiers because as their temperatures increase, their output currents decrease.

Class A Push-Pull Amplifiers

As mentioned in Chap. 5, push-pull amplifiers eliminate the even harmonic distortion present in the amplifier output. This can provide a significant improvement in the performance of linear amplifiers. Also, push-pull operation reduces the power requirements of the individual transistors. Figure 11.6 illustrates a class A transformer-coupled push-pull stage. The ac equivalent circuit is shown in Fig. 11.7. In class A operation, both transistors continuously

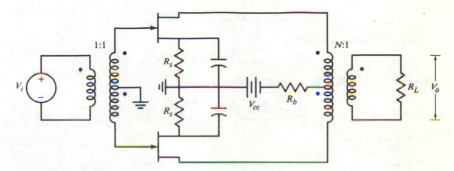

Figure 11.6 A class A transformer-coupled push-pull amplifier.

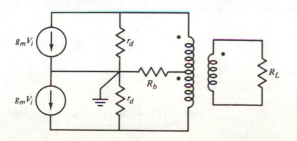

Figure 11.7 A small-signal equivalent circuit of the push-pull amplifier.

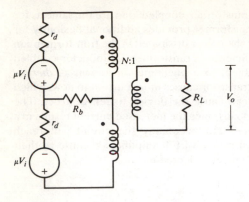

Figure 11.8 Another equivalent circuit of the push-pull amplifier shown in Fig. 11.6.

drive the output, and the transistor outputs, which are 180° out of phase, are combined in the center-tapped output transformer. The circuit of Fig. 11.7 can be redrawn as shown in Fig. 11.8 using Thévenin's theorem; the results developed for center-tapped transformers (Chap. 6) are now directly applicable. Here the amplification factor $\mu = g_m r_d$. If the dynamic drain resistances of the transistors are equal, then no current will flow through R_b and the circuit can be redrawn as shown in Fig. 11.9. The output voltage V_o is determined by

$$2\mu V_i - 2 r_d i_d = -2 N V_o$$

If the transformer is lossless,

$$2 N V_o = -i_d (2N)^2 R_L = -2 R'_L i_d$$

where $R'_L = 2 N^2 R_L$, the output voltage is

$$V_o = \frac{-N \mu R_L V_i}{r_d + R'_L} = \frac{-N g_m R_L V_i}{1 + R'_L / r_d}$$

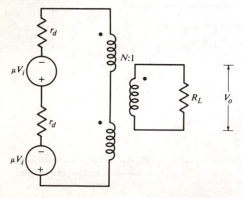

Figure 11.9 A simplified equivalent circuit of the push-pull amplifier.

Normally $r_d \gg R'_L$, so the voltage gain is

$$A_v = -g_m R_L N$$

since the load for each transistor is

$$R'_L = 2N^2 R_L$$

For maximum output power (i.e., maximum signal swing) each transistor is biased so that

$$R'_L = \frac{V_{cc}}{I_d}$$

where I_d is the dc drain current.

The maximum output power (from both transistors) is

$$P_o = \frac{2V_{cc}^2}{2R'_L} = \frac{V_{cc}^2}{(2N^2 R_L)}$$

The total output power of the class A push-pull amplifier is twice that of the single-ended class A amplifier; however, the maximum voltage drop across each transistor is the same as that of the single-ended amplifier. The power supplied to each transistor is

$$P_{cc} = I_d V_{cc} = \frac{V_{cc}^2}{R'_L}$$

The total power supplied is $2P_{cc}$, so the maximum efficiency of the class A push-pull amplifier is $\eta = 50$ percent, the same as the efficiency of the single-ended amplifier. Besides reducing even harmonic distortion, the push-pull configuration can provide twice the output power of the single-ended design.

Square-Wave Input

The efficiency of a class A amplifier depends on the input signal level and also on the signal waveshape. Consider the case of a square-wave input. If the amplifier is class A, the collector current is also a square-wave, as shown in Fig. 11.10. The dc value is I_Q and the peak current is

$$I_p \le I_Q$$

In this case

$$P_{cc} = I_Q V_{cc} = I_Q^2 R_L$$

and

$$P_o = I_p^2 R_L$$

so the efficiency

$$\eta = \frac{P_o}{P_{cc}} = \frac{I_p^2}{I_Q^2} \tag{11.9}$$

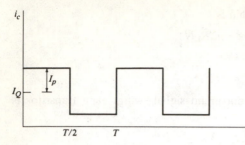

Figure 11.10 A collector-current waveform.

will approach 100 percent if I_p approaches I_Q. The efficiency of the class A amplifier depends on the waveform of the input signal.

If a tuned circuit is used as the load, as is frequently the case in order to minimize distortion, then efficiency is reduced. With a square-wave input the collector current will again be a square wave, as shown in Fig. 11.10, and the power supplied is again

$$P_{cc} = I_Q V_{cc}$$

If the tuned-circuit bandwidth is sufficiently narrow and tuned to the fundamental frequency, the output power is

$$P_o = \frac{I_1^2 R_L'}{2}$$

where I_1 is the amplitude of the fundamental frequency component of the output current

$$I_1 = \frac{4}{\pi} I_p$$

The maximum output power will be

$$(P_o)_{max} = \frac{8 I_p^2 R_L'}{\pi^2} = \frac{8 I_Q V_{cc}}{\pi^2}$$

and the maximum efficiency will be

$$\eta = \frac{(P_o)_{max}}{P_{cc}} = \frac{8}{\pi^2}$$

If the output is tuned to the nth harmonic (n-odd), the efficiency will be $\eta = 8/n^2\pi^2$, since the amplitude of the nth (odd) harmonic of a square wave is $1/n$ times that of the fundamental frequency. A tuned-circuit load decreases the efficiency, but it is frequently used since the tuned circuit reduces the output harmonic distortion.

11.3 CLASS B AMPLIFIERS

A major disadvantage of the class A amplifier is that all of the supply power is dissipated in the transistor when there is no input. It is usually advantageous to have no power supplied when there is no input signal, which is the case with a class B amplifier. A class B amplifier is biased as shown in Fig. 11.11. The quiescent collector current is zero, and the collector-to-emitter quiescent voltage is V_{cc}. It is biased just at the edge of the active region so that for a sine-wave input the transistor will conduct over 180° of the input waveform, as illustrated in Fig. 11.12.

The output current is a highly distorted sine wave, but the distortion can be removed by using a narrowband tuned circuit for the load, or what is more frequently done, by operating two class B transistor amplifiers in push-pull, as illustrated in Fig. 11.13. Ideally, each transistor conducts over alternate 180° of the input cycle, and the two outputs are summed so that an undistorted sine wave appears across the load resistor R_L. At any time, one transistor is conducting and the other is not, so the equivalent circuit can be drawn as shown in Fig. 11.14. The load seen by each transistor

$$R'_L = N^2 R_L$$

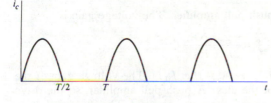

Figure 11.11 Collector-current waveform of a class B amplifier.

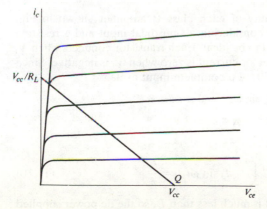

Figure 11.12 Q-point biasing for a class B amplifier.

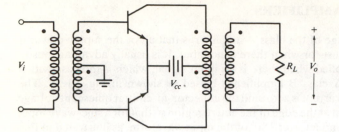

Figure 11.13 A class B push-pull amplifier.

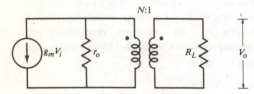

Figure 11.14 The small-signal equivalent circuit of the class B push-pull amplifier.

is one-half that of the class A push-pull amplifier. The voltage gain is

$$\frac{V_o}{V_i} = \frac{-g_m r_o N R_L}{r_o + R'_L}$$

It is usually the case that $r_o \gg R'_L$, so $A_v = -g_m R_L N$. The voltage gain of the class B stage is one-half that of the class A push-pull amplifier, so the drive requirements for the class B stage are two times as great. The increased drive requirement, however, is offset by the greater efficiency and power-handling capabilities of the class B amplifier.

The power-handling capability of each class B amplifier shown in Fig. 11.13 will now be evaluated by considering a sinusoidal input and a resistive load. The transistors are assumed to be ideal. Each transistor conducts when V_i is greater than zero, and the collector current is zero when V_i is negative. Since the load current conducts for 180° of a complete input cycle,

$$i_c(t) = I_p \sin \omega t \qquad 0 \le \omega t \le \pi$$

$$= 0 \qquad \pi \le \omega t \le 2\pi$$

The dc collector current is then

$$I_c = T^{-1} \int_0^{T/2} I_p \sin \omega t \, dt = \frac{I_p}{\pi} \qquad (11.10)$$

The dc value of the base current is much less than I_c, so the dc power supplied by the supply voltage is

$$P_{cc} = I_c V_{cc} = \frac{I_p V_{cc}}{\pi}$$

The ac power delivered to the load is

$$P_o = \frac{R_L'}{T} \int_0^{T/2} (I_p \sin \omega t)^2 \, dt = \frac{I_p^2 R_L'}{4}$$

Since $I_p \leq V_{cc}/R_L'$, the maximum output power from each transistor is

$$(P_o)_{max} = \frac{V_{cc}^2}{4 R_L} \tag{11.11}$$

The power supplied by the input circuit will normally be much less than the dc power, so the efficiency at maximum output power is

$$\eta = \frac{V_{cc}^2}{4 R_L} \frac{\pi R_L}{V_{cc}^2} = \frac{\pi}{4} = 78.5\%$$

The efficiency of a class B amplifier is much higher than that of a class A stage. In addition, the class B stage consumes no power when an input signal is not present. This is a most significant advantage in many applications (the class A amplifier has maximum power dissipation when no signal is present). The power dissipated in the transistor of the class B amplifier is

$$P_T = P_{cc} - P_o = \frac{I_p V_{cc}}{\pi} - I_p^2 \frac{R_L'}{4} \tag{11.12}$$

The maximum power dissipated in the transistor is found by differentiating Eq. (11.12) with respect to I_p. It is found that the maximum dissipation occurs for

$$I_p = \frac{2 V_{cc}}{\pi R_L'}$$

and

$$(P_T)_{max} = \frac{V_{cc}^2}{\pi^2 R_L'} = (P_o)_{max} \frac{4}{\pi^2} \tag{11.13}$$

Note, however, that the maximum transistor dissipation does not occur when the output power is a maximum. The class B stage also results in less transistor power dissipation than a class A stage. This is an expected result of greater operating efficiency. Another important difference between the two is that the maximum voltage drop across the transistor in a class B amplifier is V_{cc}; it is $2 V_{cc}$ for the class A amplifiers previously described. Class A amplifiers, therefore, require transistors with a higher collector-to-emitter breakdown voltage. Differences in class A and B power amplifiers are illustrated by the following example.

Example 11.2 Design a class B push-pull amplifier to deliver 5 W (maximum) to a 50-Ω load.

SOLUTION Assume that a transformer-coupled amplifier such as the one illustrated in Fig. 11.13 is used with a $1:1$ turns ratio. Since a push-pull amplifier is used, each class B amplifier will supply 2.5 W. The required supply voltage can be obtained from Eq. (11.11):

$$V_{cc}^2 = 4R_L(P_o)_{max} = 500 \text{ V}^2$$

A supply voltage V_{cc} of 24 V would be a suitable choice. The power-handling requirements of the transistor can be determined from Eq. (11.13):

$$(P_T)_{max} = (P_o)_{max}\frac{4}{\pi^2} = 1 \text{ W}$$

The peak output current will be

$$I = \left(\frac{4P_o}{R_L}\right)^{1/2} = 0.45 \text{ A}$$

The voltage and/or current requirements can be modified by selecting a different turns ratio for the transformer.

Example 11.1 discussed using a class A amplifier to realize these same specifications. Note that the class A amplifier transistor dissipation is over 11 times greater than that of the class B transistor, so if a class A push-pull amplifier is used, the power-handling requirements of each transistor would be $5\frac{1}{2}$ times that of the transistors used in the class B amplifier. Also, the collector-to-emitter breakdown voltage of the class A transistors must be double that of the transistors used in class B amplifiers.

While the class B amplifier also has the significant advantage of no power being dissipated when an input signal is not present, practical class A amplification does produce less distortion. All in all, if the distortion produced by class B amplification is acceptable, its benefits make it preferable to class A amplification. Large-signal distortion can be reduced with negative feedback.

Push-Pull Amplification with Complementary Transistors

Transformerless push-pull amplifiers can be realized using either bipolar or field-effect transistors with complementary symmetry. Figure 11.15 illustrates a push-pull amplifier using power MOSFETs. The current source and resistor R_b are used to bias the transistors for class B operation, and the MOSFETs do not turn on until the input voltage exceeds the threshold voltage of the device.

Power MOSFETs have several advantages over bipolar transistors. The most important is that the drain current has a negative temperature coefficient that decreases with increases in temperature, since the positive temperature

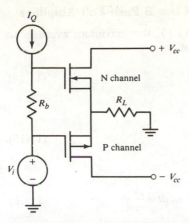

Figure 11.15 A class B push-pull amplifier using complementary MOSFETs.

coefficient of bipolar transistors can result in self-destruction unless complicated biasing circuits are used. Bipolar transistors also have an undesirable characteristic when operated at high voltages. The collector current is no longer uniform, but tends to concentrate in small areas, causing very high peak temperatures, known as *hot spots*, in these areas. The average junction temperature, measured at the transistor case, does not show the presence of hot spots, which degrade the transistor performance. The drain current of power MOSFETs is distributed uniformly, preventing the development of hot spots. The high input impedance (and hence high power gain) of the MOSFETs is another advantage. The upper frequency limits and power-handling capability of power MOSFETs have been continually increasing as manufacturing techniques have improved. Bipolar transistors are minority carrier devices that accumulate charge in the base region. This causes a problem in class B and C operation because removing the charge takes time and energy. One consequence is increasing power dissipation with increasing frequency. FETs are majority carrier devices; the charge carriers are controlled by an electric field and not by injection of minority carriers into the active region. The gate regions, therefore, contain no stored charge, and switching between the on and off states is very fast.

The first field-effect transistors were useful only at lower (less than 1 W) power levels. Their major limitation was that the FET drain-to-source channel was parallel to the chip surface, so the current density was much less than that of bipolar transistors (which utilized vertical current flow). For a given current the FET chip had to be much larger than an equivalent BJT chip, which meant a lower yield and a higher cost. Several new technologies have recently been developed that produce high-voltage, high-current FETs. The three most frequently used manufacturing technologies are VMOS (vertical MOS), V-JFET (vertical JFET) and DMOS (double-diffusion MOS).[11.2]

Power Relations in the Direct-Coupled Class B Push-Pull Amplifier

For the push-pull amplifier shown in Fig. 11.15, the maximum average ac output power in response to a sine-wave input is

$$(P_o)_{av} = \frac{V_{cc}^2}{2R_L} \qquad (11.14)$$

The power delivered by each transistor is

$$(P_o)_{av} = \frac{V_{cc}^2}{4R_L} \qquad (11.15)$$

The power delivered by each supply is

$$P_{cc} = V_{cc} T^{-1} \int_0^{T/2} \frac{V_{cc}}{R_L} \sin \omega t \, dt = \frac{V_{cc}^2}{\pi R_L} \qquad (11.16)$$

so the efficiency of this circuit is the same as that of the transformer-coupled push-pull amplifier. The maximum power dissipated in each transistor is

$$(P_T)_{max} = \frac{V_{cc}^2}{\pi^2 R_L} \qquad (11.17)$$

the same as for the transformer-coupled push-pull circuit. The only difference between the two circuits deduced from a power analysis is that the transformerless circuit requires two power supplies.

When the current requirements exceed the limitations of a single transistor, transistors can be operated in parallel. Figure 11.16 illustrates a class B amplifier containing two power MOSFETs.[11.3] It is usually easier to parallel

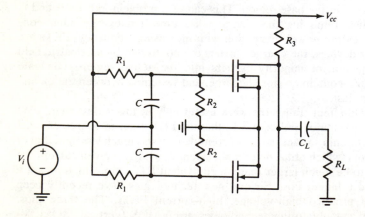

Figure 11.16 A class B amplifier using two MOSFETs in parallel.

FETs than bipolar transistors because of their larger input impedance. In this particular circuit, approximately two times as much output power could be obtained as from a class B circuit using a single transistor of the same type.

Class B amplifiers all suffer from crossover distortion, which occurs because of the nonlinear behavior of small-signal levels. The silicon BJT requires that the base-to-emitter junction be forward-biased by approximately 0.7 V before the collector current will flow. FETs also exhibit a nonlinear behavior for low signal levels. For these reasons power amplifiers are usually operated with a slight forward bias to reduce crossover distortion. This operation is often referred to as *class AB*, but it is essentially the same in that the power levels and efficiency are only slightly reduced from those of class B amplifiers.

11.4 CLASS C AMPLIFIERS

If the output current conduction angle is less than 180°, the amplifier operation is referred to as *Class C*. This mode of operation can have a greater efficiency than class B, but it creates more distortion than class A or B amplifiers. The distortion is sometimes acceptable or, in the case of frequency multiplication, desirable. Class C is often used where there is no variation in signal amplitude and the output circuit contains a tuned circuit to filter out all the harmonics of the output current. In many applications, such as the amplification of FM signals, the signal frequency, and not the amplitude, is important. Class C power amplifiers are usually used for these applications. Figure 11.17 provides examples of FET and bipolar-transistor class C amplifier circuits, and Fig. 11.18 illustrates the drain (or collector) current of a class C amplifier in which the conduction angle (2θ) is less than 180° and the drive level is sufficiently small that the output current does not saturate.

Several different models can be assumed for the current pulses. A relatively simple model is to assume that the pulses represent the tip of a sine wave. This is the model used here. That is (see Fig. 11.18),

$$i_c = I_p \sin \omega t - I_D \qquad \theta_1 \le \omega t \le \theta_2$$
$$= 0 \qquad \text{otherwise} \tag{11.18}$$

where
$$I_p > I_D$$

and
$$I_D = I_p \sin \theta_1 \tag{11.19}$$

For this model the direct current is

$$I_c = T^{-1} \int_{\theta_1/\omega}^{\theta_2/\omega} (I_p \sin \omega t - I_D)\, dt$$

$$= \frac{2 I_p \cos \theta_1 - I_D(\theta_2 - \theta_1)}{2\pi} \tag{11.20}$$

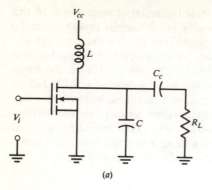

(a)

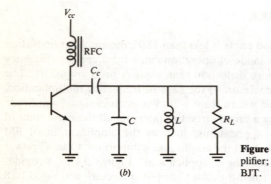

(b)

Figure 11.17 (a) A MOSFET class C amplifier; (b) a class C amplifier realized with a BJT.

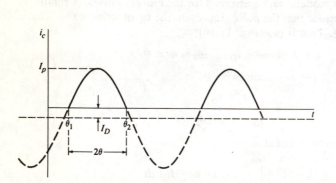

Figure 11.18 Collector-current waveform of a class C amplifier.

In order to simplify the notation we will define the conduction angle as

$$2\theta = \theta_2 - \theta_1 \qquad \text{or} \qquad \theta = \frac{\pi}{2} - \theta_1 \qquad (11.21)$$

Equation (11.20) can be rewritten as

$$I_c = \frac{I_p}{\pi}(\sin\theta - \theta\cos\theta) \qquad (11.22)$$

The direct current determines the power supplied, since the base (or gate) direct current is much smaller than the output current. That is,

$$P_{cc} = V_{cc}I_c = \frac{V_{cc}}{\pi}I_p(\sin\theta - \theta\cos\theta) \qquad (11.23)$$

If the output is a narrowband circuit tuned to the fundamental frequency of the current pulses, then the output power will be

$$P_o = \frac{I_1^2 R_L}{2} \qquad (11.24)$$

where I_1 is the amplitude of the fundamental current component

$$I_1 = \frac{4}{T}\int_0^{\theta/\omega}(I_p\cos\omega t - I_D)\cos\omega t\, dt$$

Here the time origin has been shifted to the center of the current pulse to simplify the integration. The time shifting does not alter the amplitudes of the frequency components, only their phase. The amplitude of the fundamental frequency component is

$$I_1 = \frac{I_p}{2\pi}[2\theta - \sin 2\theta] \qquad (11.25)$$

Since the conduction angle depends on the input amplitude, the fundamental current amplitude and, thus, the output voltage are a nonlinear function of the input signal amplitude.

For the class C FET amplifier shown in Fig. 11.17a, the maximum drain-to-source voltage is

$$(v_{ds})_{max} = V_{cc} + (I_1)_{max}R_L \le 2V_{cc}$$

The efficiency at maximum output power is

$$\eta = \frac{P_o}{P_{cc}} = \frac{I_1^2 R_L}{V_{cc}I_c} = \frac{V_{cc}I_1}{V_{cc}I_c} = \frac{2\theta - \sin 2\theta}{4(\sin\theta - \theta\cos\theta)}$$

where I_c is the dc value of the current [Eq. (11.20)]. The efficiency as a function of conduction angle is plotted in Fig. 11.19. The class C amplifier efficiency can be increased towards 100 percent (in an ideal amplifier) by decreasing the

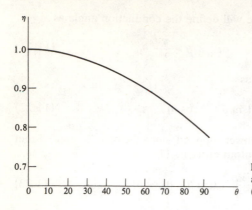

Figure 11.19 A class C amplifier efficiency as a function of current conduction angle (2θ).

conduction angle toward zero. When the conduction angle θ is 90°, the operation is class B and the efficiency is 78.5 percent. The efficiency increases monotonically as the conduction angle decreases. This high efficiency is why class C amplifiers are often used for power amplification.

Class C Power-Amplifier Design

For class C power amplifiers, as for all power amplifiers, the design parameters of greatest importance are the output power, transistor power dissipation, maximum collector-to-emitter (or drain-to-source) voltage, and maximum transistor output current I_p. For the BJT class C amplifier shown in Fig. 11.17b, the maximum collector-to-emitter voltage is

$$(v_{ce})_{max} = 2 V_{cc}$$

The maximum collector current is from Eq. (11.18):

$$I_M = (i_c)_{max} = I_p \sin \frac{\pi}{2} - I_D = I_p - I_p \sin \theta_1$$

and since $\theta = \pi/2 - \theta_1$,

$$I_M = I_p(1 - \cos \theta) \tag{11.27}$$

The peak current is related to the amplitude (I_1) of the fundamental frequency component [using Eq. (11.25)] by

$$I_M = \frac{2\pi I_1(1 - \cos \theta)}{2\theta - \sin 2\theta} \tag{11.28}$$

The ac power output of the amplifier is approximately

$$P_o = \frac{I_1^2 R_L}{2} \tag{11.29}$$

provided the Q of the tuned circuit is sufficiently high. The peak output current is a function of both the collector current and output power. The maximum output power occurs for maximum I_p, and the average (maximum) output power is

$$P_o = \frac{I_1^2 R_L}{2} = \frac{V_{cc}^2}{2R_L} \tag{11.30}$$

since

$$I_1 R_L = V_{cc} \tag{11.31}$$

for maximum output power.

The power dissipated in the transistor is

$$P_T = P_{cc} - P_o = \frac{V_{cc} I_p}{\pi} (\sin \theta - \theta \cos \theta) - \frac{V_{cc}^2}{2R_L}$$

$$= \frac{V_{cc} I_p}{\pi} \frac{\sin \theta - \theta \cos \theta}{1 - \cos \theta} - \frac{V_{cc}^2}{2R_L} \tag{11.32}$$

For a specified load resistance, Eq. (11.30) determines the required supply voltage for a specified output power. The corresponding maximum current I_M is [from Eqs. (11.28) and (11.31)]

$$I_M = \frac{2\pi V_{cc}(1 - \cos \theta)}{R_L(2\theta - \sin 2\theta)} \tag{11.33}$$

A normalized peak collector current is defined as

$$I'_M = \frac{I_M R_L}{2\pi V_{cc}} = \frac{1 - \cos \theta}{2\theta - \sin 2\theta} \tag{11.34}$$

The normalized peak collector current I'_M as a function of conduction angle is plotted in Fig. 11.20. For a fixed level of output power, the peak value of the collector current increases as the conduction angle decreases.

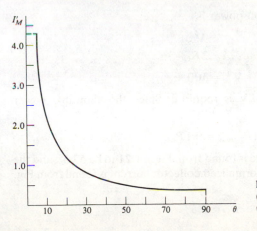

Figure 11.20 Normalized peak collector (or drain) current as a function of conduction angle (2θ).

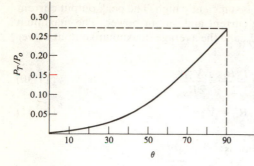

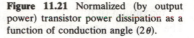

Figure 11.21 Normalized (by output power) transistor power dissipation as a function of conduction angle (2θ).

The transistor dissipation for maximum output power can be expressed as a function of the output power and conduction angle [from Eqs. (11.32) and (11.33)]:

$$P_T = P_o \left[\frac{4(\sin \theta - \theta \cos \theta)}{2\theta - \sin 2\theta} - 1 \right] \qquad (11.35)$$

The normalized transistor dissipation (P_T/P_o) is plotted as a function of the conduction angle in Fig. 11.21. As expected, the transistor dissipation increases with an increasing conduction angle. For a given maximum P_T the conduction angle must be limited to a maximum value for a specified output power. The corresponding maximum value of transistor output current is then determined from Fig. 11.20. As the conduction angle decreases the transistor dissipation decreases, but the peak output current increases.

Example 11.3 Design a class C amplifier that will deliver 5-W average power to a 50-Ω load at the frequency 1 MHz using a transistor with a safe power dissipation rating of 0.5 W.

SOLUTION The average output power is

$$P_o = \frac{V_{cc}^2}{2R_L}$$

so

$$V_{cc}^2 = 500$$

and a supply voltage of 22.4 V is required. Since the allowable power dissipation is

$$(P_T)_{max} = 0.1 P_o$$

the maximum conduction angle is found from Fig. 11.21 to be 57.5°, and the corresponding peak value of normalized collector current is found from Fig. 11.20:

$$I'_M = 0.5$$

Therefore, the peak collector current is

$$I_M = \frac{2\pi \times 22.4}{50} 0.5 = 1.4 \text{ A}$$

The selected transistor must be capable of handling this much current and the collector circuit should be tuned to resonate at 1 MHz.

An alternate design procedure for class C amplifiers is to select the power supply and transistor and then determine the maximum output power possible without exceeding the ratings of the transistor. The transistor can then be driven to its maximum allowed value of output current. From Eq. (11.30) we know that the maximum output power will occur when R_L is at a minimum. R_L can be expressed in terms of the supply voltage V_{cc}, and the peak transistor output current can be expressed as [using Eqs. (11.28) and (11.31)]

$$R_L = \frac{V_{cc}}{I_M} \times 2\pi \frac{1 - \cos\theta}{2\theta - \sin 2\theta}$$

For a specified V_{cc} and I_M, it is seen from Fig. 11.20—a plot of $(1 - \cos\theta)/(2\theta - \sin 2\theta)$—that R_L decreases with an increasing conduction angle. Actually, R_L reaches a minimum for $\theta = 122.6°$ (class AB operation). This is the conduction angle for maximum output power constrained by the peak current limitation. If the maximum transistor dissipation is exceeded at this conduction angle, the conduction angle must be reduced, and the output power will be decreased accordingly.

The transistor dissipation can also be written as [from Eqs. (11.31) and (11.32)]

$$\begin{aligned}
P_T &= \frac{V_{cc}I_M}{\pi} \frac{\sin\theta - \theta\cos\theta}{1 - \cos\theta} - V_{cc}\frac{V_{cc}}{2R_L} \\
&= \frac{V_{cc}I_M}{\pi} \left(\frac{\sin\theta - \theta\cos\theta}{1 - \cos\theta} - \frac{V_{cc}I_1}{2} \right)
\end{aligned}$$

I_1 is given by Eq. (11.28), so

$$P_T = \frac{V_{cc}I_M}{4\pi} \frac{4(\sin\theta - \theta\cos\theta) - (2\theta - \sin 2\theta)}{1 - \cos\theta} = \frac{V_{cc}I_M}{4\pi} f(\theta) \qquad (11.36)$$

The normalized transistor power dissipation

$$P'_T = \frac{4\pi P_T}{V_{cc}I_M} = f(\theta)$$

is plotted as a function of the conduction angle in Fig. 11.22.

From Eq. (11.36) it is seen that for a given V_{cc} and I_M the maximum transistor dissipation occurs for maximum $f(\theta)$.

For a specified P_T, V_{cc}, and I_M, the value of θ that satisfies this equa-

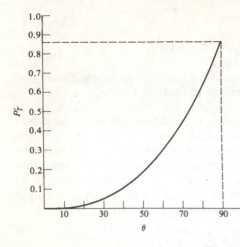

Figure 11.22 Normalized transistor power dissipation as a function of conduction angle (2θ).

tion—that is, determines $(P_T)_{\text{max}}$—is the value of θ for maximum output power. $(P_o)_{\text{max}}$ is then determined from Fig. 11.21.

Example 11.4 Determine the maximum output power and the conduction angle of a class C amplifier using a transistor with a maximum power dissipation rating of 4 W and a maximum drain current of 1.5 A. The supply voltage is 48 V.

SOLUTION The normalized maximum transistor dissipation is

$$P'_T = \frac{4\pi \times 4}{48 \times 1.5} = 0.7$$

From Fig. 11.21 it is found that the maximum possible conduction angle is $\theta = 80°$ without exceeding the maximum transistor dissipation. Figure 11.21 indicates that for this conduction angle the corresponding $P_T / P_o = 0.22$, so the output power is

$$P_o = \frac{4}{0.22} = 18.18 \text{ W}$$

The value of load resistance that results in this output power is determined from Eq. (11.30):

$$R_L = \frac{48^2}{2 \times 18.18} = 63.4 \ \Omega$$

Frequency Multiplication

Since the current pulses of a class C amplifier are rich in the harmonics of the input waveform, the class C amplifier can be used as a frequency multiplier by

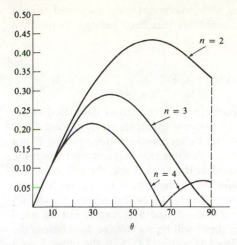

Figure 11.23 Amplitude of output current harmonics as a function of current conduction angle (2θ).

tuning the output circuit to the desired harmonic. The amplitude of the nth harmonic of the output current can be determined from a Fourier expansion of the current waveform. The collector current can be written [using Eqs. (11.18), (11.19), and (11.21)] as

$$i_c = I_p \cos \omega t - I_p \cos \theta$$

so

$$I_n = \frac{4}{T} \int_0^{\theta/\omega} (I_p \cos \omega t - I_p \cos \theta) \cos n\omega t$$

$$= \frac{I_p}{\pi} \left[\frac{\sin(n+1)\theta}{n+1} - \frac{\sin(n-1)\theta}{n-1} - \frac{2\cos\theta \sin n\theta}{n} \right] \qquad n \geq 2$$

$$= \frac{I_p}{\pi} \frac{\cos\theta \sin n\theta - n \sin\theta \cos n\theta}{n(n^2 - 1)} \tag{11.37}$$

The amplitudes of the harmonics as a function of the conduction angle are plotted in Fig. 11.23 (for $n = 2$, 3, and 4).

Example 11.5 A frequency quadrupler is to be designed. What should the conduction angle be to maximize the output signal voltage?

SOLUTION From Fig. 11.23 it is seen that the amplitude of the fourth harmonic has a maximum value for a conduction angle (2θ) of approximately 60°. The output circuit would be tuned to the fourth harmonic of the input signal.

The analysis of class C operation assumed that the current pulses could be modeled as the tip of a sine wave. In many class C applications the transistor will saturate during part of the output cycle; however, a more detailed analysis of this behavior is possible but rarely necessary. Efficiency decreases as saturation increases, therefore stable operation with maximum efficiency and output capability is achieved by driving the amplifier just hard enough to produce saturation of the transistor.

11.5 CLASS D AMPLIFICATION

The main source of power amplifier inefficiency is the power dissipated in the transistor. A class A application is the poorest example, since current flows continuously through the device and the collector-to-emitter voltage is not zero. If the collector-to-emitter (or drain-to-source) voltage is zero when the current flows, no power will be dissipated in the device, and the efficiency will approach 100 percent. This is the basic idea behind class D, E, and S power amplifiers. A class D amplifier is illustrated in Fig. 11.24. Transistors Q_1 and Q_2 operate as switches. When Q_1 is on, Q_2 is off and vice versa. For an ideal transistor with zero saturation voltage, there will be no voltage drop across the transistor, and the circuit can be modeled as in Fig. 11.25. If the input V_i is a square wave, the voltage V_a at the input to the series tuned circuit will be as shown in Fig. 11.26. Since V_a is a square wave, it can be expanded in a Fourier

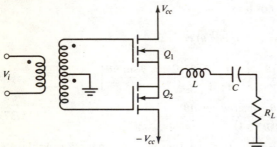

Figure 11.24 A class D amplifier.

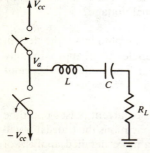

Figure 11.25 The equivalent circuit of a class D amplifier.

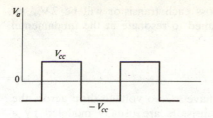

Figure 11.26 Tuned-circuit output voltage wave-form for a square-wave input to the class D amplifier.

series, and the amplitude of the fundamental frequency component is

$$V_1 = \frac{4\,V_{cc}}{\pi}$$

If the output filter is relatively high Q with a center frequency equal to the frequency of the input signal, the drain current in each transistor will be one-half of a sinusoid at the same frequency. Therefore, the direct current in each transistor is

$$I_D = (R_L T)^{-1} \int_0^{T/2} \frac{4\,V_{cc}}{\pi} \sin \omega t \, dt = \frac{4\,V_{cc}}{\pi^2 R_L} \tag{11.38}$$

and the total power supplied is

$$P_i = 2\,V_{cc} I_D = \frac{8\,V_{cc}^2}{\pi^2 R_L}$$

The output power will be

$$P_o = \left(\frac{4\,V_{cc}}{\pi}\right)^2 (2R_L)^{-1} = \frac{8\,V_{cc}^2}{\pi^2 R_L} \tag{11.39}$$

which is the same as the power supplied, so the theoretical efficiency of the ideal class D amplifier is $\eta = 100$ percent.

Example 11.6 Design a class D power amplifier to deliver 20 W to a 50-Ω load.

SOLUTION Since the output power is

$$P_o = \frac{8\,V_{cc}^2}{\pi^2 R_L} = 20\,\text{W}$$

$$V_{cc} \approx 35.1\,\text{V}$$

The direct current in each transistor will be

$$I_{dc} = \frac{4\,V_{cc}}{R_L \pi^2} = 0.285\,\text{A}$$

and the maximum voltage drop across each transistor will be $2V_{cc}$, or 70.2 V. The load circuit would be tuned to resonate at the fundamental frequency of the input signal.

Nonideal Performance

For actual transistors it is impossible to have a zero voltage drop across the device when it is saturated. Bipolar transistors are usually modeled by a saturation voltage V_{sat}, and FETs by an on-resistance R_{on}. The class D FET amplifier can then be modeled as shown in Fig. 11.27. The voltage at the input to the tuned circuit is still a square wave, as illustrated in Fig. 11.26; only the amplitude is reduced. The output voltage will be (again assuming a narrowband filter)

$$V_o = \frac{4V_{cc}R_L}{\pi(R_L + R_{on})} \sin \omega t \qquad (11.40)$$

and the direct current in each transistor will be

$$I_d = \frac{4V_{cc}}{\pi^2(R_L + R_{on})} \qquad (11.41)$$

The total power supplied is

$$P_{cc} = \frac{8V_{cc}^2}{\pi^2(R_L + R_{on})}$$

and the output power is

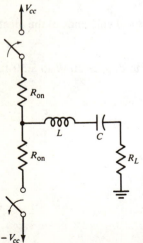

Figure 11.27 Circuit model of a class D FET amplifier.

$$P_o = \frac{8\,V_{cc}^2}{\pi^2}\,\frac{R_L}{(R_L + R_{on})^2} = P_{o_{ideal}}\left(\frac{R_L}{R_L + R_{on}}\right)^2 \qquad (11.42)$$

so the efficiency of the nonideal class D FET amplifier is

$$\eta = \frac{R_L}{R_L + R_{on}}$$

Example 11.7 Calculate the actual power output and efficiency of the class D amplifier used in the previous example if VMOS 2N6659s are used with a gate signal level of 8 V.

SOLUTION The specification sheet for the 2N6659 indicates that the on-resistance is approximately $2\,\Omega$ for this drive level. Therefore, the output power is

$$P_o = 20\left(\frac{R_L}{R_L + R_{on}}\right)^2 = 18.5\,\text{W}$$

and the efficiency is

$$\eta = \frac{50}{52} = 96.2\%$$

11.6 CLASS S POWER AMPLIFIERS

The high efficiency of the class D amplifier has encouraged the design of other switching amplifiers. Another possibility is to pulse-width modulate the information, amplifying the pulses with a high-efficiency amplifier and then demodulating the amplified signal. This is the principle of class S amplification. The classification "class S" is not universal. Some authors interchange the class S and class D classifications, while others refer to class S as "broadband class D." A class S amplifier block diagram is illustrated in Fig. 11.28. The input signal is pulse-width modulated, and the constant amplitude pulses are increased by the pulse amplifier, which is a highly efficient switching amplifier. Pulse frequency modulation, which varies the rate of constant-width pulses, can also be used. It can be shown[11.4] that for pulse-width modulation (with fixed

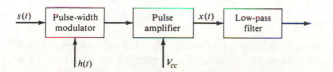

Figure 11.28 Block diagram of a class S amplifier.

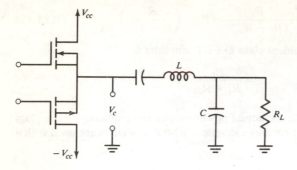

Figure 11.29 Simplified circuit for a class S amplifier.

leading edges) of a sinusoidal signal $S(t) = A \cos \omega_s t$, the modulated signal is

$$X(t) = T^{-1} \left\{ \sum_{m=-\infty}^{\infty} -1^m \frac{e^{jm\omega_o t}}{jm\omega_o} - \sum_{\substack{m=-\infty \\ |m|+|n| \neq 0}}^{\infty} \sum_{n=-\infty}^{\infty} \frac{(-j)^n J_n[A(m\omega_o + n\omega_s)]}{j(m\omega_o + n\omega_s)} + \frac{1}{2} \right\}$$

(11.43)

The modulated signal contains a term at the input signal frequency ($m = 0$, $n = 1$). As long as the modulating frequency is much greater than the signal frequency, the other significant frequency components will be at much higher frequencies and can be removed by low-pass filtering. The significant difference, then, between a class D and a class S power amplifier is that the output circuit of the class D amplifier is tuned to the fundamental frequency of the input signal, while the output circuit of the class S amplifier is a low-pass filter that recovers the input signal. It is important that the amplifier's fluctuations be large enough so that they do not become a factor in determining the pulse width of the modulated signal.

The transistor-switching characteristics of a class S amplifier limit its high-frequency performance. A simplified class S amplifier is illustrated in Fig. 11.29. If the saturation resistance of the transistors is neglected, the efficiency will approach 100 percent; however, the nonzero switching time of the transistors can also affect its efficiency.

The class D amplifier is a nonlinear power amplifier that preserves the frequency (but not the amplitude) of the input signal. The purpose of the circuit is to amplify digital data modulated on a carrier such as frequency-shift-keyed (FSK) data.

Pulse-Width Modulators

Class S amplifiers include a pulse-width modulator. A relatively simple method for obtaining pulse-width modulation consists of using a comparator as a zero-crossing detector to generate a pulse-width modulated signal. One comparator input is a triangular wave $f(t)$, and the other input is the signal to be

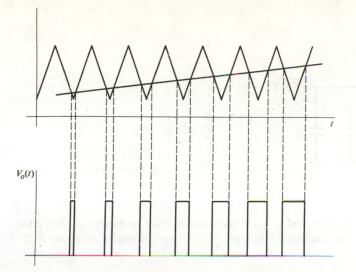

Figure 11.30 Input and output waveforms of a pulse-width modulator.

modulated $V_m(t)$. An example of the modulated signal is shown in Fig. 11.30 for the case in which the signal is a ramp waveform. The comparator output is high whenever $V_m(t) > f(t)$. This particular technique modulates both the leading and trailing edges of the waveform. Single-edge modulation can also be easily implemented with integrated circuitry.

11.7 CLASS E TUNED POWER AMPLIFIERS[11.5]

The class D amplifier utilizes transistors as switches and is a power converter. The input signal waveform is not preserved, just the frequency. In class E operation the transistor acts as a switch, as it does in class D, but only one transistor is used. A simple class E amplifier is illustrated in Fig. 11.31. Class E further differs from class D in that the output tuned circuit is designed to realize certain collector-voltage and current waveform characteristics that are selected to minimize the power dissipated while the transistor is switching from on to off, or vice versa. The transistor switching time can occupy an appreciable fraction of the input signal period during which substantial power can be dissipated in the transistor, reducing the amplifier efficiency. To prevent this problem the output tuned circuit is designed so that (1) the rise of the collector voltage is delayed until after the transistor is turned off, (2) the collector voltage is reduced to zero when the transistor is turned on, and (3) the slope of the collector voltage is zero at the time of turn-on. The analysis of the circuit is as

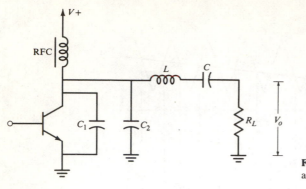

Figure 11.31 A class E amplifier.

yet not well developed, and its main parameters must be further manipulated. An analysis of the idealized amplifier with design formulas is given in the literature.[11.6]

PROBLEMS

11.1 A transistor with a peak current rating of 5 A and a maximum drain-to-source voltage of 50 V is to be used in a class A power amplifier. What is the maximum output power? If a load resistance of 50 Ω is specified, what turns ratio for a matching transformer is required?

11.2 Compare the designs of class A and class B power amplifiers to deliver 20 W to a 50-Ω load. The power supply is 24 V. Specify the maximum transistor dissipation, peak output voltage, and peak output current for each design.

11.3 Calculate the efficiency of the direct-coupled class A power amplifier shown in Fig. P11.3. What is the maximum voltage across the transistor?

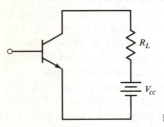

Figure P11.3 A direct-coupled class A amplifier.

11.4 Design a transformerless class B power amplifier using complementary symmetry transistors to deliver 20 W to a 50-Ω load. Specify the maximum transistor dissipation, peak output voltage, and peak output current for each transistor.

11.5 Figure P11.5 illustrates three different methods of biasing a class A amplifier. Compare the efficiency factor

$$\frac{\text{Peak output power}}{\text{Transistor dissipation with no input signal}}$$

for the three configurations.

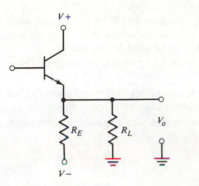
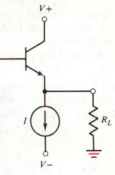
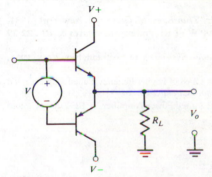

Figure P11.5 Three methods of class A biasing.

11.6 Design a power amplifier using the DV 1260T power MOSFET (see App. 1 for specification). The load resistance is 50 Ω, and is to be coupled to the transistor with a transformer (which is assumed to be 90 percent efficient). What is the maximum possible output power without exceeding the transistor maximum ratings? What are the required supply voltage and current?

11.7 Design a class A push-pull amplifier to deliver 100 W to a 50-Ω load. Use DV1260T MOSPOWER FETs. (App. 1). Specify the supply voltages, bias network, and turns ratios of any transformers used. The source resistance is 500 Ω.

11.8 Design a class C amplifier that will deliver an average power of 20 W at 1 MHz using the DV1260T MOSPOWER FET (see App. 1).

11.9 Determine the maximum output power and conduction angle of a class C amplifier using the DV1260T MOSFET and a 48-V supply voltage (see App. 1).

11.10 If a class C amplifier is to be used as a frequency tripler, what should the conduction angle be for maximum output power?

11.11 Determine the maximum output power of a frequency tripler using the DV1260T MOSFET and a 48-V supply voltage (see App. 1).

11.12 Design a complementary class D amplifier to deliver 100 W to a 50-Ω load using DV1260T transistors (see App. 1). Assume typical transistor values. A 48-V supply is to be used, and an impedance matching transformer can be used for the output.

11.13 What is the efficiency of the amplifier of Prob. 11.12? What will efficiency be if a DV1260T FET is used?

11.14 What is the maximum output power possible using DV1260T transistors in a class D amplifier? (See App. 1). What will be the maximum value of supply voltage?

11.15 Describe a technique for class S amplification that uses single-edge pulse-width modulation.

REFERENCES

11.1 Joyce, M. V., and K. Clarke: *Transistor Circuit Analysis*, Addison-Wesley, Reading, Mass., 1961.
11.2 Evans, A. D. (ed.): *Designing with Field-Effect Transistors*, McGraw-Hill, New York, 1981.
11.3 Granberg, H. O.: Four MOSFETs Deliver 600 W of RF Power, *Microwaves & RF*, **22**:89 (January 1983).
11.4 Rowe, H. E.: *Signals and Noise in Communication Systems*, Van Nostrand, New York, 1965, p. 277.
11.5 Sokal, N. O., and A. D. Sokal: Class E—A New Class of High-Efficiency Tuned Single-Ended Switching Power Amplifiers, *IEEE J. Solid-State Circuits*, **10**:168–176 (1975).
11.6 Raab, F. H.: Idealized Operation of the Class E Tuned Power Amplifier, *IEEE Trans. Circuits and Systems*, **24**:725–735 (1977).

ADDITIONAL READING

Clarke, K., and D. Hess: *Communication Circuits: Analysis and Design*, Addison-Wesley, Reading, Mass., 1971.
Furlan, J.: Preliminary Analysis of the Transistor Tuned Power Amplifier, *Proc. IEEE*, **52**:311 (1964).
Krauss, H. L., C. W. Bostian, and F. H. Raab: *Solid State Radio Engineering*, Wiley, New York, 1980.
Senak, P.: Amplitude Modulation of the Switched-Mode Tuned Power Amplifier, *Proc. IEEE*, **53**:1658–1659 (1965).

TWELVE

MODULATORS AND DEMODULATORS

12.1 INTRODUCTION

Communication systems require circuits for frequency conversion, modulation, and detection. Modulation is the modification of a high-frequency carrier signal to include the information present in a relatively low frequency signal (the modulating signal). The information is modulated onto a higher-frequency signal because radio-wave propagation is more efficient at higher frequencies and smaller antennas can be used. Also, a larger bandwidth can be obtained at the higher frequencies, enabling many information-containing signals to be multiplexed onto one carrier and sent simultaneously.

A practical illustration of this is the composite color-television signal used in the United States that was developed by the National Television Systems Committee (the NTSC system). This television signal includes a frequency-modulated sound signal with a resulting bandwidth of approximately ±25 kHz with 100 percent modulation. The sound signal is then shifted to a center frequency of 4.5 MHz so that it does not overlap with the video signal. The composite signal also includes the video information consisting of the picture signal and vertical and horizontal synchronization pulses with a bandwidth of 4 MHz. The color information, or chrominance signal, which is also contained in the composite signal, is modulated onto a 3.58-MHz subcarrier.

The total signal, then, includes time- and frequency-multiplexed components. This signal is then amplitude-modulated (AM) onto one of the standard broadcast channel carrier frequencies, which creates both upper- and lower-sideband frequency components. For standard TV broadcasting most of the lower sideband is filtered out before transmission. The removal of one sideband reduces the amount of power that must be transmitted and reduces the channel bandwidth, allowing for more channels in a given frequency spectrum.

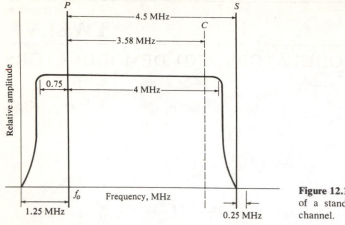

Figure 12.1 Frequency spectra of a standard broadcast TV channel.

The frequency spectra of the transmitted signal are illustrated in Fig. 12.1. The signal bandwidth is 6 MHz, which includes 4.75 MHz above the carrier frequency and 1.25 MHz (the vestigial sideband) below the carrier frequency. The color subcarrier C is 3.58 MHz above the carrier frequency, and the center frequency of the sound signal is 4.5 MHz above the carrier frequency.

Demodulation, then, separates the received information from the high-frequency carrier signal. Radio pioneers used the word "detector" for the device that detected the transmitted information. Today the terms "detector" and "demodulator" are often used interchangeably, but "detector" can have other connotations, such as when applied to an AGC detector or a frequency detector (for automatic frequency-control applications). In this chapter the terms "detector" and "demodulator" are used interchangeably, but it will be clear from the context in which the words are used which type of detector is being discussed.

In the following sections we will discuss the various types of mixers, modulators, and demodulators, and their respective applications. The differences between time-varying and time-invariant circuits are also considered.

12.2 FREQUENCY MIXERS

The most commonly used device for frequency modification is referred to as a *frequency mixer*, or simply *mixer*. The ideal mixer, represented by the schematic shown in Fig. 12.2, is a device which multiplies two input signals. If the inputs are sinusoids, the ideal mixer output is

$$V_o = A_1 \sin \omega_1 t (A_2 \sin \omega_2 t) = \frac{A_1 A_2}{2} [\cos (\omega_1 - \omega_2)t - \cos (\omega_1 + \omega_2)t] \quad (12.1)$$

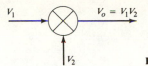

$$V_o = V_1 V_2$$

Figure 12.2 Circuit symbol of a four-quadrant multiplier.

The output consists of the sum and difference frequencies of the two input frequencies, one of which is the desired component. The other input frequency is removed with filtering—this combination of a mixer and filter to remove an output frequency is often referred to as a *single-sideband mixer*.

Although the ideal mixer does not exist, there are many different circuits for approximating the ideal mixer. There are mixer circuits which provide gain (active mixers) and passive mixers which actually have a conversion loss. (The lowest noise mixers are passive mixers.) Mixers can be classified in various ways, but in this text they are classified on the basis of the mode of operation—namely, nonlinear or switching-type mixers.

Switching-Type Mixers

In switching-type mixers, one or more switches, realized with diodes or transistors, function as time-varying circuit elements. The nonlinear or switching characteristics of diodes are often used for frequency mixing, particularly at high frequencies. Figure 12.3 provides an example of a simple switching-type mixer circuit containing diodes. If the center-tapped transformer is ideal, then the voltages will be as indicated in Fig. 12.4. The local oscillator V_L is a constant-amplitude signal. The idea is for the local oscillator signal to be much larger than V_i so that diode D_1 is on when V_L is positive and diode D_2 is on when V_L is negative. Therefore, the output voltage will be (ignoring the drop across the diode)

$$V_o = V_i + V_L \qquad V_L > 0$$

and

$$V_o = -V_i + V_L \qquad V_L < 0$$

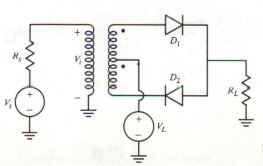

Figure 12.3 Circuit symbol of a simple two-diode switching-type mixer.

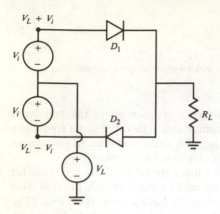

Figure 12.4 Simplified equivalent circuit of the two-diode mixer.

That is, the output consists of the local oscillator signal plus V_i switched by 180° at the frequency of the local oscillator. If we consider this switched form of V_i as V_i^*, then

$$V_o = V_L + V_i^*$$

where

$$V_i^* = V_i P(t)$$

$$P(t) = 1 \qquad V_L > 0$$

and

$$P(t) = -1 \qquad V_L < 0$$

$P(t)$ is a square wave (illustrated in Fig. 12.5) with a frequency equal to that of the local oscillator frequency ω_L. It can be expanded in a Fourier series as

$$P(t) = \frac{4}{\pi} \sum_{n=0}^{\infty} \frac{\sin(2n+1)\omega_L t}{2n+1} \tag{12.2}$$

so

$$V_i^* = V_i \left[\frac{4}{\pi} \sum_{n=0}^{\infty} \frac{\sin(2n+1)\omega_L t}{2n+1} \right]$$

If V_i is a sine wave,

$$V_i = V \sin \omega_i t$$

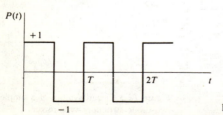

Figure 12.5 A local oscillator waveform.

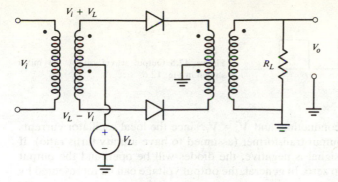

Figure 12.6 A two-diode mixer in which the load oscillator signal does not appear in the output.

then $\qquad V_i^* = \dfrac{2V}{\pi} \displaystyle\sum_{n=0}^{\infty} \dfrac{\cos\left[(2n+1)\omega_L - \omega_i\right]t - \cos\left[(2n+1)\omega_L + \omega_i\right]t}{2n+1}$ $\qquad$ (12.3)

Since $V_o = V_L + V_i^*$, the mixer output consists of the local oscillator signal plus an infinite number of additional frequencies created in the mixer. The output frequencies in addition to the upper and lower sidebands are called *spurious*. The desired mixing component is selected from the mixer by filtering the output.

The preceding analysis assumed that the local oscillator signal was much larger than the input signal and sufficiently large to instantaneously switch the diodes. Any deviation from these assumptions increases the distortion of the desired frequency component. Distortion is considered further in the discussion of double-balanced mixers. A disadvantage of the mixer circuit shown in Fig. 12.3 is that the local oscillator signal appears in the output. If the local oscillator frequency is much larger than the input frequency, then the desired mixing product $\omega_L + \omega_i$ or $\omega_L - \omega_i$ may be close to the local oscillator frequency, and it will be difficult to separate the frequencies by filtering.

The local oscillator signal does not appear in the output of the mixer circuit shown in Fig. 12.6. If the center-tapped transformer is ideal, then the voltages will be as shown in Fig. 12.7. If V_L is positive and much larger than V_i, then

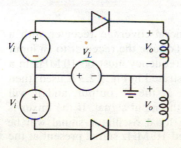

Figure 12.7 Simplified equivalent circuit of the diode mixer shown in Fig. 12.6.

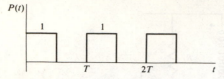

Figure 12.8 Output waveform of the mixer shown in Fig. 12.6.

both diodes will be conducting, and $V_o = V_i$, since the local oscillator currents balance out in the output transformer (assumed to have a unity turns ratio). If the local oscillator signal is negative, the diodes will be open and the output signal will be equal to zero. In general, the output voltage can be represented by

$$V_o = V_i P(t)$$

where

$$P(t) = 1 \qquad V_L > 0$$

and

$$P(t) = 0 \qquad V_L \leq 0$$

In this case $P(t)$ is a square wave (illustrated in Fig. 12.8) with a frequency equal to that of the local oscillator frequency, but it differs from the preceding case in that the dc value is no longer equal to zero. $P(t)$ can be expanded in a Fourier series as

$$P(t) = \frac{1}{2} + \frac{2}{\pi} \sum_{n=0}^{\infty} \frac{\sin (2n+1)\omega_L t}{2n+1} \qquad (12.4)$$

If V_i is a sine wave

$$V_i = V \sin \omega_i t$$

then the output voltage is

$$V_o(t) = V \frac{\sin \omega_i t}{2} + \frac{V}{\pi} \sum_{n=0}^{\infty} \frac{\cos [(2n+1)\omega_L - \omega_i] t - \cos [(2n+1)\omega_L + \omega_i] t}{2n+1} \qquad (12.5)$$

The output of this mixer differs from that shown in Fig. 12.3 in that it does not contain the local oscillator signal, but it does contain a signal at the same frequency as the input signal. In some applications this can be a problem, as illustrated by the following example.

Example 12.1 The input section of a general-coverage receiver with a 10-MHz IF filter is illustrated in Fig. 12.9. To tune the receiver to an input frequency of 20 MHz, the local oscillator frequency must be 10 MHz in a down-conversion receiver. If the mixer illustrated in Fig. 12.3 is used, then the local oscillator signal will be present at the IF output, and it will probably be so large that it will mask out the input signal. If the mixer is connected as illustrated in Fig. 12.6, the local oscillator signal will be removed from the output. Still, an unwanted 10-MHz signal present at the

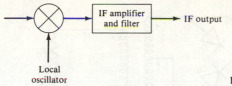

IF output

Local
oscillator

Figure 12.9 Receiver input section.

input will appear at the IF output unless it is removed by an input filter
(preselector).

A double-balanced mixer circuit, which can be used with a balanced load, is
shown in Fig. 12.10. The behavior of this circuit is the same as that of the mixer
illustrated in Fig. 12.6, and the output is given by Eq. (12.5) if one-to-one
transformers are used.

A switching-type mixer containing four diodes, in which neither the local
oscillator signal nor the input signal appears at the output, is shown in Fig.
12.11. If the local oscillator signal V_L is positive, diodes D_2 and D_3 will
conduct, and the equivalent circuit will be as shown in Fig. 12.12, with r_d
representing the equivalent diode on-resistance. The circuit is more easily
recognized if it is redrawn as shown in Fig. 12.13; the balanced output
transformer eliminates the local oscillator signal from the load. The two loop
equations are

$$V_i = (I_1 + I_2)R_L + I_1 r_d - V_L$$

and

$$V_i = (I_1 + I_2)R_L + I_2 r_d + V_L$$

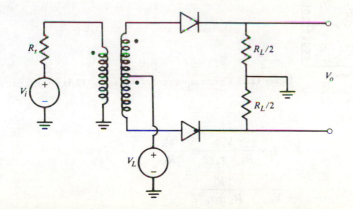

Figure 12.10 A double-balanced mixer.

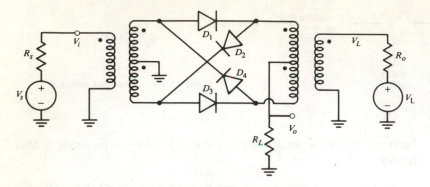

Figure 12.11 A four-diode switching-type mixer.

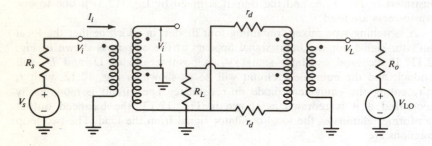

Figure 12.12 Equivalent circuit of Fig. 12.11 (for positive local oscillator voltage).

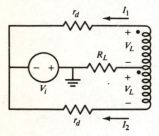

Figure 12.13 Circuit equivalent of that in Fig. 12.11.

If V_L is eliminated,

$$I_1 + I_2 = \frac{V_i}{R_L + r_d/2} = -\frac{V_o}{R_L}$$

or

$$\frac{V_o}{V_i} = -\frac{R_L}{R_L + r_d/2}$$

If the local oscillator signal is negative, diodes D_1 and D_4 conduct, and the

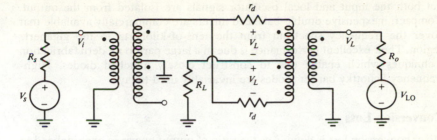

Figure 12.14 Equivalent circuit of Fig. 12.11 (for negative local oscillator voltage).

equivalent circuit is as shown in Figs. 12.14 and 12.15. The output voltage is then

$$V_o = \frac{V_i R_L}{R_L + r_d/2}$$

In this mixer the output voltage is proportional to the input voltage and is switched at the local oscillator frequency; therefore

$$V_o(t) = V_i P(t) \frac{R_L}{R_L + r_d/2}$$

where $P(t)$ is given by Eq. (12.4). If $V_i(t)$ is a sine wave

$$V_i = V \sin \omega_i t$$

then

$$V_o(t) = \frac{R_L}{R_L + r_d/2} \left\{ \frac{2V}{\pi} \sum_{n=0}^{\infty} \frac{\cos[(2n+1)\omega_L - \omega_i]t - \cos[(2n+1)\omega_L + \omega_i]t}{2n+1} \right\}$$

$$(12.6)$$

A double-balanced mixer with perfectly matched diodes and ideal transformer coupling will generate the upper and lower sidebands plus an infinite number of spurious centered on odd harmonics of the local oscillator frequency,

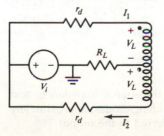

Figure 12.15 Circuit equivalent of that in Fig. 12.14.

but both the input and local oscillator signals are isolated from the output. Compact, inexpensive double-balanced mixers are commercially available that cover the frequency spectrum from the tens-of-kilohertz to the gigahertz region. Their excellent performance is due in a large part to modern fabrication techniques which enable one to construct closely matched diodes. High-frequency Schottky barrier diodes are invariably used today.

Conversion Loss

Mixer conversion loss is defined as the ratio of output power in one sideband to signal input power. It is a most important mixer parameter, particularly for the receiver input stage. To calculate the conversion loss, we will assume that the external impedances are adjusted for maximum power transfer. Consider first the double-balanced mixer circuit shown in Fig. 12.11. If the input transformer has a one-to-one turns ratio, then the equivalent circuit is as shown in Fig. 12.13, and the load impedance seen by V_i is

$$\frac{V_i}{I_1 + I_2} = \frac{V_i}{I_i} = R_L + \frac{r_d}{2}$$

Normally $R_L \gg r_d$, so the input will be matched for maximum power transfer if $R_L = R_s$. Under this condition $V_i = V_s/2$ and

$$P_i = \frac{V_s^2}{4 R_L}$$

From Eq. (12.6) we see that the output voltage in one sideband is (assuming $R_L \gg r_d$)

$$V_o\big|_{\omega_L \pm \omega_i} = \frac{V_i \times 2}{\pi} = \frac{V_s}{\pi}$$

and the output power is

$$P_o = \frac{V_s^2}{\pi^2 R_L}$$

So the conversion gain of the double-balanced mixer is

$$G = \frac{P_o}{P_i} = \frac{4 R_L}{\pi^2 R_L} = \frac{4}{\pi^2} \tag{12.7}$$

which is less than 1. The mixer has a conversion loss of

$$L = 10 \log \frac{\pi^2}{4} \approx 4 \text{ dB}$$

For an ideal double-balanced mixer matched to the source impedance, and ignoring the power lost in the transformer and switching diodes, approximately 40 percent of the signal input power will be transferred to the output.

For the single-balanced mixer shown in Fig. 12.5 the output voltage of one sideband is [from Eq. (12.5)]

$$V_o\big|_{\omega_L+\omega_i} = \frac{V_i}{\pi}$$

If the input port is matched for maximum power transfer,

$$V_o = \frac{V_s}{2\pi}$$

$$P_i = \frac{V_s^2}{4R_L}$$

and

$$P_o = \frac{V_s^2}{4\pi^2 R_L}$$

The power gain is

$$G = \frac{P_o}{P_i} = (\pi^2)^{-1} \tag{12.8}$$

so the conversion loss is $L = 10\log \pi^2 \approx 10\,\text{dB}$, and it is seen that the conversion loss of the single-balanced mixer is four times (6 dB) larger than that of the double-balanced mixer.

Distortion

As the mixer input signal power increases, it will reach the level at which it is larger than the local oscillator power. The input signal then assumes the switching role, and the output power becomes proportional to the local oscillator power. Since the local oscillator power is constant, the output power will be constant. An idealized power transfer characteristic is shown in Fig. 12.16. At low input-power levels the power transfer is linear, but as the input power increases, distortion begins and the response becomes nonlinear. At high

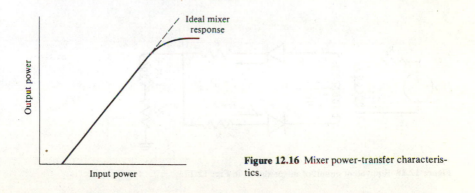

Figure 12.16 Mixer power-transfer characteristics.

input levels the output saturates at a level proportional to the local oscillator power. As the signal level further increases, the intermodulation distortion (IMD) also increases.

Intermodulation Distortion in Diode-Ring Mixers[12.1]

Consider a diode-ring mixer with a resistance R in series with each diode, as shown in Fig. 12.17. The purpose of the additional resistors will become clear once the IMD is determined. The transformers are assumed to be ideal, and the diodes all have identical characteristics. If the local oscillator power is sufficiently large, the circuit during either half-cycle is as shown in Fig. 12.18. The diode current then consists of a constant component I, due to the local oscillator, and a small component i, due to the input signal. The diode current is

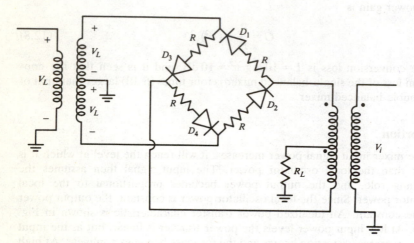

Figure 12.17 A four-diode mixer with linearizing resistors.

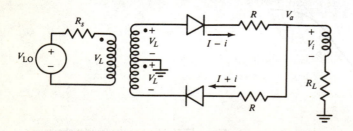

Figure 12.18 Equivalent circuit of mixer shown in Fig. 12.17.

described by

$$i_D = I_s e^{V_d/V_T}$$

where V_d is the voltage drop across the diode and $V_T = kT/q$.

The input signal V_i causes a signal current $(2i)$ to flow through the load. Because of the circuit symmetry one half flows through each diode. That is,

$$i_{D_1} = I - i$$

and

$$i_{D_2} = I + i$$

The currents are shown in Fig. 12.18. The voltage equations are

$$V_L - V_a = V_{D_1} + i_{D_1}R$$

and

$$-V_L - V_a = -V_{D_2} - i_{D_2}R$$

or

$$V_L - V_a = V_T \ln \frac{I - i}{I_s} + (I - i)R$$

and

$$-V_L - V_a = -\left[V_T \ln \frac{I + i}{I_s} + (I + i)R \right]$$

Adding the two equations, one obtains

$$-2 V_a = V_T \ln \frac{I - i}{I + i} - 2iR$$

and since

$$2 V_a = 2(V_i - 2iR_L)$$

the relation between input voltage and diode current is

$$V_i = i(2R_L + R) + \frac{V_T}{2} \ln \frac{I + i}{I - 1}$$

This can be expanded for $(i \ll I)$

$$V_i = (2R_L + R)i + \frac{V_T}{2} \left[\frac{i}{I} - \frac{1}{2}\left(\frac{i}{I}\right)^2 + \frac{1}{3}\left(\frac{i}{I}\right)^3 \left(\frac{i}{I}\right)^3 + \cdots \right.$$

$$\left. - \left[-\frac{i}{I} - \frac{1}{2}\left(\frac{i}{I}\right)^2 - \frac{1}{3}\left(\frac{i}{I}\right)^3 + \text{higher-order terms} \ldots \right] \right]$$

The even-order terms cancel, so

$$V_i = \left(2R_L + R + \frac{V_T}{I}\right)i + \frac{1}{3}\left(\frac{i}{I}\right)^3 + \text{odd-higher order terms}$$

Since the first term of the power series is not zero, the series can be inverted:[12.2]

$$i = \frac{V_i}{2R_L + R} - \frac{V_T}{3} \frac{V_i^3}{(2R_L + R)^4 I^3}$$

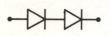

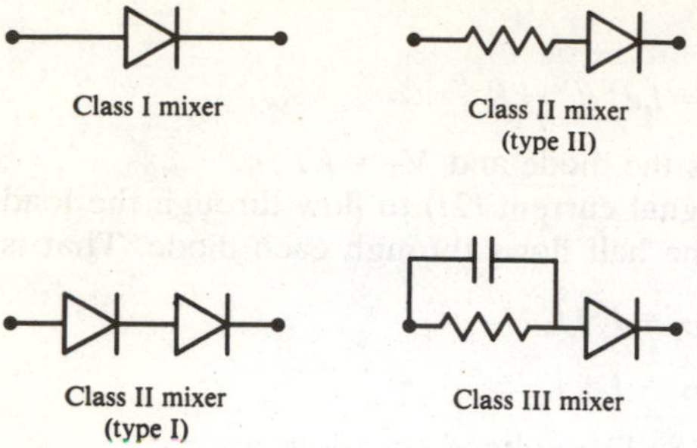

Class I mixer

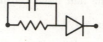

Class II mixer
(type II)

Class II mixer
(type I)

Class III mixer

Figure 12.19 Three classes of diode-ring mixers. (*After Cheadle, 1973.*)

Comparing this equation with Eq. (3.64) we see that

$$k_3 = -\frac{V_T}{3(2R_L + R)^4 I^3} \tag{12.9}$$

The third-order intermodulation distortion is proportional to k_3^2 [Eq. (2.60)], so the smaller the magnitude of k_3, the smaller will be the IMD. The IMD is inversely proportional to the sixth power of the current I_1, so the larger the local oscillator drive level, the smaller will be the IMD. The addition of a resistor R in series with the diode reduces the IMD. An additional diode can be used in place of the resistor. This permits higher local oscillator drive levels and a corresponding reduction in IMD.

Diode-ring mixers are classified by the different elements in each leg. Figure 12.19 illustrates three classes of diode-ring mixers. The class II mixers have the highest power-handling capabilities, but because of the additional matched components, they are also the most expensive to manufacture. The capacitor in parallel with the series resistor in the class III diodes acts to reduce the distortion that occurs when the local oscillator signal switches state; it also reduces conversion loss by acting as a high-frequency bypass for the signal voltage. At high frequencies the effects of finite switching time also cause distortion. It can be shown that for the same power levels, a square-wave signal will cause less distortion than will be the case if a sine-wave signal is used for the local oscillator.[12.3]

Square-Law Mixers

The square-law characteristic is approximated by several electronic devices. That a square-law device can function as a mixer is readily seen by squaring the sum of two sine waves:

$$(A_1 \sin \omega_1 t + A_2 \sin \omega_2 t)^2 = (A_1^2 \sin \omega_1 t)^2 + (A_2 \sin \omega_2 t)^2 + 2A_1 A_2 \sin \omega_1 t \sin \omega_2 t$$

$$= \frac{A_1^2(1 - \cos 2\omega_1 t)}{2} + \frac{A_2^2(1 - \cos 2\omega_2 t)}{2}$$

$$+ \frac{2A_1 A_2[\cos (\omega_1 - \omega_2)t - \cos (\omega_1 + \omega_2)t]}{2} \tag{12.10}$$

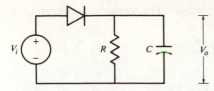

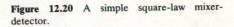

Figure 12.20 A simple square-law mixer-detector.

An ideal square-law device will provide the upper and lower sidebands, together with a dc. component and the second harmonic of both input waveforms. The circuit is frequently used at microwave frequencies for down conversion to the lower sideband, which is at a lower frequency than either of the input signals. A simple square-law mixer, which has been used since the earliest radio receivers were built, is shown in Fig. 12.20. Crystal diodes were originally used, but today Schottky barrier diodes are the best choice.

At lower frequencies this form of the diode mixing is normally not used because of the large conversion loss. Transistor mixers are preferred because they can provide conversion gain. Transistors are often used to approximate the square-law characteristic. The input and local oscillator signal voltages are applied to the transistor so that they effectively add to the dc bias voltage to produce the total gate-source or base-emitter voltage. The composite signal is then passed through the device nonlinearity to create the sum and difference frequencies.

BJT Mixers

Figure 12.21 illustrates a simplified bipolar transistor mixer. For this circuit the base-to-emitter voltage is

$$V_{be} = V_{DC} + V_i - V_L$$

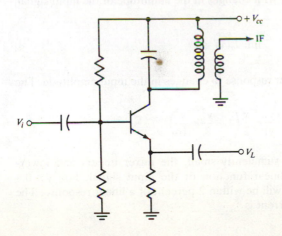

Figure 12.21 A bipolar transistor fixer.

where V_{DC} is the base-to-emitter bias voltage. The collector current in a bipolar transistor is described by ($V_{be} > 0$)

$$i_C = I_S e^{V_{be}/V_T}$$

and since

$$V_{be} = V_{DC} + V_i - V_L$$

the current is

$$i_C = I_S e^{V_{DC}/V_T} e^{V_i/V_T} e^{-V_L/V_T}$$

If $V_i = V_1 \cos \omega_i t$ and $V_L = V_L \cos \omega_o t$, then the current can be expanded in a series of modified Bessel functions[12.4] as

$$\begin{aligned}
i_C(t) = I_S e^{V_{DC}/V_T} [& I_o(y) I_o(x) - 2 I_o(y) I_1(x) \cos \omega_o t \\
& + 2 I_1(y) I_o(x) \cos \omega_i t - 4 I_1(y) I_1(x) \cos \omega_i t \cos \omega_o t \\
& + \text{higher-order terms}]
\end{aligned} \qquad (12.11)$$

where $y = V_1/V_T$, $x = V_L/V_T$, and I_n is the nth order modified Bessel function.

The collector current consists of a dc component I_C, components at both the input and local oscillator frequencies, components at the frequencies $\omega_o \pm \omega_i$, and an infinite number of high-frequency components. The amplitude of either the upper- or lower-sideband component is

$$\begin{aligned}
I &= I_S e^{V_{DC}/V_T} 2 I_1'(y) I_1(x) \\
&= 2 I_C \frac{I_1(y) I_1(x)}{I_o(y) I_o(x)}
\end{aligned} \qquad (12.12)$$

The local oscillator voltage amplitude is constant, and if $V_L \gg V_1$, then the dc collector current will not vary with changes in the amplitude of the input signal, since

$$\lim_{y \to 0} I_o(y) = 1$$

The mixer should have a linear response to changes in the input amplitude. The ratio is given as

$$\frac{I_1(y)}{I_o(y)} \approx \frac{y}{2} \left(1 - \frac{y^2}{8} + \frac{y^4}{16} \right)$$

so if the input amplitude is sufficiently small, the mixer upper- and lower-sideband outputs will be a linear function of the input signal. For $y \leq 0.4$ ($V_1 \leq 10.5$ mV) the response will be within 2 percent of a linear response. The amplitude of the sideband current is

$$I = 2I_C \frac{I_1(y)}{I_o(y)} \frac{I_1(x)}{I_o(x)}$$

$$\approx I_C y \frac{I_1(x)}{I_o(x)} = g_m V_1 \frac{I_1(x)}{I_o(x)} \tag{12.13}$$

provided the input signal is sufficiently small. g_m is the transconductance I_C/V_T. The ratio $I_1(x)/I_o(x)$ rapidly approaches 1 as x increases:

$$\lim_{x \to \infty} \frac{I_1(x)}{I_o(x)} = 1$$

and

$$\frac{I_1(x)}{I_o(x)} \approx 0.86 \qquad \text{for } x = 4$$

So if the local oscillator drive level is approximately 100 mV ($x \approx 4$) or larger, the amplitude of the upper- and lower-sideband current components is $I \approx g_m V_1$. If the collector circuit is tuned to either of these frequencies the mixer conversion gain $G = g_m R_L$, where R_L is the equivalent load resistance at the frequency of interest. The BJT mixer has the advantage over diode mixers in that it provides conversion gain, although it will be much noisier than a properly designed diode-ring-type mixer. Another advantage of the BJT mixer over diode mixers is that the local oscillator drive level required is much smaller. This reduces oscillator design and system-shielding requirements.

FET Mixers

If a FET is operated in its "constant-current" region, the idealized FET current transfer characteristic is the square-law relation

$$i_D = I_{DSS} \left(1 - \frac{V_{gs}}{V_p}\right)^2 \tag{12.14}$$

where V_{gs} is the gate-to-source voltage and V_p is the transistor pinch-off voltage. Because of the square-law characteristic, the FET will not generate any harmonics higher than second-order. Since the third-order intermodulation distortion arises from the cubic term in the transfer characteristic, an ideal FET mixer will not produce any third-order intermodulation distortion. However, in reality the transfer characteristic deviates from the idealized version, and some intermodulation distortion will be produced. Still, a properly biased and operated FET mixer will produce much smaller high-order mixing products than a bipolar transistor. This is one reason why a FET mixer is usually preferred over a bipolar transistor mixer. The FET also provides at least 10 times as great an input voltage range as the BJT. Figure 12.22 illustrates a FET mixer circuit. The drain current is (in the constant-current region)

$$i_D = I_{DSS} \left(1 - \frac{v_i - v_L + V_{DC}}{V_p}\right)^2 \tag{12.15}$$

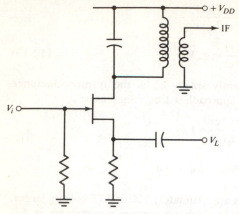

Figure 12.22 A FET mixer circuit.

where V_{DC} is the gate-to-source bias voltage (or $V_{GS} - V_T$ for a MOSFET). If the applied signals are sine waves

$$v_i = V_i \sin \omega_i t$$

and

$$v_L = V_L \sin \omega_L t$$

the output current is [obtained by expanding Eq. (12.15)]

$$i_D = I_{DC} + K_1(V_i \sin \omega_i t - V_L \sin \omega_L t) + K_2(V_i \sin 2\omega_i t + V_L \sin 2\omega_L t)$$

$$- K_3 \sin (\omega_i - \omega_L)t + K_3 \sin (\omega_i + \omega_L)t \qquad (12.16)$$

The amplitude of the sum and difference frequencies is

$$K_3 = \frac{I_{DSS} V_i V_L}{V_p^2}$$

The term

$$\frac{K_3}{V_i} = \frac{I_{DSS} V_L}{V_p^2} \qquad (12.17)$$

is referred to as the *conversion transconductance* g_c. In general, the device with the lowest pinch-off voltage has the highest gain, and the conversion transconductance is directly proportional to the amplitude of the local oscillator signal. It would also appear that FETs with high I_{DSS} are preferred, but I_{DSS} and V_p are related. It is usually the case that devices selected for high I_{DSS} also have a high V_p and a lower conversion transconductance than low-I_{DSS} devices. Since the device is to be operated in the constant-current region, V_L must be less than the magnitude of the pinch-off voltage. If

$$V_L = \frac{|V_p|}{2}$$

then
$$K_3 = \frac{I_{DSS}}{2\,V_p}\,V_i$$

and the sideband current is

$$i_D = K_3 \sin\left(\omega_i \pm \omega_2\right)t = V_1 \frac{I_{DSS}}{2\,V_p}\left[\sin\left(\omega_i \pm \omega_1\right)t\right]$$

Since for a JFET the transconductance is

$$g_m = \frac{\partial i_D}{\partial V_{gs}} = -2\,\frac{I_{DSS}}{V_p}\left(1 - \frac{V_{gs}}{V_p}\right)$$

$$= -2\,\frac{I_{DSS}}{V_p}\bigg|_{V_{gs}=0} \tag{12.18}$$

the conversion transconductance is one-fourth the small-signal transconductance evaluated at $V_{gs} = 0$ (provided $V_L = V_p/2$). For a MOSFET it can be shown that the conversion conductance cannot exceed one-half of the transconductance of the device when it is used as a small-signal amplifier.

Example 12.2 A FET with $I_{DSS} = 40$ mA and transconductance $g_m = 14 \times 10^{-3}$ S at $V_{gs} = 0$ is to be used in a mixer. Estimate the conversion gain if a 50-Ω load is used.

SOLUTION Since

$$g_m = \frac{-2I_{DSS}}{V_p}\left(1 - \frac{V_{gs}}{V_p}\right)$$

the transistor pinch-off voltage is

$$|V_p| = \frac{2 \times 40 \times 10^{-3}}{14 \times 10^{-3}} = 5.7 \text{ V}$$

The peak local oscillator voltage V_L should be kept to approximately 50 percent of this value in order to minimize distortion. If $V_L \approx 2.85$ V, then the conversion transconductance is found from Eq. (12.18) to be

$$g_c = \frac{40 \times 10^{-3}}{2 \times 5.7} = 3.5 \times 10^{-3} \text{ S}$$

The magnitude of the voltage gain is

$$A_v \approx g_c R_L = 1.75$$

If the device is operated in the common-gate configuration, the current gain will be close to unity, so the voltage gain will also be equal to the power gain.

Although the conversion transconductance is smaller than the small-signal

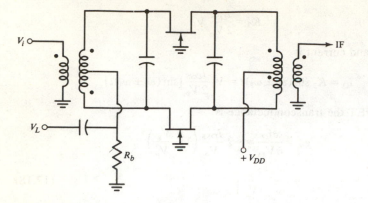

Figure 12.23 A FET double-balanced mixer.

transconductance, it is large enough that the circuit can be operated as a mixer with power and voltage gains. This is an important difference from that of the diode-switching mixer. A FET mixer is capable of producing lower intermodulation and harmonic products than a comparable bipolar or diode mixer. Also, a FET mixer operating at a high level has a larger dynamic range and greater signal-handling capacity than does a diode mixer operated at the same local oscillator level. However, the noise figure of FET mixers is currently higher than that of diode mixers. The best intermodulation and cross-modulation performance is obtained with the FET operated in the common-gate configuration, where the input impedance is much lower than for the common-source configuration. Figure 12.23 illustrates a double-balanced mixer in which the FET transistors are operated in the common-gate configuration. The push-pull output cancels the even-order output harmonics.

The dual-gate MOSFET is often used as a mixer. A typical dual-gate MOSFET mixer circuit is illustrated in Fig. 12.24. If the input signals are sinusoidal, the output will contain frequency components at both the sum and difference frequencies. Several other frequency components are also present in the output. The magnitude of either the sum or difference frequency is proportional to

$$A = K|V_{g_2}|$$

so the conversion gain is proportional to the magnitude of the local oscillator voltage. For maximum conversion gain the local oscillator amplitude should be selected so that it drives the gate just to the point of transistor saturation.

The input signal is normally connected to the lower (closest to the ground) input gate terminal and the local oscillator signal to the upper gate. If the input is connected to the upper terminal then the drain resistance of the lower

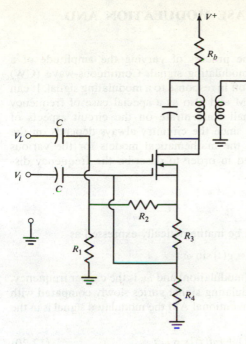

Figure 12.24 A dual-gate MOSFET mixer circuit.

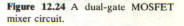

transistor section appears as a source resistance to the input signal. It has been shown in Chap. 2 that the source resistance will reduce the voltage gain at the collector. Also, the connection has a larger drain-to-gate capacitance with a lower bandwidth than is attainable when the input signal is connected to the lower gate. The device is usually biased so that both transistors are operating in their triode (nonsaturated) region. In this region it can be shown[12.5] that the small-signal drain current is

$$i_d = g_{m_1} V_{g_1} + g_{m_2} V_{g_2}$$

where

$$g_{m_1} = a_o + a_1 V_{g_1} - a_2 V_{g_2}$$

and

$$g_{m_2} = b_o + b_1 V_{g_1} + b_2 V_{g_2}$$

The drain current can be written as

$$i_d = a_1 V_{g_1}^2 + a_o V_{g_1} + (a_2 + b_1) V_{g_1} V_{g_2} + b_o V_{g_2} + b_2 V_{g_2}^2 \qquad (12.19)$$

Since the drain current contains the product of the two signals, the dual-gate MOSFET can be used as a mixer when both transistors are operated in the nonsaturated region.

12.3 AMPLITUDE AND PHASE MODULATION AND DEMODULATION

Amplitude modulation (AM) is the process of varying the amplitude of a constant frequency signal with a modulating signal. Continuous-wave (CW) modulations turn the carrier on and off in response to a modulating signal. It can be considered a special case of AM and also as a special case of frequency modulation. In this section we shall concentrate on the circuit aspects of modulators and demodulators, but since the circuitry always depends on the frequency spectrum of the signal, the mathematical models for the various modulation methods are introduced in order to describe the frequency distributions of the modulated signals.

Amplitude Modulation

An amplitude-modulated wave can be mathematically expressed as

$$S(t) = g(t) \sin \omega_c t$$

where $g(t)$ is the modulating signal (modulation) and ω_c is the carrier frequency. It is normally the case that the modulating signal varies slowly compared with the carrier signal frequency. For conventional AM the modulated signal is in the form of

$$S(t) = A[1 + mf(t)] \sin \omega_c t \tag{12.20}$$

m is called the *modulation factor* and is normally less than 1; $100m$ is the *percent modulation*. Consider a simple modulating signal:

$$f(t) = \cos \omega_m t$$

then
$$S(t) = A \left\{ \sin \omega_c t + \frac{m}{2} [\sin (\omega_c + \omega_m)t + \sin (\omega_c - \omega_m)t] \right\} \tag{12.21}$$

The frequency spectrum of the modulated signal is shown in Fig. 12.25. It consists of the carrier frequency plus upper- and lower-sideband components

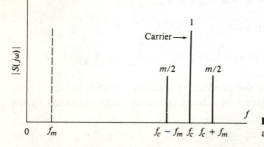

Figure 12.25 Frequency spectrum of an amplitude-modulated signal.

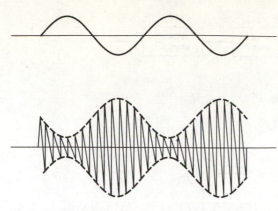

Figure 12.26 Amplitude-modulating and -modulated signals.

centered about the carrier frequency. This signal is often referred to as the *full-carrier double-sideband signal*. Figure 12.26 illustrates a sinusoidal modulating signal and the carrier amplitude modulated by the sinusoid. The signal envelope is of the same form as the modulating signal ($m < 1$).

Equation (12.21) shows that for $m < 1$ the amplitude of the carrier is at least twice as large as the amplitude of either sideband component, so at least two-thirds of the signal power will be in the carrier and at most one-third in the two sidebands. Because the carrier does not contain any modulating information, it is often removed or suppressed, resulting in the signal

$$S(t) = \frac{Am}{2}[\sin(\omega_c + \omega_m)t + \sin(\omega_c - \omega_m)t] \qquad (12.22)$$

which is referred to as a *double-sideband suppressed-carrier* (*DSB*) *signal*. The carrier component is not present in the DSB signal. It is the case, however, that as the waveform gets more efficient in terms of power-to-information content the detection method gets more complex. Some means of recovering the carrier component is needed for the detector to recover the amplitude and frequency of the modulating signal. The DSB signal, although more efficient in terms of transmitted power, still occupies the same bandwidth as normal AM signal. Since both sidebands contain the same information, one sideband can be removed, resulting in a *single-sideband-signal* (*SSB*). SSB is the most efficient form of an amplitude-modulated signal. Circuitry for the three types of AM will now be described.

Amplitude Modulators: Standard AM

Full-carrier double-sideband amplitude modulation is achieved either by modulating the oscillator signal at a relatively low power level and amplifying the modulated signal with a cascade of amplifiers (including an output power

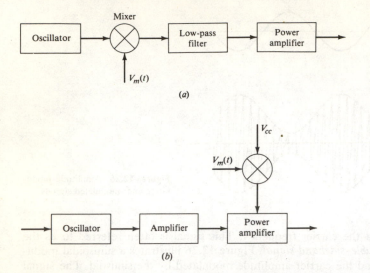

Figure 12.27 (a) A low-level amplitude-modulation circuit; (b) amplitude modulation at high power levels.

amplifier) or by using the modulating signal to control the supply voltage of the power amplifier. Both methods are illustrated in Fig. 12.27. The power requirements of the modulator and modulating signal can be estimated by considering the power in an amplitude-modulated waveform:

$$S(t) = A[1 + m(t)] \sin \omega_c t$$

The peak output power is

$$(P_o)_{\text{peak}} = \frac{A^2}{2}(1 + m)^2 \tag{12.23}$$

so if the maximum modulation index is unity,

$$(P_o)_{\text{peak}} = 2A^2 = 4(P_o)_{\text{car}}$$

The modulator must be designed to handle four times the average carrier power with 100 percent modulation; the output power will be four times the carrier power.

Several of the previously analyzed mixer circuits with output filtering can be used to realize the low-level modulation illustrated in Fig. 12.27a. For example, the output of the diode mixer shown in Fig. 12.3 is given by Eq. (12.3). If V_L is a sine wave

$$V_L = V_1 \sin \omega_L t$$

and if a low-pass filter is added to the output with a bandwidth of

$$B = \omega_L + \omega_i$$

then the output will be

$$S(t) = V_1 \left(1 + \frac{4}{\pi} \frac{V}{V_1} \sin \omega_i t\right) \sin \omega_L t$$

Since the low-pass filter removes the higher frequency component, the modulation index of the resulting AM waveform is $m = (4/\pi) V/V_1$. This particular amplitude modulator only functions well for low indices of modulation.

Both the FET and BJT mixers described earlier can function as amplitude modulators with a relatively high modulation index. The final amplifier will need to be linear. Linear class C can be obtained with the output circuit tuned to the carrier frequency. The output will then be linearly related to the input, provided the amplifier output circuit is not current-limited. (The peak value of the current pulses must be proportional to the peak value of the input drive current.)

The most frequently used method of amplitude modulation at high power levels is to modulate the supply voltage to the power amplifier, as shown in Fig. 12.27b. Figure 12.28 illustrates a collector-modulated circuit. The transistor can be operated as either class B or C; the output circuit is tuned to the carrier frequency and has a bandwidth equal to twice that of the modulating signal. The modulating signal is applied in series with the dc supply voltage, so the total low-frequency supply voltage for the transistor is

$$V = V_{cc} + V_m(t)$$
$$= V_{cc}(1 + m \cos \omega_m t)$$

provided

$$V_m(t) = mV_{cc} \cos \omega_m t$$

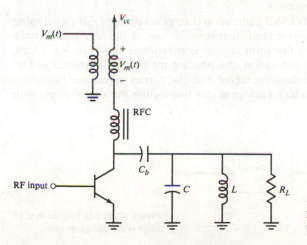

Figure 12.28 A collector-modulated circuit.

It was shown in the discussion of class C power amplifiers that the amplitude of the output signal under saturation-limited conditions equals the power supply voltage. Therefore, changing the transistor supply voltage modulates the output signal amplitude proportionally, and the output voltage becomes

$$V_o = V_{cc}(1 + m \cos \omega_m t) \cos \omega_c t$$

For 100 percent modulation the peak value of the voltage $V_m(t)$ must equal V_{cc}. At full modulation the sidebands each contain one-fourth the power of the carrier, so $P_o = \frac{3}{2} P_c$. The carrier signal has an amplitude of V_{cc}, and the amplitude of each sideband component is one-half that of the carrier amplitude. The total output power is then

$$P_o = \frac{3}{2} \frac{V_{cc}^2}{2 R_L}$$

The unmodulated carrier power is supplied by the power supply. The remaining power must be furnished by the modulator. One reason that output modulation has been the most frequently used method is that collector modulation results in less intermodulation distortion than existing low-level modulation circuits.

All of the information in an AM wave is contained in one sideband. It is possible to eliminate the other sideband without loss of information; thus the required transmitter power is reduced to one-third of that previously required, since the power in each sideband is one-fourth the carrier power for sine-wave modulation. Double-sideband suppressed-carrier (DSB) modulation can easily be realized with the double-balanced mixers previously described. Although transmitter complexity is not markedly increased for DSB signals, the receiver is more complex. Single-sideband suppressed-carrier (SSB) transmission results in an even more complex receiver, and the quality of the demodulated signal is generally not as good. However, SSB is extensively used because of the reduced power and bandwidth requirements. The signal quality is adequate for applications of voice transmission.

The simplest method of SSB generation is to generate the DSB signal using a double-balanced modulator, and then remove one of the sidebands with a filter. A block diagram of this form of SSB generation is shown in Fig. 12.29. Another method of SSB generation, the phasing method, is illustrated in Fig. 12.30. Here both the modulating signal and the carrier signal are processed through phase splitters, which each generate two signals 90° out of phase with

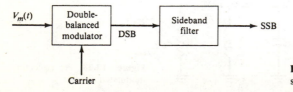

Figure 12.29 Filtering method of single-sideband generation.

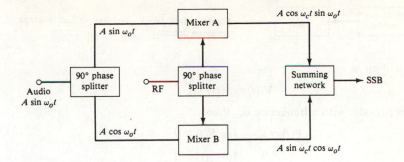

Figure 12.30 Phasing method of single-sideband generation.

each other. The summing network output,

$$S(t) = A \cos \omega_c t \sin \omega_o t + A \sin \omega_c t \cos \omega_o t$$

$$= A \sin (\omega_c + \omega_o) t$$

is the desired SSB signal. The phasing method has the advantage of not requiring the sharp cutoff filters of the filtering method of SSB generation, but it is difficult to realize a broadband phase-shifting network for the lower-frequency modulating signal. The television signal described at the beginning of the chapter uses vestigial sideband plus the low-frequency portion (the part closest to the carrier) of the other sideband. The technique is used primarily to conserve bandwidth. It is suitable for television because the low-frequency components are sufficient for reconstructing an acceptable picture.

Demodulators

AM detection can be divided into synchronous and asynchronous detection. Synchronous detection employs a time-varying or nonlinear element synchronized with the incoming carrier frequency. Otherwise the detection is asynchronous. The simplest asynchronous detector, the average envelope detector, is described here.

Average Envelope Detectors

A block diagram of the average envelope detector is shown in Fig. 12.31. The rectifier output

$$V_r(t) = S(t) \qquad S(t) > 0$$

$$= 0 \qquad S(t) < 0$$

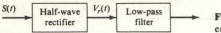

Figure 12.31 Block diagram of an average envelope detector.

can be written as

$$V_r(t) = S(t)P(t)$$

If $S(t)$ is periodic with a frequency ω_c, then

$$P(t) = 1 \qquad \text{for } S(t) \geq 0$$

$$= 0 \qquad \text{for } S(t) \leq 0$$

$P(t)$ is a rectangular wave (identical to that illustrated in Fig. 12.8) with the same frequency ω_c, and so [from Eq. (12.4)]

$$P(t) = \frac{1}{2} + \frac{2}{\pi} \sum_{n=0}^{\infty} \frac{\sin(2n+1)}{2n+1} \omega_c t$$

If $S(t)$ is the AM wave described by Eq. (12.20), then

$$V_r(t) = A[1 + mf(t)] \left(\frac{\sin \omega_c t}{2} + \pi^{-1} + \frac{\cos 2\omega_c t}{\pi} + \text{higher harmonics of } \omega_c \right)$$

$$(12.24)$$

If the low-pass filter bandwidth is chosen to filter out the component at ω_c and all higher harmonics, the output will be

$$V_o(t) = \frac{A[1 + mf(t)]}{\pi}$$

which is a dc term plus the modulating information.

Two additional points will be made to further describe the operation of the envelope detector. First of all, consider the case where

$$f(t) = \sin \omega_m t$$

Then Eq. (12.24) can be written as

$$V_r(t) = A \left[\frac{\sin \omega_c t}{2} + \frac{m}{\pi} (\sin \omega_m t) + \pi^{-1} + \pi^{-1} \left(\frac{\cos(\omega_c - \omega_m)t - \cos(\omega_c + \omega_m)t}{2} \right) \right.$$

$$\left. + \text{higher-frequency terms} \right] \qquad (12.25)$$

The output will contain a term at the frequency $\omega_c - \omega_m$, which must also be removed by the low-pass filter. This is not possible if ω_m is close to ω_c. To ensure that this distortion does not occur, the maximum modulating frequency should be constrained so that

$$(\omega_m)_{max} \leq \frac{\omega_c}{2}$$

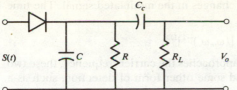

Figure 12.32 A diode envelope detector.

and the corresponding low-pass filter bandwidth B must be selected so that

$$V_r(t) > 0 \qquad \text{if } S(t) > 0$$

This is only possible if m is not greater than 1 (the maximum modulation is 100 percent) and the carrier term is present. Average envelope detection will only work for normal AM with a modulation index less than 1. However, if a large carrier component $A \cos \omega_c t$ is added to the SSB signal, the resultant signal can also be detected with an envelope detector (see Prob. 12.10).

A simple diode envelope-detector circuit is shown in Fig. 12.32. It is assumed here that the input signal amplitude is large enough that the diode can be considered either on or off, depending upon the input signal polarity. The diode can then be replaced by an open circuit when it is reverse-biased and by a constant resistance when it is forward-biased. The series capacitor C_c is included to remove the dc component [Eq. (12.24)]. The purpose of the load capacitor C in the circuit is to eliminate the high-frequency component from the output and to increase the average value of the output voltage. The effect of the load capacitor can be seen from Fig. 12.33, which illustrates the input and output signal waveforms of a diode detector. As the input signal is applied, the capacitor charges up until the input waveform begins to decrease. At this time the diode becomes open-circuited and the capacitor discharges through the load resistance R_L as

$$V_L = V_p e^{-t/R_L C}$$

V_p is the peak value of the input signal, and the diode opens at the time $t = 0$. The larger the value of capacitance used, the smaller will be the output ripple (which means the smaller will be the undesired high-frequency components and the larger will be the dc value of the output voltage). However, C cannot be too

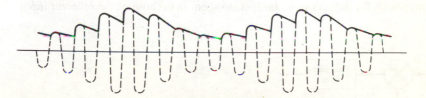

Figure 12.33 Envelope detector input (dashed line) and output waveforms.

large, or it will not be able to follow changes in the modulated signal. The time constant is often selected as

$$RC = [(\omega_m\omega_c)^{-1}]^{1/2}$$

If the highest modulation frequency approaches the carrier frequency, these two constraints become incompatible and some other form of detection, such as a synchronous detector, must be used.

Synchronous Detection

Envelope detectors will not detect some AM signals such as DSB. If it is possible to obtain a signal synchronized in frequency and phase with the original carrier then it is possible to readily detect DSB signals. Some communications systems transmit a small pilot-carrier signal synchronized with the original carrier signal. FM stereo uses this technique. If a local oscillator signal synchronized with the carrier signal is available, then demodulation can be accomplished, as shown in Fig. 12.34. Consider, for example, the DSB signal given by Eq. (12.22). If the local oscillator signal is

$$V_L = V \sin \omega_c t$$

then the output signal is

$$V_o = V_L S(t)$$

$$= \frac{Am}{4} V[2 \cos \omega_m t - \cos (2\omega_c + \omega_m)t - \cos (2\omega_c - \omega_m)t]$$

If this signal is low-pass filtered ($\omega_m < B \le \omega_c$), the output is

$$V'_o = \frac{AVm}{2} \cos \omega_m t$$

which is proportional to the original modulating signal. The same form of synchronous detection can be used to detect normal AM and also SSB signals. Therefore, if a signal phase-synchronized with the carrier is available, any of the previously described mixer circuits together with a low-pass filter can be used as a detector. If the local oscillator signal is a square wave, the same results apply. As previously shown, the square wave will result in additional high-frequency mixing products, but these can all be removed by the low-pass filter. The mixer requirements for detection are far less stringent than those of the receiver input

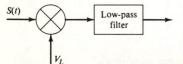

Figure 12.34 Synchronous amplitude demodulator.

mixer. For detection the signal level is larger and mixer noise is less of a problem. Also the mixer-detector usually follows a narrowband filter, and intermodulation distortion is not a significant problem. Phase-locked loops can also be used for AM detection and pilot-carrier recovery. These applications are discussed in Chap. 8.

Angle Modulation

Amplitude modulation alters the amplitude of a sine-wave carrier. Information can also be transmitted by modulating the phase and frequency. Such modulation is, of course, often used. Angle modulation has the disadvantage, compared with amplitude modulation, of occupying a wider bandwidth, but it can provide better discrimination against noise and other interfering signals.

An angle-modulated waveform can be written as

$$S(t) = A(t) \cos \left[\omega_c t + \theta(t) \right] \qquad (12.26)$$

where ω_c is the carrier frequency and $A(t)$ is the amplitude modulation. Ideally, $A(t)$ will be constant, with the phase $\theta(t)$ representing the angle modulation. Angle modulation can be further subdivided into phase and frequency modulation, depending on whether it is the phase or the derivative of phase (frequency) that is modulated. Frequency and phase modulation are not distinct, since changing the frequency will result in a change in phase, and modulating the phase also modulates the frequency.

Angle Modulators

Frequency modulation can be achieved directly by modulating a VCO (direct FM) or, indirectly, by phase-modulating the RF waveform by the integrated audio input signal (indirect FM). Another method of FM is to use a phase-locked loop, as shown in Fig. 12.35. The output in response to the modulating signal V_M (assuming a linear model for the loop) is

$$\theta_o(s) = \frac{(K_o/s) V_M(s)}{1 + K_o K_d F(s)/sN} \qquad (12.27)$$

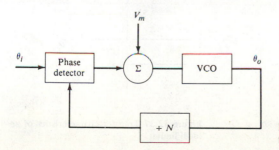

Figure 12.35 A phase-locked-loop frequency demodulator.

where K_d is the phase-detector gain constant and K_o is the VCO sensitivity (hertz per volt). In the steady state the output phase will be proportional to the modulating voltage, so the phase-locked loop can serve either as a phase modulator or, if V_M is the integral of the modulating signal of interest, as a phase modulator. The phase-locked loop bandwidth (described in Chap. 9) needs to be wider than the bandwidth of the modulating signal in order for the loop not to distort the signal.

FM Demodulators

The same type circuitry is used for detecting both types of angle modulation, and we will refer to either process as *FM detection*. In this section a few of the many different methods that have been used for the detection of frequency-modulated waves are evaluated. FM detector circuits are often referred to as *frequency discriminators*.

The ideal FM detector produces an output voltage that changes linearly with changes in the input frequency, as illustrated in Fig. 12.36. The output voltage will usually be zero at the carrier frequency. Any deviation from the linear characteristic distorts the detected waveform. Amplitude modulation caused by noise, signal fading, etc., can also cause the recovered signal to be distorted. Limiting circuitry is usually included in an FM detector to reduce the amount of amplitude modulation.

The transfer characteristic of an ideal limiter is illustrated in Fig. 12.37. The limiter output is restricted to the values that depend only on the sign of the input waveform. A single-stage differential-pair limiter is shown in Fig. 12.38. This circuit gives a close approximation to the ideal limiter characteristic. The transfer characteristic is symmetrical (about $V_i = 0$) so there are no even harmonics (or direct current) in the output waveform. The input signal must be large enough to drive the differential amplifier into saturation. If the input signal is too small, several differential-pair stages can be cascaded in order for the output to be saturated. Integrated-circuit limiters frequently contain three cascaded differential sections.

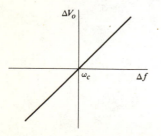

Figure 12.36 Ideal characteristics of a frequency-to-voltage converter.

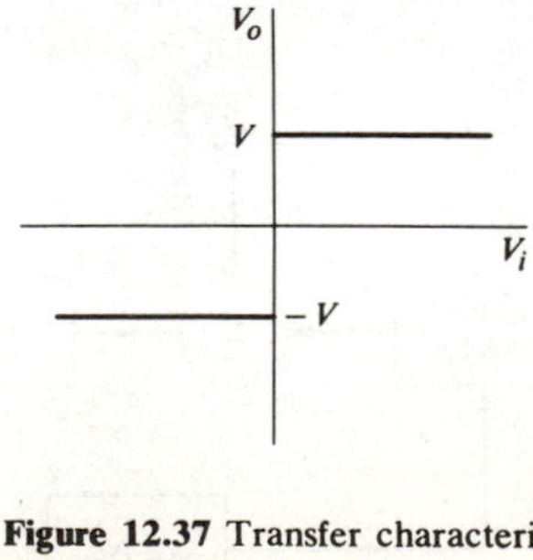

Figure 12.37 Transfer characteristics of an ideal limiter.

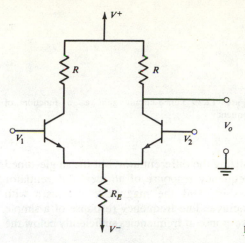

Figure 12.38 A differential-pair limiter.

Example 12.3 A differential pair has a voltage gain of 60 dB into a 2-kΩ load and a differential input impedance of 2 kΩ. If 75-mV input voltage drives the stage into saturation, how many stages will be needed to limit a 5-μV input signal?

SOLUTION A voltage gain of 60 dB ($A_v = 1000$) means that the output of the first stage will be 5 mV, which is not sufficient to saturate the output of the second stage. Therefore, the complete limiter consists of three cascaded differential pairs.

FM Detectors

An analytical basis for FM detection is obtained by considering the derivative of the FM signal:

$$\frac{d}{dt}[A\cos(\omega_c t + \theta(t))] = -\left(\omega_c + \frac{d\theta}{dt}\right)A\sin[\omega_c t + \theta(t)] \qquad (12.28)$$

The derivative of an angle-modulated signal is an amplitude-modulated FM waveform; all of the modulating information is contained in the amplitude of the differentiated waveform. It will normally be the case that $\omega_c \gg d\theta/dt$. If so, the amplitude modulation can be removed with an envelope detector. Of course, any amplitude modulation on the original signal must first be removed by limiting.

The output of the envelope detector will be proportional to $\omega_c + d\theta/dt$, which is $\omega_c + KV_m(t)$ for a frequency-modulated waveform. If the output is then high-pass-filtered to remove the constant term ω_c, the remainder will be proportional to the modulating signal. This approach does have the disadvantage that any dc component in the modulating signal is lost.

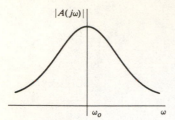

Figure 12.39 Tuned-circuit gain as a function of frequency.

There are many circuits for realizing the differentiator, but the single-tuned circuit is used most often. The frequency response of an ideal differentiator $H(j\omega) = j\omega K$ has a $+90°$ phase shift, and the magnitude increases with increasing frequency at 6 dB per octave. The frequency response of a simple tuned circuit will approximate this response at frequencies sufficiently below the circuit's resonant frequency.

The frequency response magnitude of the parallel tuned circuit is obtained from Eq. (4.7) and Table 4.1:

$$|A(j\omega)| = \frac{R}{[1 + Q^2(\omega/\omega_o - \omega_o/\omega)^2]^{1/2}}$$

Values for Q and ω_o of the parallel tuned circuit are given in Table 4.1.

The magnitude of the frequency response of this circuit is plotted in Fig. 12.39. The carrier frequency must be sufficiently below the resonant frequency of the circuit so that the magnitude response is approximately linear. If the carrier frequency is too far below the resonant frequency, there will be too much attenuation, with a corresponding degradation of the signal-to-noise ratio.

At the frequency $\omega_c + \Delta\omega$,

$$|A(j\omega)| = \frac{R}{\{1 + Q^2[(\omega_c + \Delta\omega)/\omega_o - \omega_o/(\omega_c + \Delta\omega)]^2\}^{1/2}}$$

$$\approx \frac{R\omega_o(\Delta\omega + \omega_c)}{Q[\omega_o^2 - (\omega_c + \Delta\omega)^2]}$$

provided ω_c is close enough to ω_o that

$$Q\left[\frac{(\omega_c + \Delta\omega) - \omega_o^2}{\omega_o(\omega_c + \Delta\omega)}\right] \gg 1$$

Also, if

$$\omega_c + \Delta\omega \ll \omega_o$$

then

$$|A(j\omega)| \approx \frac{R(\omega_c + \Delta\omega)}{Q\omega_o}$$

The output consists of a constant term corresponding to ω_c, plus a

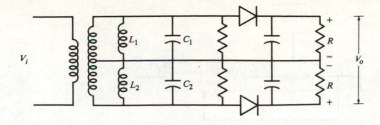

Figure 12.40 A balanced frequency discriminator.

component proportional to the frequency deviation $\Delta\omega$. Balanced discriminators are often used to eliminate the constant term. A simplified balanced discriminator is illustrated in Fig. 12.40. The upper resonant circuit is tuned to the frequency $\omega_o - \omega_c$ and the output is proportional to $\omega_c + \Delta\omega$. The lower resonant circuit is tuned to $\omega_o + \omega_c$ and the output is proportional to $\omega_c - \Delta\omega$. The differential output is then

$$V_o = K[\omega_c + \Delta\omega - (\omega_c - \Delta\omega)] = 2K\Delta\omega$$

which is proportional to the frequency deviation from the carrier frequency.

Pulse Discriminators

A different approach to FM discrimination, one becoming increasingly popular as advances in digital integrated circuits continue, is shown in Fig. 12.41. The zero-crossing detector outputs a pulse on every other zero-crossing of the input signal corresponding to a zero-crossing for every full cycle of the input waveform. The period discriminator then determines the period between alternate zero-crossings and converts the period information to an analog voltage with the period-to-voltage converter.

Figure 12.42 shows one method of implementing this frequency discriminator. The voltage comparator functions as an amplitude-limiter. Whenever the input signal goes positive the contents of the counter are strobed into the D/A converter and the counter is reset. The counter increments at a constant rate equal to the clock frequency. Therefore, the contents of the counter are proportional to the time lapse since the last positive-going zero-crossing. A D/A converter connected to the counter output provides an analog signal proportional to the period or frequency. The circuit can also be

Figure 12.41 A pulse frequency discriminator.

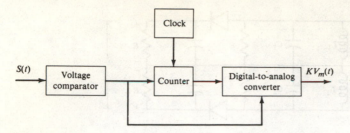

Figure 12.42 Block diagram implementation of a pulse frequency discriminator.

Figure 12.43 Another pulse frequency discriminator.

implemented as shown in Fig. 12.43. For each trigger signal the multivibrator outputs a positive pulse of constant duration. For this circuit the monostable multivibrator outputs a positive pulse of constant duration and a square wave when the input is triggered at a rate equal to the carrier frequency. Since the average value of square wave is zero, the output voltage will also be zero. If the input frequency increases, the multivibrator output will be as shown in Fig. 12.44. Since the pulse length is constant, the multivibrator output is no longer

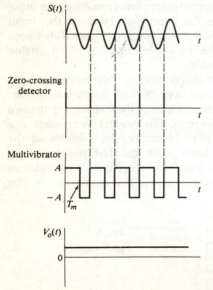

Figure 12.44 Pulse-frequency-discriminator wave-forms.

symmetrical, and the average output voltage will be positive and proportional to the increase in frequency. If the input frequency decreases below the carrier frequency, the output voltage will become negative.

The pulse duration discriminator is very insensitive to amplitude modulation. Compared with ratio discriminators, it is primarily limited by the speed of the digital circuitry and the increased complexity of its hardware. Another popular FM discriminator, the phase-locked-loop frequency discriminator, is described in Chap. 8.

12.4 DIGITAL MODULATION

Today more and more information is being transmitted in digital form. This is true from low-frequency voice transmission to an all-digital television system to satellite communications at microwave frequencies. The many reasons for the increased use of digital modulation include the advances in modern technology, which made available small, compact, and inexpensive digital circuits; the ability to transmit at low signal-to-noise interference ratios; and the ease of time-multiplexing the data. Also, digital signals allow for the use of very efficient power amplifiers. Digital modulation can refer to the modulation of an analog signal by a digital waveform or the digital modulation of an analog signal. An analog signal is one that takes on a continuum of values, such as (0 to 10 V) and it can be either amplitude- or angle-modulated. A digital waveform assumes only the values corresponding to logical 0s and 1s. An analog signal can be modulated with digital data by varying either the amplitude (such as CW modulation), frequency (as in frequency-shift keying), or phase (as in phase-shift keying). Digital data can also be transmitted directly using various encoding methods—such as BCD—and data formats—such as nonreturn to zero (NRZ) or return to zero (RZ).

Figure 12.45 illustrates the components required to convert an analog-modulated signal $V_m(t)$ to a digitally encoded waveform. The signal is first sampled (see Chap. 8 for an analysis of the sampling process), providing a pulse-amplitude-modulated (PAM) signal, which can be transmitted directly but which is usually converted to a pulse-code-modulated (PCM) signal in the A/D converter. The Nyquist sampling criterion states that the sampling frequency f_s must be greater than twice the bandwidth B of the input signal in order to

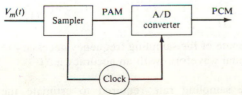

Figure 12.45 Digital encoding of a signal.

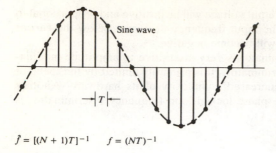

Figure 12.46 A uniformly sampled sine wave.

$\hat{f} = [(N + 1)T]^{-1}$ $f = (NT)^{-1}$

preserve all information. That is, $f_s > 2B$. If the sampling rate is less than twice the bandwidth, the frequency spectrum centered about the sampling frequency overlaps the original spectrum and cannot be separated from the original spectrum by filtering. (This is known as *foldover distortion or aliasing*.)

Thus, a PAM system must include a band-limiting filter before the sampler to ensure that foldover distortion does not occur. The Nyquist sampling rate may need to be significantly exceeded if precise resolution of the signal is required in a short time. The Nyquist criterion states the minimum sampling rate required, but a large number of samples are required before the signal can be accurately reconstructed. Consider the problem of measuring the period T of a sine wave (as illustrated in Fig. 12.46) from its PAM values. The signal period could be estimated by counting the number of samples between successive zero-crossings or between zero-crossings of the same sign (corresponding to a full input cycle). If there are N samples between zero-crossings of the same sign, the estimated period of the signal is $\hat{T} = NT_s$, where T_s is the sampling period. The maximum error in the estimation of the period is one sample period. That is,

$$NT_s \leq T \leq (N+1)T_s$$

The uncertainty in the frequency estimate is

$$\Delta f = (NT_s)^{-1} - [(N+1)T_s]^{-1} = [N(N+1)T_s]^{-1} \approx (N^2 T_s)^{-1}$$

Therefore, for small errors

$$\Delta f \approx \frac{T_s}{N^2 T_s^2} \approx \frac{f^2}{f_s}$$

or

$$f_s = \frac{f^2}{\Delta f} \tag{12.29}$$

This equation provides a good estimate of the sampling frequency necessary to estimate the frequency of a sinusoidal waveform with an accuracy $\pm \Delta f$.

Example 12.4 Determine the sampling rate required to estimate the frequency of a sine wave with a maximum error of 1 Hz. The maximum frequency of the unknown signal is 1 kHz.

SOLUTION If the estimate is to be made from one cycle of data, Eq. (12.29) can be used. The required sampling frequency is

$$f_s = \frac{(10^3)^2}{1} = 1 \text{ MHz}$$

The sampling rate requirements can be reduced if more samples (at the same rate) are used to estimate the period. For example, if P full cycles of the sampled signal are used, the sampling uncertainty is still one sample period, so the required sampling frequency is

$$f_s = \frac{f^2}{P\Delta f} \qquad (12.30)$$

The PAM signal is usually converted to a PCM signal before transmission, using an analog-to-digital converter. In uniformly encoded PCM, all signals falling within a prescribed amplitude interval are represented by a single discrete value. The process is illustrated in Fig. 12.47, where eight equal-size-quantization levels and a binary representation of each level are presented. If the signal level lies anywhere within the lowest level (the lowest 12.5 percent), it is assigned the binary code 000. If the signal lies in the highest level (the highest 12.5 percent), it is assigned the binary code 111. Signal amplitudes must be confined to the maximum range of the encoder or overload distortion will occur. The quantization process introduces quantization noise; the more quantization levels, the smaller will be the quantization noise. It was mentioned in Chap. 10 that for sampled data the signal-to-noise ratio after uniform quantization of a

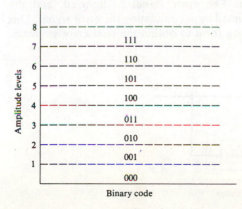

Figure 12.47 Amplitude levels of a PCM signal.

noiseless signal is approximately

$$\frac{S}{N} = +6n \quad \text{dB} \tag{12.31}$$

where $(n + 1)$ is the number of bits in the digital word (including the sign bit). The uniform quantization process results in each level having the same amount of noise. Therefore, the small signals have a small signal-to-noise ratio, and the large signals have a large signal-to-noise ratio. In many applications such as voice transmission, the large signals are the least likely to occur; uniform PCM is not efficient for such applications. A more efficient encoding procedure occurs if smaller quantization levels are used for small signals and large quantization levels for large signals. The process is known as *companding*; it is possible to achieve the result with a linear A/D converter preceded by a nonlinear amplifier whose gain decreases with increasing signal level. Various nonlinear characteristics can be used to implement the amplifier (*compandor*). A widely used characteristic is the u law, defined as

$$F_u(x) = sgn(x)\frac{\ln(1 + u|x|)}{\ln(1 + u)} \tag{12.32}$$

where x is the input signal amplitude ($-1 \le x \le 1$), $sgn(x)$ is the polarity of x, and u is a parameter used to define the amount of compression.[12.6] The gain compression functions are now included directly in the design of A/D converters (encoders), and the modulation is frequently referred to as *log-PCM*.

Many signals, such as voice signals, contain a large amount of sample-to-sample redundancy, which permits a reduction in the digital bit rate (and thus the bandwidth). If the sampling rate is fast enough, the next sample value is probably going to be close to the previous sample in amplitude. Differential pulse code modulation (DPCM) encodes the difference between samples. As in PCM, the analog-to-digital conversion process can be conventional or companded. DPCM is ideally suited for digital implementation and large-scale integration. Figure 12.48 contains a block-diagram representation of one system for DPCM implementation. The error signal is digitized, and the estimate of the input signal is constructed by accumulating the error signal. This estimate is then reconverted to analog form to obtain the next error estimate.

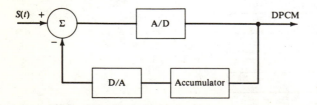

Figure 12.48 A differential pulse code modulator.

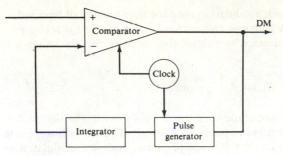

Figure 12.49 A delta modulator.

There are many implementations of this technique, including many which do not reconstruct the analog signal but rather generate the error estimate in the digital domain. DPCM provides approximately a one-bit reduction in word length for the digital modulation of speech signals.

Delta modulation (DM) is another method for the digital modulation of analog signals. DM is a special case of DPCM. It uses only one bit per sample to represent the difference signal. The signal bit specifies whether the signal has increased or decreased since the last sample. Delta modulation requires an appreciably higher sampling rate than PCM or DPCM, but the hardware required is simpler. Figure 12.49 illustrates a delta modulator consisting of a clock, comparator, pulse generator, and integrator (to reconstruct an estimate of the input signal). The comparator output consists of digital bits, and the pulse generator output consists of two equal amplitude pulses of opposite polarity. The polarity depends on the output bit.

Digital Modulation and Demodulation of Analog Data

The digitized data, often referred to as the *baseband data*, can be easily modulated onto a radio frequency carrier since the baseband has only two amplitude levels. The circuit components previously discussed are directly applicable to this problem. The following are the most frequently used techniques.

Amplitude Modulation

For digital or continuous-wave (CW) modulation, the oscillator can be directly keyed on an "off", but this can create frequency distortion referred to as *chirps*. A better method is to operate the oscillator continuously and then switch the output on and off with a buffer amplifier (switch). On/off keying is a special case of amplitude or frequency modulation. It is the most economical method of digital communications, particularly if envelope detection is used, but it gives the poorest error performance. Some form of angle modulation gives better performance.

Another type of amplitude modulation used for transmission of binary data is quadrature AM (QAM), which uses two carriers in quadrature (offset by 90°). If a carrier is amplitude-modulated by a signal that assumes only the values ±1, then the modulated signal is

$$S(t) = A \frac{1 \pm m}{2} \sin \omega_o t$$

The envelope will have an amplitude of $1 + m/2$ or $1 - m/2$. The digital data contributes a portion that is either in phase or 180° out of phase with respect to the carrier. QAM uses two carriers 90° out of phase; both are modulated by digital data, and then the modulated signals are summed. The relative phase of the signal portion is then 45, 135, 225, and 315°. QAM is able to transmit twice as much data as AM modulation.

Frequency Modulation

Frequency-shift keying (FSK) is a special case of frequency modulation that can be achieved by pulling a VCO plus or minus a standard frequency deviation $\Delta\omega$. FSK is often used for low-cost, low-data rate communication over analog telephone networks. The channel capacity can be increased by using multiple frequencies. A FSK system having four frequencies is denoted as $(FSK)_4$. FSK detection is readily achieved with phase-locked loops, as described in Chap. 8.

Phase Modulation

It can be shown that for two-level line coding the maximum use of transmitted power is achieved when one signal is the negative of the other. Phase-shift keying (PSK) has become the most popular digital modulation method for high-performance applications. A relatively simple method of PSK modulation is to generate the carrier signals 180° apart in phase and then select between the two signals, depending on the data values. The data rate can be increased by transmitting multiple-phase signals. $(PSK)_8$, for example, will assume one of eight values spaced 45° apart.

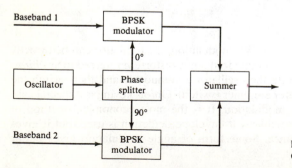

Figure 12.50 Block diagram of a QPSK modulator.

Quadrature-phase-shift keying [QPSK or (PSK)$_4$] uses four phase values spaced 90° apart. QPSK is achieved by using two binary phase-shift-keyed (BPSK) modulators with the two carriers 90° apart in phase, as shown in Fig. 12.50. The two BPSK signals are then summed. The phase splitting can be achieved with a PLL, as described in Chap. 8. It can be shown that an ideal QPSK modem can transmit two bits per second per unit bandwidth. To transmit 50×10^6 bits per second, for example, will require a transmission bandwidth of 25 MHz if QPSK is used. Coherent demodulation is normally used for QPSK demodulation by recovering the carrier as described in Chap. 8. Many other phase-modulation methods are possible[12.7]—such as staggered QPSK (S PSK) and minimal shift keying (MSK)—and can be realized with the basic circuitry described in this text.

12.5 INTEGRATED CIRCUIT TECHNIQUES[12.8]

Most integrated-circuit modulators and demodulators use some form of a variable transconductance multiplier that relies on the dependence of the transistor's transconductance upon its bias current. The technique is illustrated by the differential amplifier illustrated in Fig. 12.51. It was shown in Chap. 2 that for small values of differential input voltage, the differential output voltage is

$$V_o \approx g_m R_L V_1$$

and the transconductance is

$$g_m = \frac{I_E}{V_T} \approx \frac{V_2}{R_E V_T}$$

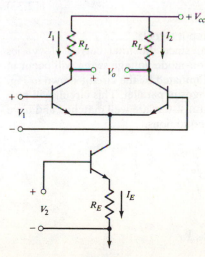

Figure 12.51 Balanced-modulator stage commonly used in integrated circuits.

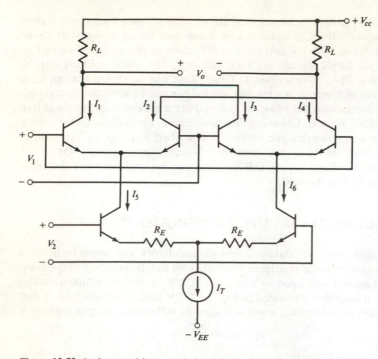

Figure 12.52 An improved integrated-circuit balanced modulator.

Therefore, the output

$$V_o \approx \frac{R_L V_1 V_2}{R_E V_T} \qquad (12.33)$$

is proportional to the product of the two input signals.

One difficulty with this circuit is that since the total current I_E varies directly as a function of V_2, a large common-mode voltage swing will occur in the circuit. This common-mode swing is eliminated by the circuit shown in Fig. 12.52, which consists of two differential stages in parallel. This circuit will now be analyzed with the assumptions that the transistors are well matched and have a large current gain ($\beta \gg 1$). Under these conditions,

$$I_1 + I_2 = I_5$$

$$I_3 + I_4 = I_6$$

$$I_5 + I_6 = I_T$$

If the differential input voltage is small,

$$I_1 - I_2 = g_{m_1} V_1$$

and
$$I_3 - I_4 = g_{m_3} V_1$$

where g_{m_1} is the transconductance of the transistor pair Q_1 and Q_2, and g_{m_3} is the transconductance of transistor pair Q_3 and Q_4. Also,

$$g_{m_1} = \frac{I_5}{V_T}$$

and
$$g_{m_3} = \frac{I_6}{V_T}$$

The output voltage is

$$V_o = R_L[(I_1 - I_2) + (I_3 - I_4)]$$

$$= R_L(g_{m_1} - g_{m_3}]V_1$$

$$= \frac{V_1 R_L}{V_T}(I_5 - I_6)$$

Also
$$V_2 = V_{be_5} + I_5 R_e - I_6 R_E - V_{be_6}$$

$$= V_T \ln \frac{I_5}{I_6} + R_E(I_5 - I_6) \approx R_E(I_5 - I_6)$$

provided R_E is sufficiently large. Under these conditions

$$V_o = \frac{R_L}{R_E V_T} V_1 V_2 \qquad (12.34)$$

The circuit operates as a multiplier, provided the input voltages are sufficiently small. If the voltages are comparable to, or larger than, the thermal voltage V_T, the transistor pairs function as synchronous switches, and the circuit functions as a balanced modulator. Improved circuits, which provide four-quadrant multiplication, have been developed, but the balanced modulator circuit suffices for most communications applications. In integrated-circuit terminology a balanced modulator is a multiplier circuit in which the output is a linear function of only one of the inputs. This input is known as the *modulating input*, and the other input is referred to as the *carrier input*. When differential inputs are used (such as shown in Fig. 12.52), the carrier signal appearing at the output is greatly suppressed, and the circuit is referred to as a *balanced modulator*.

PROBLEMS

12.1 Design a passive double-balanced mixer (including the specification of the transformer turns ratio) that will provide maximum power transfer between a 300-Ω source resistance and a 50-Ω load resistance.

12.2 If a passive double-balanced mixer is used to realize the mixing product $n\omega_L - \omega_i$, what is the conversion loss?

12.3 In the circuit illustrated in Fig. P12.3 the local oscillator signal is fed to the primary of the center-tapped transformer. Derive an expression for the output voltage as a function of the input voltage, assuming the diodes and transformer are ideal.

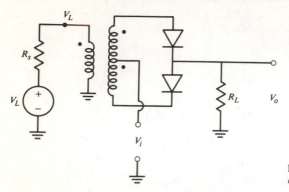

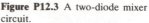

Figure P12.3 A two-diode mixer circuit.

12.4 In synchronous detection of the DSB signal given by Eq. (12.22), what happens if the local oscillator signal is of the form

$$V_L = V \cos \omega_c t$$

What does this imply in terms of phase synchronization for the receiver?

12.5 Show that synchronous detection can be used for SSB detection. What must be the phase relation between the carrier and local oscillator signal?

12.6 Determine the sampling rate required to measure the period of a sine wave using the successive zero-crossings to estimate the half-cycle period of the signal. The maximum signal frequency is 1000 Hz and the maximum tolerable error is 1 Hz. What is the required rate if five consecutive zero-crossings are used to estimate the period?

12.7 Verify Eq. (12.30).

12.8 Calculate the conversion loss of the simple double-balanced mixer shown in Fig. 12.10. The diode on-resistance is much less than the load resistance R_L. What value of R_L is required for maximum power transfer if the circuit employs a 1:1 transformer?

12.9 If a 2:1 voltage transformer is used on the input of the double-balanced mixer

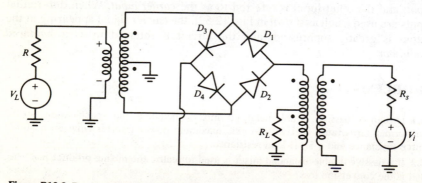

Figure P12.9 Double-balanced mixer circuit.

shown in Fig. P12.9, what is the optimum value of R_L, in terms of R_s, for maximum power transfer?

12.10 Show that an amplitude-modulated wave consisting of one sideband plus a carrier component can be demodulated with envelope detection, provided the carrier amplitude is sufficiently large.

12.11 Calculate the conversion loss of the single-balanced mixer shown in Fig. 12.3 for the condition $R_L = 2R_s$, and also for the case $R_s = 2R_L$.

12.12 Determine the output voltage of the circuit shown in Fig. P12.12. Compare the performance of this circuit with that of the circuit illustrated in Fig. 12.10.

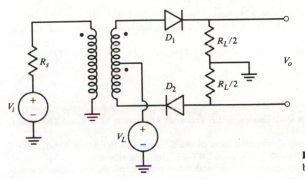

Figure P12.12 Mixer with balanced load.

12.13 Verify that the differential output of the balanced discriminator illustrated in Fig. 12.40 is $2K\Delta\omega$, provided one tuned circuit is tuned to $\omega_c + \Delta\omega$ and the other is tuned to $\omega_c - \Delta\omega$.

12.14 Design a demodulator for QAM signals using phase-locked-loop techniques. Describe the system in block diagram form.

12.15 Design a $(FSK)_4$ demodulator using PLL techniques. Include circuitry to indicate which signal is present at the input.

12.16 A system is to measure the period of a sine wave. Describe how the sampling rate can be reduced if single-point data interpolation midway between the sampling points is used.

12.17 Describe a block diagram for QPSK detection.

12.18 Calculate the common-mode voltage swing of the double-balanced modulator shown in Fig. 12.52 as a function of the input signal V_2.

REFERENCES

12.1 Walker, H. P.: Sources of Intermodulation in Diode-Ring Mixers, *Radio and Electronic Engineer*, **46**:247–255 (1976).

12.2 Kaplan, W.: *Advanced Calculus*, Addison-Wesley, Reading, Mass., 1952, p. 362.

12.3 Cheadle, D.: Selecting Mixers for Best Intermod Performance, *Microwaves*, 1973, pp. 48–52.

12.4 Clarke, K. K., and D. T. Hess: *Communication Circuits: Analysis and Design*, Addison-Wesley, Reading, Mass., 1971.

12.5 Ulfers, B. G.: *Characteristics and Applications of Dual Gate, Insulated Gate Field Effect Transistors*, M.S. thesis, University of Florida, Gainesville, 1969.
12.6 Bellamy, J.: *Digital Telephony*, Wiley, New York, 1982.
12.7 Spilker, J. J., Jr.: *Digital Communications by Satellite*, Prentice-Hall, Englewood Cliffs, N.J., 1977.
12.8 Grebene, A. B.: *Analog Integrated Circuit Design*, Van Nostrand Reinhold, New York, 1972.

ADDITIONAL READING

Hayward, W. H.: *Introduction to Radio Frequency Design*, Prentice-Hall, Englewood Cliffs, N.J., 1982.
Jayant, N. S.: Digital Coding of Speech Waveforms, *Proc. IEEE*, 611–632, 1974.
Pappenfus, E. W., W. B. Brune, and E. O. Schoenke: *Single Sideband Principles and Circuits*, McGraw-Hill, New York, 1964.
Wheeler, H. A.: Design Formulas for Diode Detectors, *Proc. IRE*, **26**:745–780 (1938).

Mixers

Abdullah, F., and F. M. Clayton: A Large-Signal Theory for Broad-Band Frequency Converters Using Abrupt Junction Varactor Diodes, *IEEE Trans. on Microwave Theory and Techniques*, MTT-25, **101**:117–136 (1977).
Barber, M. R.: Noise Figure and Conversion Loss of the Schottky Barrier Mixer Diode, *IEEE Trans. on Microwave Theory and Techniques*, MIT-15, 1967, pp. 629–635.
Gardiner, J. G.: The Relationship between Cross-Modulation and Intermodulation Distortions in the Double-Balanced Modulator, *Proc. IEEE*, **56**:2069–2971 (1968).
Howson, D. P., and J. G. Gardiner: High-Frequency Mixers Using Square-Law Diodes, *Radio and Electronic Engineer*, November 1968, pp. 311–316.
Huang, M. Y., R. L. Buskirk, and D. E. Carlile: Select Mixer Frequencies Painlessly, *Electronic Design* **8**:104–109 (April 12, 1976).
Kurtz, S. R.: Specifying Mixers as Phase Detectors, *Microwaves*, January 1978, pp. 80–89.
Mouw, R. B., and S. M. Fukuchi: Broadband Double Balanced Mixer/Modulators, Part I, *Microwave J.*, March 1969, pp. 131–134.
——— and ———: Broadband Double Balanced Mixer/Modulators, Part II, *Microwave J.*, May 1969, pp. 71–76.
Stevens, R. T.: Linear Scales Show Mixer Harmonics, *Electronics*, 10, 1964, pp. 37–39.

APPENDIX 1. SILICONIX 1260T

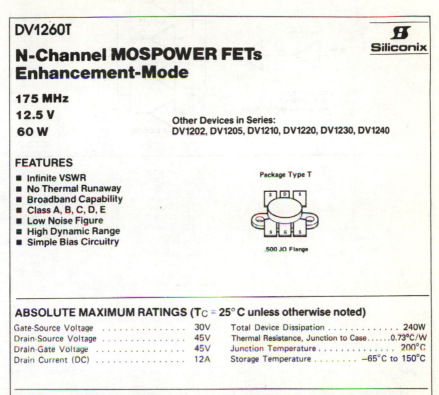

DV1260T

Siliconix

N-Channel MOSPOWER FETs
Enhancement-Mode

175 MHz
12.5 V
60 W

Other Devices in Series:
DV1202, DV1205, DV1210, DV1220, DV1230, DV1240

FEATURES

- Infinite VSWR
- No Thermal Runaway
- Broadband Capability
- Class A, B, C, D, E
- Low Noise Figure
- High Dynamic Range
- Simple Bias Circuitry

Package Type T

.500 JO Flange

ABSOLUTE MAXIMUM RATINGS (T_C = 25°C unless otherwise noted)

Gate-Source Voltage	30V	Total Device Dissipation	240W
Drain-Source Voltage	45V	Thermal Resistance, Junction to Case	0.73°C/W
Drain-Gate Voltage	45V	Junction Temperature	200°C
Drain Current (DC)	12A	Storage Temperature	−65°C to 150°C

ELECTRICAL CHARACTERISTICS (T_C = 25°C unless otherwise noted)

Symbol	Characteristics	Min	Typ	Max	Unit	Test Condition
BV_{DSS}	Drain-Source Breakdown Voltage	45			V	V_{GS} = 0V, I_D = 30 mA
I_{DSS}	Drain-Source Leakage Current			3.0	mA	V_{GS} = 0V, V_{DS} = 15V
I_{GSS}	Gate-Source Leakage Current			100	nA	V_{GS} = 30V, V_{DS} = 0V
g_m	D.C. Forward Transconductance[1]	1.5	2.4		Mho	V_{DS} = 10V, I_D = 6A, ΔV_{GS} = 1.0V
$I_{D(on)}$	On-State Drain Current[1]		20		A	V_{DS} = 12V, V_{GS} = 10V
$V_{GS(th)}$	Gate Threshold Voltage	2		6	V	V_{GS} = V_{DS}, I_D = 600 mA
C_{iss}	Common-Source Input Capacitance		285		pF	V_{GS} = 0V, V_{DS} = 12.5V, f = 1.0 MHz
C_{oss}	Common-Source Output Capacitance		340		pF	V_{GS} = 0V, V_{DS} = 12.5V, f = 1.0 MHz
C_{rss}	Reverse Transfer Capacitance		60		pF	V_{GS} = 0V, V_{DS} = 12.5V, f = 1.0 MHz
G_{ps}	Common-Source Power Gain	8.0			dB	V_{DD} = 12.5V, Po = 60W, f = 175 MHz, I_{DQ} = 6A
η	Drain Efficiency		60		%	V_{DD} = 12.5, Po = 60W, f = 175 MHz, I_{DQ} = 6A
V_{SWR}	Load Mismatch Tolerance	30:1				V_{DD} = 12.5, Po = 60W, f = 175 MHz, I_{DQ} = 6A

Note 1: Pulse Test — 80µs to 300µs, 1% duty cycle

TYPICAL AMPLIFIER LINE-UP

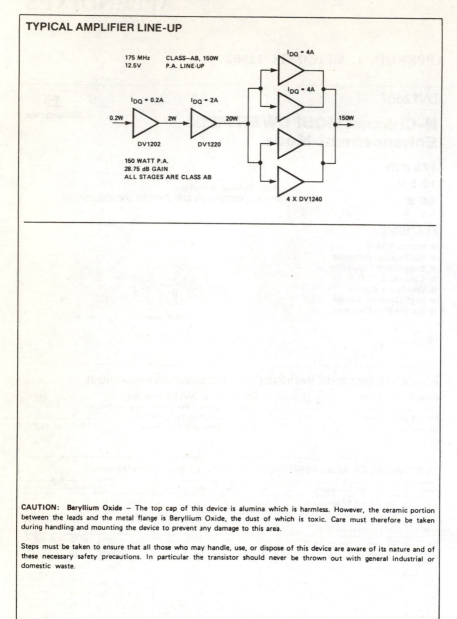

175 MHz   CLASS—AB, 150W
12.5V   P.A. LINE-UP

150 WATT P.A.
28.75 dB GAIN
ALL STAGES ARE CLASS AB

CAUTION: Beryllium Oxide — The top cap of this device is alumina which is harmless. However, the ceramic portion between the leads and the metal flange is Beryllium Oxide, the dust of which is toxic. Care must therefore be taken during handling and mounting the device to prevent any damage to this area.

Steps must be taken to ensure that all those who may handle, use, or dispose of this device are aware of its nature and of these necessary safety precautions. In particular the transistor should never be thrown out with general industrial or domestic waste.

TYPICAL PERFORMANCE CURVES (T$_C$ = 25°C unless otherwise noted)

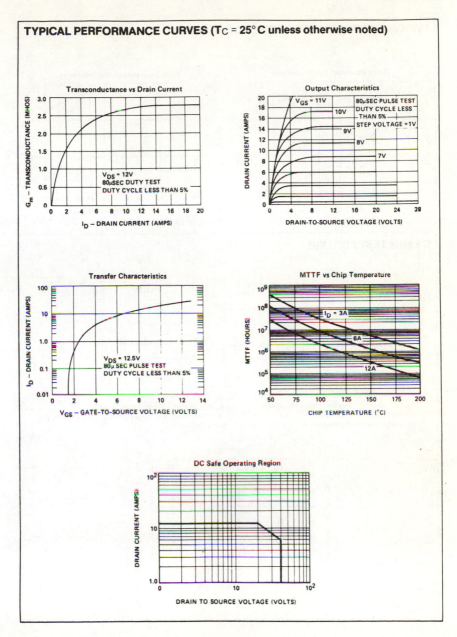

TYPICAL PERFORMANCE CURVES-CONTINUED

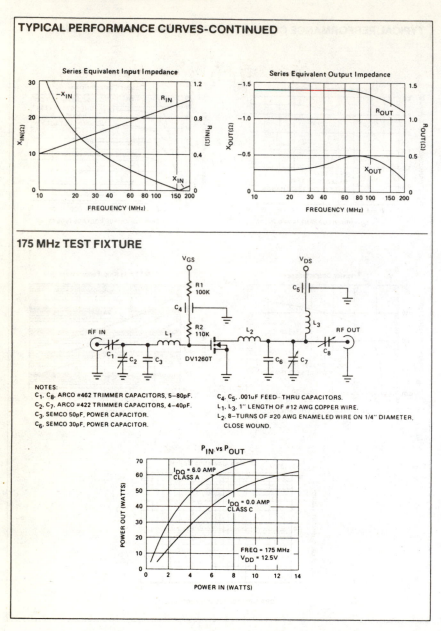

Series Equivalent Input Impedance

Series Equivalent Output Impedance

175 MHz TEST FIXTURE

NOTES:
C_1, C_8. ARCO #462 TRIMMER CAPACITORS, 5–80pF.
C_2, C_7. ARCO #422 TRIMMER CAPACITORS, 4–40pF.
C_3. SEMCO 50pF, POWER CAPACITOR.
C_6. SEMCO 30pF, POWER CAPACITOR.

C_4, C_5. .001uF FEED–THRU CAPACITORS.
L_1, L_3. 1" LENGTH OF #12 AWG COPPER WIRE.
L_2. 8–TURNS OF #20 AWG ENAMELED WIRE ON 1/4" DIAMETER, CLOSE WOUND.

P_{IN} vs P_{OUT}

APPENDIX 2. U310-SILICONIX

n-channel JFETs
designed for . . .

𝕤 Siliconix

- ■ **VHF Amplifiers**
- ■ **Front End High Sensitivity Amplifiers**
- ■ **Oscillators**
- ■ **Mixers**

BENEFITS
- ● Industry Standard
- ● High Power Gain
  16 dB at 105 MHz, Common-Gate
  11 dB at 450 MHz, Common-Gate
- ● Low Noise
  2.7 dB Noise Figure at 450 MHz
- ● Wide Dynamic Range
  Greater than 100 dB
- ● 75 Ω Input Match Common Gate

ABSOLUTE MAXIMUM RATINGS (25°C)

Gate-Drain or Gate-Source Voltage . . . . –25 V
Gate Current . . . . 20 mA
Total Power Dissipation at T_A = 25°C . . . . 500 mW
Power Derating to 150°C . . . . 4.0 mW/°C
Storage Temperature Range. . . . . –65 to + 200°C
Lead Temperature
  (1/16″ from case for 10 seconds) . . . . 300°C

TO-52
See Section 6

ELECTRICAL CHARACTERISTICS (25°C unless otherwise noted)

		Characteristic	U308 Min	U308 Typ	U308 Max	U309 Min	U309 Typ	U309 Max	U310 Min	U310 Typ	U310 Max	Unit	Test Conditions	
1		I_{GSS} Gate Reverse Current			–150			–150			–150	pA	V_{GS} = –15 V,	
2					–150			–150			–150	nA	V_{GS} = 0	T_A = 125°C
3	S T A T I C	BV_{GSS} Gate-Source Breakdown Voltage	–25			–25			–25			V	I_G = –1 µA, V_{DS} = 0	
4		$V_{GS(off)}$ Gate-Source Cutoff Voltage	–1.0		–6.0	–1.0		–4.0	–2.5		–6.0		V_{DS} = 10 V, I_D = 1 nA	
5		I_{DSS} Saturation Drain Current (Note 1)	12		60	12		30	24		60	mA	V_{DS} = 10 V, V_{GS} = 0	
6		$V_{GS(f)}$ Gate-Source Forward Voltage		1.0			1.0			1.0		V	I_G = 10 mA, V_{DS} = 0	
7	D Y N A M I C	g_{fg} Common-Gate Forward Transconductance (Note 1)	10	17		10	17		10	17		mmho	V_{DS} = 10 V, I_D = 10 mA	f = 1 kHz
8		g_{og} Common-Gate Output Conductance			250			250			250	µmho		
9		C_{gd} Drain-Gate Capacitance			2.5			2.5			2.5	pF	V_{GS} = –10 V, V_{DS} = 10 V	f = 1 MHz
10		C_{gs} Gate-Source Capacitance			5.0			5.0			5.0			
11		$\overline{e_n}$ Equivalent Short Circuit Input Noise Voltage		10			10			10		$\frac{nV}{\sqrt{Hz}}$	V_{DS} = 10 V, I_D = 10 mA	f = 100 Hz
12	H I F R E Q	g_{fg} Common-Gate Forward Transconductance		15			15			15		mmho	V_{DS} = 10 V, I_D = 10 mA	f = 105 MHz
13				14			14			14				f = 450 MHz
14		g_{og} Common-Gate Output Conductance		0.18			0.18			0.18				f = 105 MHz
15				0.32			0.32			0.32				f = 450 MHz
16		G_{pg} Common-Gate Power Gain (Note 2)	14	16		14	16		14	16				f = 105 MHz
17			10	11		10	11		10	11				f = 450 MHz
18		NF Noise Figure		1.5	2.0		1.5	2.0		1.5	2.0	dB		f = 105 MHz
19				2.7	3.5		2.7	3.5		2.7	3.5			f = 450 MHz

NOTES:
1. Pulse test duration = 2 ms.
2. Gain (G_{pg}) measured at optimum input noise n

NZA

2-89

APPENDIX 3. 2N5394 NOISE CHARACTERISTICS

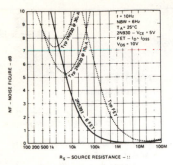

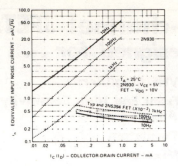

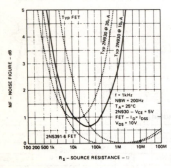

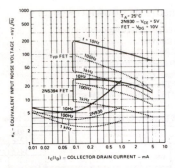

The 2N5391-6 field effect transistor optimum source resistance is 1 to 10M. However, even at the optimum source resistance for a typical 2N930 type bi-polar transistor, the noise figure is higher than the 2N5391-6 series FET's.

Substituting equation (10) into (8) gives

$$NF_{min} = 10 \log \left[1 + \frac{(e_n + \gamma e_n) 2 i_n}{4KT} \right] \qquad (11)$$

Assuming worst case $\tau = 1$,

$$NF_{min} = 10 \log \left[1 + \frac{e_n i_n}{KT} \right]$$

Bi-Polar Versus FET Noise Comparison

The significant difference between the bi-polar and FET is in the "e_n" and "i_n" values as plotted in Figure 5, 8, and 9. Bi-polars are extremely sensitive to collector current variations and the circuit designer must beware. FET's are relatively insensitive to drain current variations, particularly "i_n". Notice that the FET "i_n" is 2-3 orders of magnitude lower than the bi-polar.

FET devices perform particularly well at low frequencies (<100Hz) i.e., they do not show a significant I/f noise slope. Naturally at low frequencies, the noise figure drops rapidly as source resistance increases.

Most impressive about the FET, particularly the 2N5391-6 series, is its low noise over many decades of source resistance value. This flexibility is advantageous in strain-gauges, light sensors, thermistors, and other transducers. Designers of standard amplifier circuits who will be using varying source resistance might well use the 2N5391-6 series FET's to provide the low noise flexibility.

If low noise applications restrict source resistances to less than 500 ohms, the bi-polar will probably continue to be the best.

Noise Figure Nomograph

The Noise Nomograph is a useful method of solving the various noise equations when it is assumed that the correlation factor (γ) is unity, a close approximation for most devices.

The more important noise equations are:

$$NF = 10 \log \left[1 + \frac{(e_n + i_n R_s)^2}{4KTRg} \right] \qquad (1)$$

$$NF_{min} = 10 \log \left[1 + \frac{e_n i_n}{KT} \right] \qquad (2)$$

$$R_{opt} = \frac{e_n}{i_n} \qquad (3)$$

☀ TELEDYNE SEMICONDUCTOR

APPENDIX 4. HARRIS 5190

FEATURES

- FAST SETTLING TIME 70ns
- VERY HIGH SLEW RATE 200V/μs
- WIDE GAIN-BANDWIDTH 150MHz
- POWER BANDWIDTH 6.5MHz
- LOW OFFSET VOLTAGE 5mV
- INPUT VOLTAGE NOISE .... 15nV/√Hz
- MONOLITHIC BIPOLAR CONSTRUCTION

APPLICATIONS

- FAST, PRECISE D/A CONVERTERS
- HIGH SPEED SAMPLE-HOLD CIRCUITS
- PULSE AND VIDEO AMPLIFIERS
- WIDEBAND AMPLIFIERS

SCHEMATIC

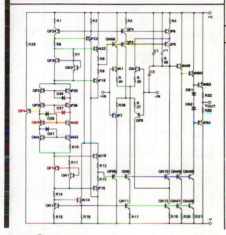

Copyright © Harris Corporation 1979

GENERAL DESCRIPTION

HA-5190/5195 are monolithic operational amplifiers featuring an ultimate combination of speed, precision, and bandwidth. Employing monolithic bipolar construction coupled with dielectric isolation, these devices are capable of delivering an unparalleled 200V/μs slew rate with a settling time of 70ns (0.1%, 5V output step.) These truly differential amplifiers are designed to operate at gains ≥ 5 without the need for external compensation. Other outstanding HA-5190/5195 features are 150MHz gain-bandwidth-product and 6.5MHz full power bandwidth. In addition to these dynamic characteristics, these amplifiers also have excellent input characteristics such as 5mV offset voltage and 15nV input voltage noise (at 1kHz).

With 200V/μs slew rate and 70ns settling time, these devices make ideal output amplifiers for accurate, high speed D/A converters or the main components in high speed sample/hold circuits. 150MHz gain-bandwidth-product, 6.5MHz power bandwidth, and 5mV offset voltage make HA-5190/5195 ideally suited for a variety of pulse and wideband video amplifier applications.

At temperatures above +75°C, a heat sink is required for HA-5190. (See note 2.) HA-5190 is specified over the –55°C to +125°C range while HA-5195 is specified from 0°C to +75°C.

PINOUTS

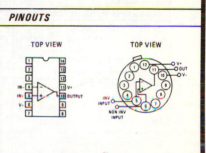

CAUTION:
These devices are sensitive to electrostatic discharge.

SPECIFICATIONS

ABSOLUTE MAXIMUM RATINGS (Note 1)

Voltage between V+ and V− Terminals	35V
Differential Input Voltage	6V
Output Current	50mA (Peak)
Internal Power Dissipation (Note 2)	870mW (Cerdip); 1W (TO−8) Free Air
Operating Temperature Range: (HA-5190)	$-55^\circ C \leq T_A \leq +125^\circ C$
(HA-5195)	$0^\circ C \leq T_A \leq +75^\circ C$
Storage Temperature Range	$-65^\circ C \leq T_A \leq +150^\circ C$

ELECTRICAL CHARACTERISTICS $V_{SUPPLY} = \pm 15$ Volts; $R_L = 200$ ohms, unless otherwise specified.

PARAMETER	TEMP	HA-5190 −55°C to +125°C			HA-5195 0°C to +75°C			UNITS
		MIN	TYP	MAX	MIN	TYP	MAX	
INPUT CHARACTERISTICS								
Offset Voltage	+25°C		3.0	5.0		3.0	6	mV
	FULL			10.0			10.0	mV
Average Offset Voltage Drift	FULL		20			20		µV/°C
Bias Current	+25°C		5	15		5	15	µA
	FULL			20			20	µA
Offset Current	+25°C		1	4		1	4	µA
	FULL			6			6	µA
Input Resistance	+25°C		10			10		Kohms
Input Capacitance	+25°C		1.0			1.0		pF
Common Mode Range	FULL	±5			±5			V
Input Noise Voltage (f = 1kHz, $R_g = 0\Omega$)	+25°C		15			15		nV/$\sqrt{Hz}$
TRANSFER CHARACTERISTICS								
Large Signal Voltage Gain (Note 3)	+25°C	15K	30K		10K	30K		V/V
	FULL	5K			5K			V/V
Common-Mode Rejection Ratio (Note 4)	FULL	74			74			dB
Gain-Bandwidth-Product (Notes 5 & 6)	+25°C		150			150		MHz
OUTPUT CHARACTERISTICS								
Output Voltage Swing (Note 3)	FULL	±5	±8		±5	±8		V
Output Current (Note 3)	+25°C	25	30		25	30		mA
Output Resistance	+25°C		TBD			TBD		Ohms
Full Power Bandwidth (Note 3 & 7)	+25°C	5	6.5		5	6.5		MHz
TRANSIENT RESPONSE (Note 8)								
Rise Time	+25°C		13	18		13	18	ns
Overshoot	+25°C		8			8		%
Slew Rate	+25°C	160	200		160	200		V/µs
Settling Time:								
5V Step to 0.1%	+25°C		70			70		ns
5V Step to 0.01%	+25°C		100			100		ns
2.5V Step to 0.1%	+25°C		50			50		ns
2.5V Step to 0.01%	+25°C		80			80		ns
POWER REQUIREMENTS								
Supply Current	FULL		19	25		19	25	mA
Power Supply Rejection Ratio (Note 9)	FULL	70	90		70	90		dB

* 100% tested for DASH 8. All other parameters for design information only.

NOTES

1. Absolute maximum ratings are limiting values, applied individually, beyond which the serviceability of the circuit may be impaired. Functional operability under any of these conditions is not necessarily implied.

2. Derate at 8.7mW/°C for operation at ambient temperatures above +75°C. Heat sinking required at temperatures above +75°C. $T_{JA} = 115°C/W$; $T_{JC} = 35°C/W$. Thermalloy model 6007 heat sink recommended.

3. $R_L = 200\,\Omega$, $C_L < 10pF$, $VO = \pm 5V$.

4. $V_{CM} = \pm 5V$.

5. $V_0 = 90mV$.

6. $A_V = 10$.

7. Full power bandwidth guaranteed based on slew rate measurement using $FPBW = \dfrac{Slew\ Rate}{2\pi V_{peak}}$.

8. Refer to Test Circuits section of data sheet.

9. $V_{SUPPLY} = \pm 10\ V.D.C.\ to\ \pm 15\ V.D.C.$

TEST CIRCUITS

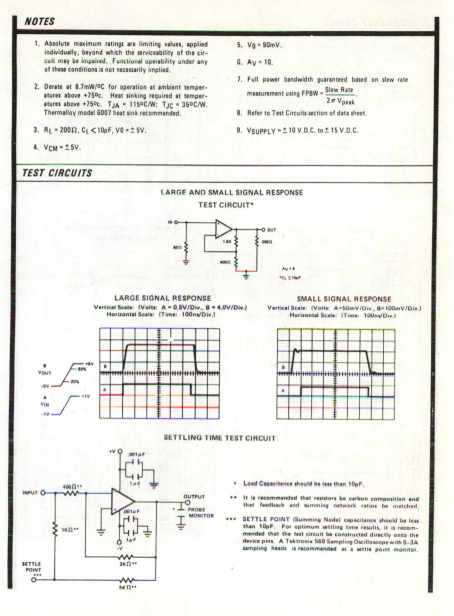

LARGE AND SMALL SIGNAL RESPONSE
TEST CIRCUIT*

LARGE SIGNAL RESPONSE
Vertical Scale: (Volts: A = 0.5V/Div., B = 4.0V/Div.)
Horizontal Scale: (Time: 100ns/Div.)

SMALL SIGNAL RESPONSE
Vertical Scale: (Volts: A=50mV/Div., B=100mV/Div.)
Horizontal Scale: (Time: 100ns/Div.)

SETTLING TIME TEST CIRCUIT

* Load Capacitance should be less than 10pF.

** It is recommended that resistors be carbon composition and that feedback and summing network ratios be matched.

*** SETTLE POINT (Summing Node) capacitance should be less than 10pF. For optimum settling time results, it is recommended that the test circuit be constructed directly onto the device pins. A Tektronix 568 Sampling Oscilloscope with S-3A sampling heads is recommended as a settle point monitor.

PERFORMANCE CURVES

V+ = +15V, V– = –15V, T$_A$ = +25°C unless otherwise stated.

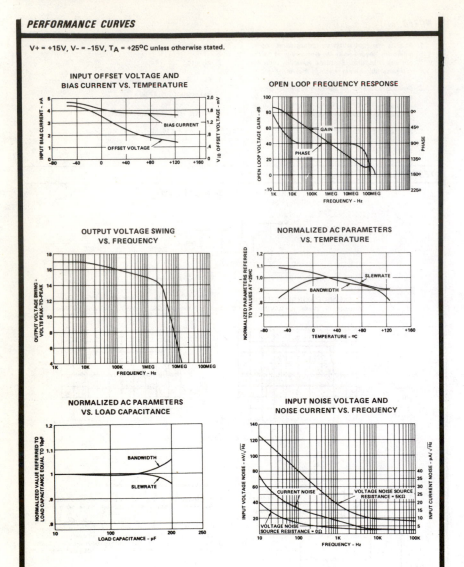

PERFORMANCE CURVES (Cont'd)

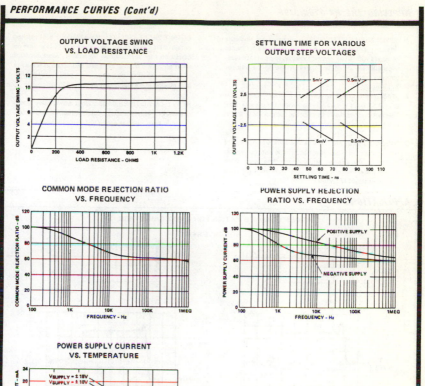

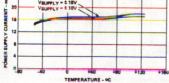

APPLYING THE HA-5190/5195

1. POWER SUPPLY DECOUPLING: Although not absolutely necessary, it is recommended that all power supply lines be decoupled with .01µF ceramic capacitors to ground. Decoupling capacitors should be located as near to the amplifier terminals as possible.

2. STABILITY CONSIDERATIONS: HA-5190/5195 is stable at gains ≥ 5. Gains < 5 are covered elsewhere in this data sheet. Feedback resistors should be of carbon composition located as near to the input terminals as possible.

3. WIRING CONSIDERATIONS: Video pulse circuits should be built on a ground plane. Minimum point to point connections directly to the amplifier terminals should be used. When ground planes cannot be used, good single point grounding techniques should be applied.

4. OUTPUT SHORT CIRCUIT: HA-5190/5195 does not have output short circuit protection. Short circuits to ground can be tolerated for approximately 10 seconds. Short circuits to either supply will result in immediate destruction of the device. In applications where short circuiting is possible, current limiting resistors in the supply lines are recommended.

5. HEAVY CAPACITIVE LOADS: When driving heavy capacitive loads (≥ 100pF) a small resistor (≈100Ω) should be connected in series with the output and inside the feedback loop.

APPLICATIONS

SUGGESTED COMPENSATION FOR UNITY GAIN STABILITY

NON-INVERTING

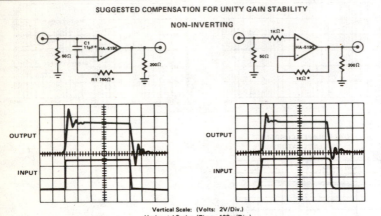

Vertical Scale: (Volts: 2V/Div.)
Horizontal Scale: (Time: 100ns/Div.)

* Values were determined experimentally for optimum speed and settling time.

R1 and C1 should be optimized for each particular application to ensure best overall frequency response.

INVERTING

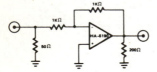

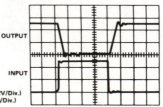

Vertical Scale: (Volts: 2V/Div.)
Horizontal Scale: (50ns/Div.)

APPLICATIONS (Cont'd)

VIDEO PULSE AMPLIFIER/75Ω COAXIAL DRIVER

VIDEO PULSE AMPLIFIER COAXIAL LINE DRIVER

FAST DAC OUTPUT BUFFER

Vertical Scale: (Volts: 2V/Div.)
Horizontal Scale: (Time: 50ns/Div.)

B = V$_{OUT}$ C = DIGITAL INPUT

* Time delay between B and C represents total time delay for 0V to +5V full scale coded change.

PACKAGING

CHIP

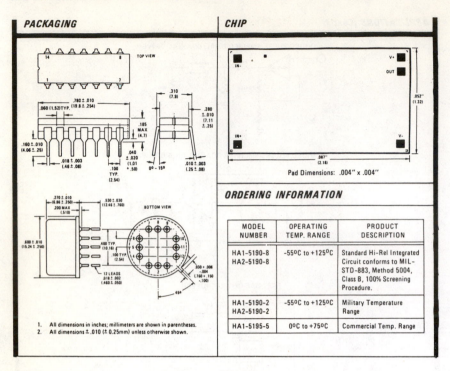

Pad Dimensions: .004" x .004"

1. All dimensions in inches; millimeters are shown in parentheses.
2. All dimensions ± .010 (± 0.25mm) unless otherwise shown.

ORDERING INFORMATION

MODEL NUMBER	OPERATING TEMP. RANGE	PRODUCT DESCRIPTION
HA1-5190-8 HA2-5190-8	-55°C to +125°C	Standard Hi-Rel Integrated Circuit conforms to MIL-STD-883, Method 5004, Class B, 100% Screening Procedure.
HA1-5190-2 HA2-5190-2	-55°C to +125°C	Military Temperature Range
HA1-5195-5	0°C to +75°C	Commercial Temp. Range

APPENDIX 5. BURR-BROWN 3554

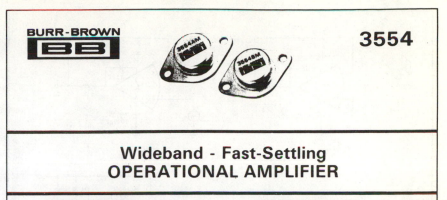

BURR-BROWN
BB

3554

Wideband - Fast-Settling
OPERATIONAL AMPLIFIER

FEATURES

- SLEW RATE, 1000V/μsec
- FAST SETTLING, 150nsec, max (to ±.05%)
- GAIN-BANDWIDTH PRODUCT, 1.7GHz
- FULL DIFFERENTIAL INPUT

APPLICATIONS

- PULSE AMPLIFIERS
- TEST EQUIPMENT
- WAVEFORM GENERATORS
- FAST D/A CONVERTERS

DESCRIPTION

The 3554 is a full differential input, wideband operational amplifier. It is designed specifically for the amplification or conditioning of wideband data signals and fast pulses. It features an unbeatable combination of gain-bandwidth product, settling time and slew rate. It uses hybrid construction. On the beryllia substrate are matched input FETs, thin-film resistors and high speed silicon dice. Active laser trimming and complete testing provide superior performance at a very moderate price.

The 3554 has a slew rate of 1000V/μsec and will output ±10V and ±100mA. When used as a fast

settling amplifier, the 3554 will settle to ±0.05% of the final value within 150nsec. A single external compensation capacitor allows the user to optimize the bandwidth, slew rate or settling time in the particular application.

The 3554 is reliable and rugged and addresses almost any application when speed and bandwidth are serious considerations. It is particularly a good choice for use in fast settling circuits, fast D/A converters, multiplexer buffers, comparators, waveform generators, integrators, and fast current amplifiers. It is available in several grades to allow selection of just the performance required.

International Airport Industrial Park - P.O. Box 11400 - Tucson, Arizona 85734 - Tel. (602) 746-1111 - Twx: 910-952-1111 - Cable: BBRCORP - Telex: 66-6491

TYPICAL CIRCUITS

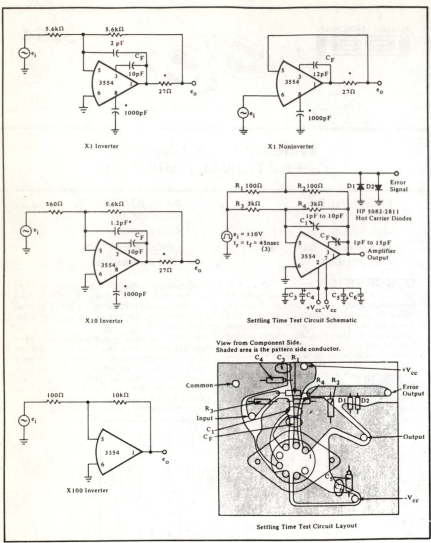

X1 Inverter

X1 Noninverter

X10 Inverter

Settling Time Test Circuit Schematic

X100 Inverter

View from Component Side.
Shaded area is the pattern side conductor.

Settling Time Test Circuit Layout

NOTES:
1. These circuits are optimized for driving large capacitive loads (to 470pF).
2. The 3554 is stable at gains of greater than 55 ($C_L \leq 100pF$) without any frequency compensation.
3. 45nsec is optimum. Very fast rise times (10-20nsec) may saturate the input stage causing less than optimum settling time performance.
*Indicates component that may be eliminated when large capacitive loads are not being driven by the device.

2

ELECTRICAL SPECIFICATIONS

At $T_{CASE} = 25°C$ and $±15VDC$, unless otherwise noted.

PARAMETERS	CONDITIONS	3554AM MIN	TYP	MAX	3554BM MIN	TYP	MAX	3554SM MIN	TYP	MAX	UNITS
OPEN-LOOP GAIN,DC					•	•	•	•	•	•	
No Load		100	106								dB
Rated Load	$R_L = 100Ω$	90	96								dB
RATED OUTPUT					•	•	•	•	•	•	
Voltage	$I_O = ±100mA$	±10	±11								V
Current	$V_O = ±10V$	±100	±125								mA
Output Resistance, open loop	$f = 10MHz$		20								•Ω
DYNAMIC RESPONSE					•	•	•	•	•	•	
Bandwidth (0dB, small signal)	$C_F = 0$	70†	90								MHz
Gain-bandwidth Product	$C_F = 0, G = 10 V/V$	150	225								MHz
	$C_F = 0, G = 100 V/V$	425	725								MHz
	$C_F = 0, G = 1000 V/V$	1000	1700								MHz
Full Power Bandwidth	$C_F = 0, V_o = 20Vp-p, R_L = 100Ω$	16	19								MHz
Slew Rate	$C_F = 0, V_o = 20Vp-p, R_L = 100Ω$	1000	1200								V/µsec
Settling Time to ±1%	$A = 1$		60								nsec
to ±.1%	$A = 1$		120								nsec
to ±.05%	$A = 1$		140	150							nsec
to ±.01%	$A = 1$		200	250							nsec
INPUT OFFSET VOLTAGE											
Initial offset, $T_A = 25°C$			±0.5	±2		±0.2	±1		±0.2	±1	mV
vs. Temp ($T_A = -25°C$ to $+85°C$)			±20	±50		±8	±15				µV/°C
vs. Temp ($T_A = -55°C$ to $+125°C$)									±12	±25	µV/°C
vs. Supply Voltage			±80	±300		•	•		•	•	µV/V
INPUT BIAS CURRENT					•	•	•	•	•	•	
Initial bias, 25°C		0	-10	-50							pA
vs. Temp			**								
vs. Supply Voltage			±1								pA/V
INPUT DIFFERENCE CURRENT					•	•	•	•	•	•	
Initial difference, 25°C			±2	±10							pA
INPUT IMPEDANCE					•			•			
Differential			$10^{11} \| 2$								Ω | pF
Common-mode			$10^{11} \| 2$								Ω | pF
INPUT NOISE					•	•	•.	•	•	•	
Voltage, $f_o = 1Hz$	$R_s = 100Ω$		125	450†							nV/√Hz
$f_o = 10Hz$	$R_s = 100Ω$		50	160†							nV/√Hz
$f_o = 100Hz$	$R_s = 100Ω$		25	90†							nV/√Hz
$f_o = 1kHz$	$R_s = 100Ω$		15	50†							nV/√Hz
$f_o = 10kHz$	$R_s = 100Ω$		10	35†							nV/√Hz
$f_o = 100kHz$	$R_s = 100Ω$		8	25†							nV/√Hz
$f_o = 1MHz$	$R_s = 100Ω$		7	25†							nV/√Hz
$f_B = .3Hz$ to $10Hz$	$R_s = 100Ω$		2	7†							µV, p-p
$f_B = 10Hz$ to $1MHz$	$R_s = 100Ω$		8	25							µV, rms
Current, $f_B = .3Hz$ to $10Hz$	$R_s = 100Ω$		45								fA, p-p
$f_B = 10Hz$ to $1MHz$	$R_s = 100Ω$		2								pA, rms
INPUT VOLTAGE RANGE					•	•	•	•	•	•	
Common-mode Voltage Range	Linear Operation		$±(|V_{cc}|-4)$								V
Common-mode Rejection	$f = DC, V_{CM} = +7V, -10V$	44	78								dB
Max. Safe Input Voltage			±Supply								V
POWER SUPPLY					•	•	•	•	•	•	
Rated Voltage			±15								VDC
Voltage Range, derated performance		±5		±18							VDC
Current, quiescent		±17	±35	±45							mA
TEMPERATURE RANGE (ambient)											
Specification		-25		+85	-25		+85	-55		+125	°C
Operating, derated performance		-55		+125	-55		+125	-55		+125	°C
Storage		-65		+150	-65		+150	-65		+150	°C
θ junction-case			15			15			15		°C/W
θ junction-ambient			45			45			45		°C/W

* Specifications same as for 3554AM
** Doubles every +10°C
† This parameter is untested and is not guaranteed. This specification is established to a 90% confidence level.
The information in this publication has been carefully checked and is believed to be reliable; however, no responsibility is assumed for possible inaccuracies or omissions. Prices and specifications are subject to change without notice. No patent rights are granted to any of the circuits described herein.

3

TYPICAL PERFORMANCE CURVES

at $T_C = +25^{\circ}C$ and $\pm15VDC$ unless otherwise noted.

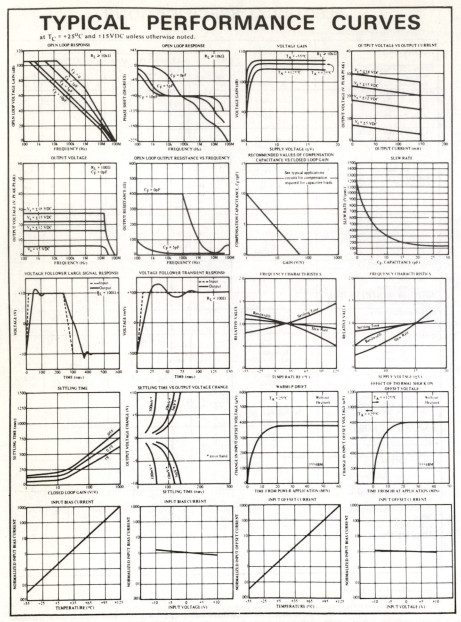

4

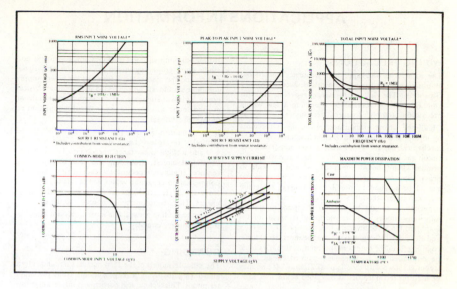

MECHANICAL

Seating Plane

DIM	INCHES		MILLIMETERS	
	MIN	MAX	MIN	MAX
A	1.510	1.550	38.35	39.37
B	.745	.770	18.92	19.56
C	.300	.400	7.62	10.16
D	.038	.042	0.97	1.07
E	.080	.105	2.03	2.67
F	40° BASIC		40° BASIC	
G	.500 BASIC		12.7 BASIC	
H	1.186 BASIC		30.12 BASIC	
J	.593 BASIC		15.06 BASIC	
K	.400	.500	10.16	12.70
Q	.151	.161	3.84	4.09
R	.980	1.020	24.89	25.91

Pin material and plating composition conform to Method 2003 (solderability) of Mil-Std-883 [except paragraph 3.2].

NOTE

Leads in true position within .010 (.25mm) R at MMC at seating plane
Pin numbers shown for reference only
Numbers may not be marked on package

AMPLIFIER CONNECTIONS

Offset Potentiometer (Optional)

$+V_S$

$20k\Omega$

$+V$

$-V_S$

C_F Frequency Compensation

There is no internal case connection.

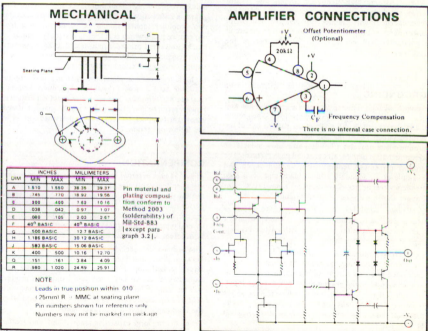

APPLICATIONS INFORMATION

WIRING PRECAUTIONS

The 3554 is a wideband, high frequency operational amplifier that has a gain-bandwidth product exceeding 1 Gigahertz. The full performance capability of this amplifier will be realized by observing a few wiring precautions and high frequency techniques.

Of all the wiring precautions, grounding is the most important and is described in an individual section. The mechanical circuit layout also is very important. All circuit element leads should be as short as possible. All printed circuit board conductors should be wide to provide low resistance, low inductance connections and should be as short as possible. In general, the entire physical circuit should be as small as practical. Stray capacitances should be minimized especially at high impedance nodes such as the input terminals of the amplifier. Pin 5, the inverting input, is especially sensitive and all associated connections must be short. Stray signal coupling from the output to the input or to pin 8 should be minimized. A recommended printed circuit board layout is shown with the "Typical Circuits." It also may be used for test purposes as described below.

When designing high frequency circuits low resistor values should be used; resistor values less than 5.6 kΩ are recommended. This practice will give the best circuit performance as the time constants formed with the circuit capacitances will not limit the performance of the amplifier.

GROUNDING

As with all high frequency circuits a ground plane and good grounding techniques should be used. The ground plane should connect all areas of the pattern side of the printed circuit board that are not otherwise used. The ground plane provides a low resistance, low inductance common return path for all signal and power returns. The ground plane also reduces stray signal pick up. An example of an adequate ground plane and good high frequency techniques is the Settling Time Test Circuit Layout shown with the "Typical Circuits."

Each power supply lead should be bypassed to ground as near as possible to the amplifier pins. A combination of a 1μF tantalum capacitor in parallel with a 470pF ceramic capacitor is a suitable bypass.

In inverting applications it is recommended that pin 6, the noninverting input, be grounded rather than being connected to a bias current compensating resistor. This assures a good signal ground at the noninverting input. A slight offset error will result; however, because the resistor values normally used in high frequency circuits are small and the bias current is small, the offset error will be minimal.

If point-to-point wiring is used or a ground plane is not, single point grounding should be used. The input signal return and the load signal return and the power supply common should all be connected at the same physical point. This will eliminate any common current paths or ground loops which could cause signal modulation or unwanted feedback.

It is recommended that the case of the 3554 not be grounded during use (it may, if desired). A grounded case will add a slight capacitance to each pin. To an already functional circuit, grounding the case will probably require slight compensation readjustment and the compensation capacitor values will be slightly different from those recommended in the typical performance curves. There is no internal connection to the case.

Proper grounding is the single most important aspect of high frequency circuitry.

GUARDING

The input terminals of the 3554 may be surrounded by a guard ring to divert leakage currents from the input terminals. This technique is particularly important in low bias current and high input impedance applications. The guard, a conductive path that completely surrounds the two amplifier inputs, should be connected to a low impedance point which is at the input signal potential. It blocks unwanted printed circuit board leakage currents from reaching the input terminals. The guard also will reduce stray signal coupling to the input.

In high frequency applications guarding may not be desirable as it increases the input capacitance and can degrade performance. The effects of input capacitance, however, can be compensated by a small capacitor placed across the feedback resistor. This is described further in the following section.

COMPENSATION

The 3554 uses external frequency compensation so that the user may optimize the bandwidth or slew rate or settling time for his particular application. Several typical performance curves are provided to aid in the selection of the correct compensation capacitance value. In addition several typical circuits show recommended compensation in different applications.

The primary compensation capacitor, C_F, is connected between pins 1 and 3. As the performance curves show, larger closed-loop gain configurations require less capacitance and an improved gain-bandwidth product will be realized. Note that no compensation capacitor is required for closed-loop gains above 55V/V and when the load capacitance is less than 100pF.

When driving large capacitive loads, 470pF and greater,

an additional capacitor, C_s, is connected between pin 8 and ground. This capacitor is typically 1000pF. It is particularly necessary in low closed loop voltage gain configurations. The value may be varied to optimize performance and will depend upon the load capacitance value. In addition, the performance may be optimized by connecting a small resistance in series with the output and a small capacitor from pin 1 to 5. See the "Typical Circuits" for the X10 Inverter.

The flat high frequency response of the 3554 may be preserved and any high frequency peaking avoided by connecting a small capacitor in parallel with the feedback resistor. This capacitor will compensate for the closed-loop, high frequency, transfer function zero that results from the time constant formed by the input capacitance of the amplifier, typically 2pF, and the input and feedback resistors. Using small resistor values will keep the break frequency of this zero sufficiently high, avoiding peaking and preserving the phase margin. Resistor values less than 5.6kΩ are recommended. The selected compensation capacitor may be a trimmer, a fixed capacitor or a planned PC board capacitance. The capacitance value is strongly dependent on circuit layout and closed-loop gain. It will typically be 2pF for a clean layout using low resistances (1kΩ) and up to 10pF for circuits using larger resistances.

SETTLING TIME

Settling time is truly a complete dynamic measure of the 3554's total performance. It includes the slew rate time, a large signal dynamic parameter, and the time to accurately reach the final value, a small signal parameter that is a function of bandwidth and open loop-gain. The settling time may be optimized for the particular application by selection of the closed-loop gain and the compensation capacitance. The best settling time is observed in low closed-loop gain circuits. A performance curve shows the settling time to three different error bands.

Settling time is defined as the total time required, from the signal input step, for the output to settle to within the specified error band around the final value. This error band is expressed as a percentage of the magnitude of the output transition.

SLEW RATE

Slew rate is primarily an output, large signal parameter. It has virtually no dependence upon the closed-loop gain or the bandwidth, per se. It is dependent upon compensation. Decreasing the compensation capacitor value will increase the available slew rate as shown in the performance curve. Stray capacitances may appear to the amplifier as compensation. To avoid limiting the slew rate performance, stray capacitances should be minimized.

CAPACITIVE LOADS

The 3554 will drive large capacitive loads (up to 1000pF) when properly compensated. See the section on "Compensation." The effect of a capacitive load is to decrease the phase margin of the amplifier. With compensation the amplifier will provide stable operation even with large capacitive loads.

The 3554 is particularly well suited for driving 50Ω loads connected via coaxial cables due to its ±100mA output drive capability. The capacitance of the coaxial cable, 29pF/foot of length for RG-58, does not load the amplifier when the coaxial cable or transmission line is terminated in the characteristic impedance of the transmission line.

OFFSET VOLTAGE ADJUSTMENT

The offset voltage of the 3554 may be adjusted to zero by connecting a 20kΩ linear potentiometer between pins 4 and 8 with the wiper connected to the positive supply. A small, noninductive potentiometer is recommended. The leads connecting the potentiometer to pins 4 and 8 should be no longer than 6 inches to avoid stray capacitance and stray signal pickup. Stray coupling from the output, pin 1, to pin 4 (negative feedback) or to pin 8 (positive feedback) should be avoided.

The potentiometer is optional and may be omitted when the guaranteed offset voltage is considered sufficiently low for the particular application.

For each microvolt of offset voltage adjusted, the offset voltage temperature drift will change by ±0.004μV/°C.

HEAT SINKING

The 3554 does not require a heat sink for operation in most environments. The use of a heat sink, however, will reduce the internal thermal rise and will result in cooler operating temperatures. At extreme temperature and under full load conditions a heat sink will be necessary as indicated in the "Maximum Power Dissipation" curve. A heat sink with 8 holes for the 8 amplifier pins should be used. Burr-Brown has heat sinks available in three sizes - 3°C/W, 4.2°C/W and 12°C/W. A separate product data sheet is available upon request.

When heat sinking the 3554, it is recommended that the heat sink be connected to the amplifier case and the combination not connected to the ground plane. For a single-sided printed circuit board, the heat sink may be mounted between the 3554 and the nonconductive side of the PC board, and insulating washers, etc., will not be required. The addition of a heat sink to an already functional circuit will probably require slight compensation readjustment for optimum performance due to the change in stray capacitances. The added stray capacitance from the heat sink to each pin will depend on the thickness and type of heat sink used.

SHORT CIRCUIT PROTECTION

The 3554 is short circuit protected for continuous output shorts to common. Output shorts to either supply will destroy the device, even for momentary connections. Output shorts to other potential sources are not recommended as they may cause permanent damage.

TESTING

The 3554 may be tested in conventional operational amplifier test circuits; however, to realize the full performance capabilities of the 3554, the test fixture must not limit the full dynamic performance capability of the amplifier. High frequency techniques must be employed. The most critical dynamic test is for settling time. The 3554 Settling Time Test Circuit Schematic and a test circuit layout is shown with the "Typical Circuits." The input pulse generator must have a flat topped, fast settling pulse to measure the true settling time of the amplifier. The layout exemplifies the high frequency considerations that must be observed. The layout also may be used as a guide for other test circuits. Good grounding, truly square drive signals, minimum stray coupling and small physical size are important.

Every 3554 is thoroughly tested prior to shipment assuring the user that all parameters equal or exceed their specifications.

APPENDIX 6. MOTOROLA 390

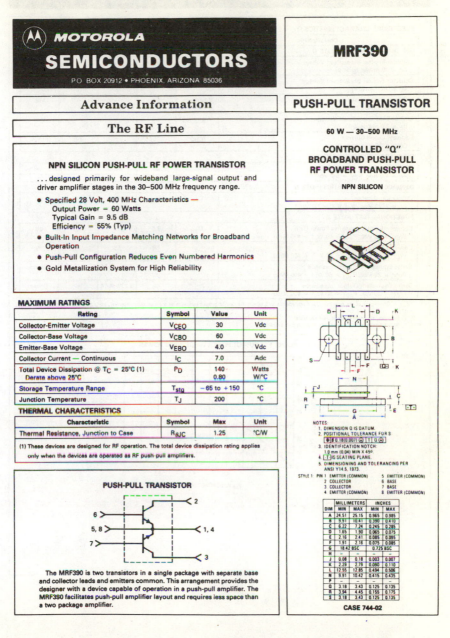

MOTOROLA
SEMICONDUCTORS
P.O. BOX 20912 • PHOENIX, ARIZONA 85036

MRF390

| Advance Information | PUSH-PULL TRANSISTOR |

The RF Line

60 W — 30–500 MHz

CONTROLLED "Q"
BROADBAND PUSH-PULL
RF POWER TRANSISTOR

NPN SILICON

NPN SILICON PUSH-PULL RF POWER TRANSISTOR

. . . designed primarily for wideband large-signal output and driver amplifier stages in the 30–500 MHz frequency range.

- Specified 28 Volt, 400 MHz Characteristics —
  Output Power = 60 Watts
  Typical Gain = 9.5 dB
  Efficiency = 55% (Typ)
- Built-In Input Impedance Matching Networks for Broadband Operation
- Push-Pull Configuration Reduces Even Numbered Harmonics
- Gold Metallization System for High Reliability

MAXIMUM RATINGS

Rating	Symbol	Value	Unit
Collector-Emitter Voltage	V_{CEO}	30	Vdc
Collector-Base Voltage	V_{CBO}	60	Vdc
Emitter-Base Voltage	V_{EBO}	4.0	Vdc
Collector Current — Continuous	I_C	7.0	Adc
Total Device Dissipation @ T_C = 25°C (1) Derate above 25°C	P_D	140 0.80	Watts W/°C
Storage Temperature Range	T_{stg}	−65 to +150	°C
Junction Temperature	T_J	200	°C

THERMAL CHARACTERISTICS

Characteristic	Symbol	Max	Unit
Thermal Resistance, Junction to Case	$R_{\theta JC}$	1.25	°C/W

(1) These devices are designed for RF operation. The total device dissipation rating applies only when the devices are operated as RF push-pull amplifiers.

PUSH-PULL TRANSISTOR

The MRF390 is two transistors in a single package with separate base and collector leads and emitters common. This arrangement provides the designer with a device capable of operation in a push-pull amplifier. The MRF390 facilitates push-pull amplifier layout and requires less space than a two package amplifier.

NOTES:
1. DIMENSION Q IS DATUM.
2. POSITIONAL TOLERANCE FOR S:
3. IDENTIFICATION NOTCH:
   1.0 mm (0.04) MIN X 45°
4. IS SEATING PLANE.
5. DIMENSIONING AND TOLERANCING PER ANSI Y14.5, 1973.

STYLE 1: PIN 1. EMITTER (COMMON)   5. EMITTER (COMMON)
2. COLLECTOR   6. BASE
3. COLLECTOR   7. BASE
4. EMITTER (COMMON)   8. EMITTER (COMMON)

DIM	MILLIMETERS MIN	MAX	INCHES MIN	MAX
A	24.51	25.15	0.965	0.985
B	9.91	10.41	0.390	0.410
C	6.22	7.24	0.245	0.285
D	1.65	1.90	0.065	0.075
E	2.16	2.41	0.085	0.095
F	1.91	2.16	0.075	0.085
G	18.42 BSC		0.725 BSC	
H	—	—	—	—
J	0.08	0.18	0.003	0.007
K	2.29	2.79	0.090	0.110
L	12.55	12.85	0.494	0.506
N	9.91	10.42	0.415	0.435
P	—	—	—	—
Q	3.18	3.43	0.125	0.135
R	3.94	4.45	0.155	0.175
S	3.18	3.43	0.125	0.135

CASE 744-02

MRF390

ELECTRICAL CHARACTERISTICS (T_C = 25°C unless otherwise noted.)

Characteristic	Symbol	Min	Typ	Max	Unit
OFF CHARACTERISTICS (NOTE 1)					
Collector-Emitter Breakdown Voltage (I_C = 30 mAdc, I_B = 0)	$V_{(BR)CEO}$	30	—	—	Vdc
Collector-Emitter Breakdown Voltage (I_C = 30 mAdc, V_{BE} = 0)	$V_{(BR)CES}$	60	—	—	Vdc
Collector-Base Breakdown Voltage (I_C = 30 mAdc, I_E = 0)	$V_{(BR)CBO}$	60	—	—	Vdc
Emitter-Base Breakdown Voltage (I_E = 3.0 mAdc, I_C = 0)	$V_{(BR)EBO}$	4.0	—	—	Vdc
Collector Cutoff Current (V_{CB} = 30 Vdc, I_E = 0)	I_{CBO}	—	—	3.0	mAdc
ON CHARACTERISTICS (NOTE 1)					
DC Current Gain (I_C = 1.0 Adc, V_{CE} = 5.0 Vdc)	h_{FE}	20	—	100	—
DYNAMIC CHARACTERISTICS (NOTE 1)					
Output Capacitance (V_{CB} = 28 Vdc, I_E = 0, f = 1.0 MHz)	C_{ob}	—	37	50	pF
FUNCTIONAL TEST (NOTE 2)					
Common-Emitter Amplifier Power Gain (V_{CC} = 28 Vdc, P_{out} = 60 W, f = 400 MHz)	G_{pe}	7.0	9.5	—	dB
Collector Efficiency (V_{CC} = 28 Vdc, P_{out} = 60 W, f = 400 MHz)	η	50	55	—	%
Series Equivalent Input Impedance (V_{CC} = 28 Vdc, P_{out} = 60 W, f = 400 MHz)	Z_{in}	—	7.22 + j5.15	—	Ohms
Series Equivalent Output Impedance (V_{CC} = 28 Vdc, P_{out} = 60 W, f = 400 MHz)	$Z_{OL}*$	—	9.17 − j4.14	—	Ohms

*Z_{OL} = Conjugate of the optimum load impedance into which the device output operates at a given output power, voltage and frequency.
NOTES:
1. Each transistor chip measured separately.
2. Both transistor chips operating in push-pull amplifier.

FIGURE 1 — OUTPUT POWER versus INPUT POWER

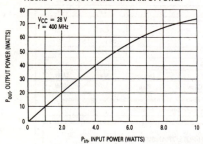

APPENDIX 7. RCA CD4046

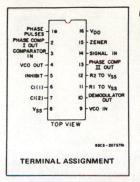

PHASE PULSES	1
PHASE COMP I OUT	2
COMPARATOR IN	3
VCO OUT	4
INHIBIT	5
CI(1)	6
CI(2)	7
V$_{SS}$	8

TOP VIEW

98CS - 20757RI

TERMINAL ASSIGNMENT

16 - V$_{DD}$
15 - ZENER
14 - SIGNAL IN
13 - PHASE COMP II OUT
12 - R2 TO V$_{SS}$
11 - R1 TO V$_{SS}$
10 - DEMODULATOR OUT
9 - VCO IN

COS/MOS Micropower Phase-Locked Loop

Features:

- Very low power consumption:
  70 μW (typ.) at VCO f$_o$ = 10 kHz, V$_{DD}$ = 5 V
- Operating frequency range up to 1.2 MHz (typ.)
  at V$_{DD}$ = 10 V
- Wide supply-voltage range: V$_{DD}$ − V$_{SS}$ = 5 to 15 V
- Low frequency drift: 0.06%/°C (typ.) at V$_{DD}$ = 10 V
- Choice of two phase comparators:
  1. Exclusive-OR network
  2. Edge-controlled memory network with phase-pulse output for lock indication

- High VCO linearity: 1% (typ.)
- VCO inhibit control for ON-OFF keying and ultra-low standby power consumption
- Source-follower output of VCO control input (Demod. output)
- Zener diode to assist supply regulation
- Quiescent current specified to 15 V
- Maximum input leakage current of 1 μA at 15 V (full package-temperature range)

Applications:

- FM demodulator and modulator
- Frequency synthesis and multiplication
- Frequency discriminator
- Data synchronization
- Voltage-to-frequency conversion
- Tone decoding
- FSK — Modems
- Signal conditioning
- (See ICAN-6101) "RCA COS/MOS Phase-Locked Loop — A Versatile Building Block for Micropower Digital and Analog Applications"

The RCA-CD4046A COS/MOS Micropower Phase-Locked Loop (PLL) consists of a low-power, linear voltage-controlled oscillator (VCO) and two different phase comparators having a common signal-input amplifier and a common comparator input. A 5.2-V zener diode is provided for supply regulation if necessary. The CD4046A is supplied in a 16-lead dual-in-line ceramic package (CD-4046AD), a 16-lead dual-in-line plastic package (CD4046AE), and a 16-lead flat pack (CD4046AK). It is also available in chip form (CD4046AH).

VCO Section

The VCO requires one external capacitor C1 and one or two external resistors (R1 or R1 and R2). Resistor R1 and capacitor C1 determine the frequency range of the VCO and resistor R2 enables the VCO to have a frequency offset if required. The high input impedance ($10^{12}\Omega$) of the VCO simplifies the design of low-pass filters by permitting the designer a wide choice of resistor-to-capacitor ratios. In order not to load the low-pass filter, a source-follower output of the VCO input voltage is provided at terminal 10 (DEMODULATED OUTPUT). If this terminal is used, a load resistor (R$_S$) of 10 kΩ or more should be connected from this terminal to V$_{SS}$. If unused this terminal should be left open. The VCO can be connected either directly or through frequency dividers to the comparator input of the phase comparators. A full COS/MOS logic swing is available at the output of the VCO and allows direct coupling to COS/MOS frequency dividers such as the RCA-CD4024, CD4018, CD4020, CD4022, CD4029, and CD4059. One or more CD4018 (Preset-table Divide-by-N Counter) or CD4029 (Pre-settable Up/Down Counter), or CD4059A (Programmable Divide by "N" Counter), together with the CD4046A (Phase-Locked Loop) can be used to build a micropower low-frequency synthesizer. A logic 0 on the INHIBIT input "enables" the VCO and the source follower, while a logic 1 "turns off" both to minimize stand-by power consumption.

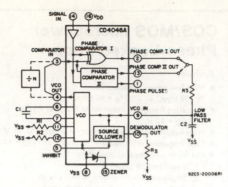

Fig.1 — COS/MOS phase-locked loop block diagram.

Phase Comparators

The phase-comparator signal input (terminal 14) can be direct-coupled provided the signal swing is within COS/MOS logic levels [logic "0" $\leq 30\%$ $(V_{DD}-V_{SS})$, logic "1" $\geq 70\%$ $(V_{DD}-V_{SS})$]. For smaller swings the signal must be capacitively coupled to the self-biasing amplifier at the signal input.

Phase comparator I is an exclusive-OR network; it operates analagously to an over-driven balanced mixer. To maximize the lock range, the signal- and comparator-input frequencies must have a 50% duty cycle. With no signal or noise on the signal input, this phase comparator has an average output voltage equal to $V_{DD}/2$. The low-pass filter connected to the output of phase comparator I supplies the averaged voltage to the VCO input, and causes the VCO to oscillate at the center frequency (f_O).

The frequency range of input signals on which the PLL will lock if it was initially out of lock is defined as the frequency capture range $(2f_C)$.

The frequency range of input signals on which the loop will stay locked if it was initially in lock is defined as the frequency lock range $(2f_L)$. The capture range is $\leq$ the lock range.

With phase comparator I the range of frequencies over which the PLL can acquire lock (capture range) is dependent on the low-pass-filter characteristics, and can be made as large as the lock range. Phase-comparator I enables a PLL system to remain in lock in spite of high amounts of noise in the input signal.

One characteristic of this type of phase comparator is that it may lock onto input frequencies that are close to harmonics of the VCO center-frequency. A second characteristic is that the phase angle between the signal and the comparator input varies between 0° and 180°, and is 90° at the center frequency. Fig. 2 shows the typical, triangular, phase-to-output response characteristic

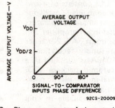

Fig.2 — Phase-comparator I characteristics at low-pass filter output.

of phase-comparator I. Typical waveforms for a COS/MOS phase-locked-loop employing phase comparator I in locked condition of f_O is shown in Fig. 3.

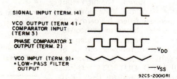
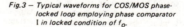

Fig.3 — Typical waveforms for COS/MOS phase-locked loop employing phase comparator I in locked condition of f_O.

Phase-comparator II is an edge-controlled digital memory network. It consists of four flip-flop stages, control gating, and a three-state output circuit comprising p- and n-type drivers having a common output node. When the p-MOS or n-MOS drivers are ON they pull the output up to V_{DD} or down to V_{SS}, respectively. This type of phase comparator acts only on the positive edges of the signal and comparator inputs. The duty cycles of the signal and comparator inputs are not important since positive transitions control the PLL system utilizing this type of comparator. If the signal-input frequency is higher than the comparator-input

CD4046A Types ———————————————— **File No. 637**

frequency, the p-type output driver is maintained ON most of the time, and both the n and p drivers OFF (3 state) the remainder of the time. If the signal-input frequency is lower than the comparator-input frequency, the n-type output driver is maintained ON most of the time, and both the n and p drivers OFF (3 state) the remainder of the time. If the signal- and comparator-input frequencies are the same, but the signal input lags the comparator input in phase, the n-type output driver is maintained ON for a time corresponding to the phase difference. If the signal- and comparator-input frequencies are the same, but the comparator input lags the signal in phase, the p-type output driver is maintained ON for a time corresponding to the phase difference. Subsequently, the capacitor voltage of the low-pass filter connected to this phase comparator is adjusted until the signal and comparator inputs are equal in both phase and frequency. At this stable point both p-

and n-type output drivers remain OFF and thus the phase comparator output becomes an open circuit and holds the voltage on the capacitor of the low-pass filter constant. Moreover the signal at the "phase pulses" output is a high level which can be used for indicating a locked condition. Thus, for phase comparator II, no phase difference exists between signal and comparator input over the full VCO frequency range. Moreover, the power dissipation due to the low-pass filter is reduced when this type of phase comparator is used because both the p- and n-type output drivers are OFF for most of the signal input cycle. It should be noted that the PLL lock range for this type of phase comparator is equal to the capture range, independent of the low-pass filter. With no signal present at the signal input, the VCO is adjusted to its lowest frequency for phase comparator II. Fig. 4 shows typical waveforms for a COS/MOS PLL employing phase comparator II in a locked condition.

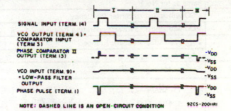

Fig.4 — Typical waveforms for COS/MOS phase-locked loop employing phase comparator II in locked condition.

RECOMMENDED OPERATING CONDITIONS

For maximum reliability, nominal operating conditions should be selected so that operation is always within the following range:

CHARACTERISTIC	LIMITS		UNITS
	Min.	Max.	
Supply Voltage Range (For T_A = Full Package Temperature Range	3	12	V

MAXIMUM RATINGS, *Absolute-Maximum Values:*

STORAGE-TEMPERATURE RANGE (T_{stg}) .... −65 to +150°C
OPERATING-TEMPERATURE RANGE (T_A):
  PACKAGE TYPES D, F, K, H .... −55 to +125°C
  PACKAGE TYPES E, Y .... −40 to +85°C
DC SUPPLY-VOLTAGE RANGE, (V_{DD})
  (Voltages referenced to V_{SS} Terminal) .... −0.5 to +15 V
POWER DISSIPATION PER PACKAGE (P_D):
  FOR T_A = −40 to +60°C (PACKAGE TYPES E, Y) .... 500 mW
  FOR T_A = +60 to +85°C (PACKAGE TYPES E, Y) .... Derate Linearly at 12 mW/°C to 200 mW
  FOR T_A = −55 to +100°C (PACKAGE TYPES D, F, K) .... 500 mW
  FOR T_A = +100 to +125°C (PACKAGE TYPES D, F, K) ...... Derate Linearly at 12 mW/°C to 200 mW
DEVICE DISSIPATION PER OUTPUT TRANSISTOR
  FOR T_A = FULL PACKAGE TEMPERATURE RANGE (ALL PACKAGE TYPES) .... 100 mW
INPUT VOLTAGE RANGE, ALL INPUTS .... −0.5 to V_{DD} +0.5 V
LEAD TEMPERATURE (DURING SOLDERING):
  At distance 1/16 ± 1/32 inch (1.59 ± 0.79 mm) from case for 10 s max. .... +265°C

DESIGN INFORMATION

This information is a guide for approximating the values of external components for the CD4046A in a Phase-Locked-Loop system. The selected external components must be within the following ranges:

$10 \text{ k}\Omega \leqslant R1, R2, R_S \leqslant 1 \text{ M}\Omega$
$C1 \geqslant 100 \text{ pF at } V_{DD} \geqslant 5 \text{ V};$
$C1 \geqslant 50 \text{ pF at } V_{DD} \geqslant 10 \text{ V}$

In addition to the given design information refer to Fig.5 for R1, R2, and C1 component selections.

Characteristics	Phase Comparator Used	Design Information	
		VCO WITHOUT OFFSET $R_2 = \infty$	**VCO WITH OFFSET**
VCO Frequency	1	(graph: VCO INPUT VOLTAGE, f_{MAX}, f_0, $2f_L$, f_{MIN}, $V_{DD}/2$, V_{DD})	(graph: VCO INPUT VOLTAGE, f_0, f_{MAX}, f_{MIN}, $2f_L$, $V_{DD}/2$, V_{DD}) 92CS-20012RI
	2	Same as for No.1	
For No Signal Input	1	VCO will adjust to center frequency, f_O	
	2	VCO will adjust to lowest operating frequency, f_{min}	
Frequency Lock Range, $2 f_L$	1	$2 f_L$ = full VCO frequency range	
		$2 f_L = f_{max} - f_{min}$	
	2	Same as for No.1	
Frequency Capture Range, $2 f_C$ Loop Filter Component Selection	1	(circuit: IN R3 OUT, $\tau 1 = R3 C2$, C2) $(1), (2)$ $2 f_C \approx \frac{1}{\pi}\sqrt{\frac{2\pi f_L}{\tau 1}}$ (circuit: IN R3 OUT, R4, C2) For $2 f_C$, see Ref. (2) 92CS-21901	
	2	$f_C = f_L$	
Phase Angle Between Signal and Comparator	1	90^0 at center frequency (f_0) approximating 0^0 and 180^0 at ends of lock range ($2 f_L$)	
	2	Always 0^0 in lock	

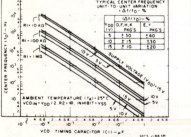

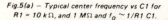

Fig.5(a) — Typical center frequency vs C1 for R1 = 10 kΩ, and 1 MΩ and f₀ ∼ 1/R1 C1.

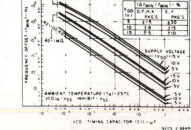

Fig.5(b) — Typical frequency offset vs C1 for R2 = 10 kΩ, 100 kΩ, and 1 MΩ.

NOTE: Lower frequency values are obtainable if larger values of C1 than shown in Figs. 5(a) and 5(b) are used.

CD4046A Types ———————————————— File No. 637

DESIGN INFORMATION (Cont'd):

Characteristics	Phase Comparator Used	Design Information	
Locks On Harmonic of Center Frequency	1	Yes	
	2	No	
Signal Input Noise Rejection	1	High	
	2	Low	
VCO Component Selection		**VCO WITHOUT OFFSET** $R_2 = \infty$	**VCO WITH OFFSET**
	1	– Given: f_o – Use f_o with Fig.5a to determine R1 and C1	– Given: f_o and f_L – Calculate f_{min} from the equation $f_{min} = f_o - f_L$ – Use f_{min} with Fig.5b to determine R2 and C1 – Calculate $\dfrac{f_{max}}{f_{min}}$ from the equation $\dfrac{f_{max}}{f_{min}} = \dfrac{f_o + f_L}{f_o - f_L}$ – Use $\dfrac{f_{max}}{f_{min}}$ with Fig.5c to determine ratio R2/R1 to obtain R1
	2	– Given: f_{max} – Calculate f_o from the equation $f_o = \dfrac{f_{max}}{2}$ –Use f_o with Fig.5a to determine R1 and C1	– Given: f_{min} & f_{max} – Use f_{min} with Fig.5b to determine R2 and C1 – Calculate $\dfrac{f_{max}}{f_{min}}$ – Use $\dfrac{f_{max}}{f_{min}}$ with Fig.5c to determine ratio R2/R1 to obtain R1

For further information, see
(1) F. Gardner, "Phase-Lock Techniques" John Wiley and Sons, New York, 1966
(2) G. S. Moschytz, "Miniaturized RC Filters Using Phase-Locked Loop", BSTJ, May, 1965.

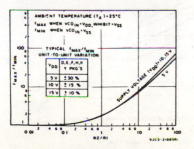

Fig.5(c) – Typical f_{max}/f_{min} vs R2/R1.

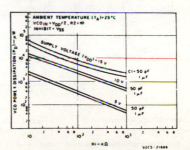
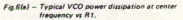

Fig.6(a) – Typical VCO power dissipation at center frequency vs R1.

File No. 637 _____ **CD4046A Types**

ELECTRICAL CHARACTERISTICS at $T_A = 25°C$

Characteristic	Test Conditions		V_O Volts	V_{DD} Volts	Limits All Package Types D,E,F,H,K,Y			Units
					Min.	Typ.	Max.	
Phase Comparator Section								
Operating Supply Voltage, $V_{DD} - V_{SS}$	VCO Operation		–		5	–	15	V
	Comparators only		–		3	–	15	
Total Quiescent Device Current, I_L: Term. 14 Open	Term. 15 open Term. 5 at V_{DD} Terms 3 & 9 at V_{SS}		5		–	25	55	μA
			10		–	200	410	
Term. 14 at V_{SS} or V_{DD}			5		–	5	15	
			10		–	25	60	
			15		–	50	500	
Term. 14 (SIGNAL IN) Input Impedance, Z_{14}			5		1	2	–	MΩ
			10		0.2	0.4	–	
			15		–	0.2		
AC-Coupled Signal Input Voltage Sensitivity* (peak-to-peak)	See Fig.7		5		–	200	400	mV
			10		–	400	800	
			15		–	700	–	
DC-Coupled Signal Input and Comparator Input Voltage Sensitivity Low Level			5		1.5	2.25	–	V
			10		3	4.5	–	
			15		4.5	6.75	–	
High Level		V_O Volts	5		–	2.75	3.5	
			10		–	5.5	7	
			15		–	8.25	–	
Output Drive Current: n-Channel (Sink), I_{DN}	Phase Comparator I & II Term. 2 & 13		0.5	5	0.43	0.86	–	mA
			0.5	10	1.3	2.5	–	
	Phase Pulses		0.5	5	0.23	0.47	–	
			0.5	10	0.7	1.4	–	
p-Channel (Source), I_{DP}	Phase Comparator I & II Term. 2 & 13		4.5	5	–0.3	–0.6	–	
			9.5	10	–0.9	–1.8	–	
	Phase Pulses		4.5	5	–0.08	–0.16	–	
			9.5	10	–0.25	–0.5	–	
Input Leakage Current, I_{IL}, I_{IH} Max.	Any Input			15	–	$\pm10^{-5}$	±1	μA

* For sine wave, the frequency must be greater than 1 kHz for Phase Comparator II.

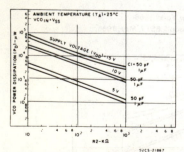
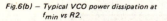

Fig.6(b) — Typical VCO power dissipation at f_{min} vs R2.

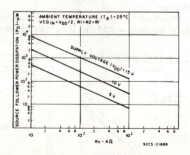
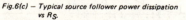

Fig.6(c) — Typical source follower power dissipation vs R_S.

NOTE: To obtain approximate total power dissipation of PLL system for no-signal input
P_D (Total) = P_D (f_o) + P_D (f_{MIN}) + P_D (R_S) – Phase Comparator I

P_D (Total) = P_D (f_{MIN}) – Phase Comparator II

CD4046A Types ———————————————— **File No. 637**

ELECTRICAL CHARACTERISTICS at $T_A = 25^\circ C$

Characteristic	Test Conditions		V_O Volts	V_{DD} Volts	Limits All Package Types D,E,F,H,K,Y			Units
					Min.	Typ.	Max.	
VCO Section								
Operating Supply Voltage $V_{DD}-V_{SS}$	As fixed oscillator only				3	–	15	V
	Phase-lock-loop operation				5	–	15	
Operating Power Dissipation, P_D	$f_o = 10$ kHz $R_2 = \infty$	$R_1 = 1$ MΩ $VCO_{IN} = \dfrac{V_{DD}}{2}$		5	–	70	–	μW
				10	–	600	–	
				15	–	2400	–	
Maximum Operating Frequency, f_{max}	$R1 = 10$ kΩ $R_2 = \infty$ $VCO_{IN} = V_{DD}$	$C1 = 100$ pF $C1 = 50$ pF		5	0.25	0.5	–	MHz
				10	0.6	1.2	–	
				15		1.5	–	
Center Frequency (f_o) and Frequency Range, $f_{max}-f_{min}$	Programmable with external components R1, R2, and C1 *See Design Information*							
Linearity	$VCO_{IN} = 2.5$ V ± 0.3 V, R1 > 10 kΩ			5	–	1	–	%
	$= 5$ V ± 2.5 V, R1 > 400 kΩ			10	–	1	–	
	$= 7.5$ V ± 5 V, R1 $= 1$ MΩ			15	–	1	–	
Temperature-Frequency Stability[*]: No Frequency Offset $f_{MIN} = 0$	%/$^\circ$C $\propto \dfrac{1}{f \cdot V_{DD}}$ $R2 = \infty$			5	–	0.12–0.24	–	%/$^\circ$C
				10	–	0.04–0.08	–	
				15	–	0.015–0.03	–	
Frequency Offset $f_{MIN} \neq 0$	%/$^\circ$C $\propto \dfrac{1}{f \cdot V_{DD}}$			5	–	0.06–0.12	–	
				10	–	0.05–0.1	–	
				15	–	0.03–0.06	–	
Input Resistance of VCO_{IN} (Term 9), R_I				5,10,15	–	10^{12}	–	Ω
VCO Output Voltage (Term 4) Low Level, V_{OL}				5,10,15	–	–	0.01	V
High Level, V_{OH}	Driving COS/MOS-Type Load (e.g. Term 3 Phase Comparator Input)			5	4.99	–	–	
				10	9.99	–	–	
				15	14.99	–	–	
VCO Output Duty Cycle				5,10,15	–	50	–	%
VCO Output Transition Times, t_{THL}, t_{TLH}			V_O Volts	5	–	75	150	ns
				10	–	50	100	
				15	–	40	–	
VCO Output Drive Current: n-Channel (Sink), I_{DN}			0.5	5	0.43	0.86	–	mA
			0.5	10	1.3	2.6	–	
p-Channel (Source), I_{DP}			4.5	5	–0.3	–0.6	–	
			9.5	10	–0.9	–1.8	–	
Source-Follower Output (Demodulated Output): Offset Voltage $(VCO_{IN}-V_{DEM})$	$R_S > 10$ kΩ			5,10	–	1.5	2.2	V
				15	–	1.5	–	
Linearity	$R_S > 50$kΩ	$VCO_{IN} = 2.5 \pm 0.3$ V		5	–	0.1	–	%
		$= 5 \pm 2.5$ V		10	–	0.6	–	
		$= 7.5 \pm 5$ V		15	–	0.8	–	
Zener Diode Voltage (V_Z): CD4046AD,AF,AK	$I_Z = 50$ μA				4.7	5.2	5.7	V
CD4046AE,AY					4.5	5.2	6.1	
Zener Dynamic Resistance, R_Z	$I_Z = 1$ mA				–	100	–	Ω

[*] Positive coefficient.

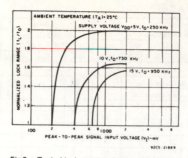

Fig.7 — Typical lock range vs signal input amplitude.

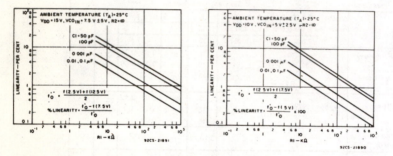

Fig.8(a) and (b) — Typical VCO linearity vs R1 and C1.

OPERATING & HANDLING CONSIDERATIONS

1. Handling

All inputs and outputs of this device have a network for electrostatic protection during handling. Recommended handling practices for COS/MOS devices are described in ICAN-6000 "Handling and Operating Considerations for MOS Integrated Circuits".

2. Operating

Operating Voltage

During operation near the maximum supply voltage limit, care should be taken to avoid or suppress power supply turn-on and turn-off transients, power supply ripple, or ground noise; any of these conditions must not cause $V_{DD} - V_{SS}$ to exceed the absolute maximum rating.

Input Signals

To prevent damage to the input protection circuit, input signals should never be greater than V_{DD} nor less than V_{SS}. Input currents must not exceed 10 mA even when the power supply is off.

Unused Inputs

A connection must be provided at every input terminal. All unused input terminals must be connected to either V_{DD} or V_{SS}, whichever is appropriate.

Output Short Circuits

Shorting of outputs to V_{DD} or V_{SS} may damage COS/MOS devices by exceeding the maximum device dissipation.

CD4046A Types ———————————————————— File No. 637

Dimensions and pad layout for CD4046A

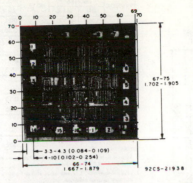

*Dimensions in parentheses are in millimeters and are
derived from the basic inch dimensions as indicated.
Grid graduations are in mils (10^{-3} inch).*

ORDERING INFORMATION

RCA COS/MOS device packages are identified by letters indicated in the following
chart. When ordering a COS/MOS device, it
is important that the appropriate suffix letter
be affixed to the type number of the device.

Package	Suffix Letter
Dual-In-Line White Ceramic	D
Dual-In-Line Frit-Seal Ceramic:	
Commercial Type	Y
Premium Type	F
Dual-In-Line Plastic	E
Ceramic Flat Package	K
Chip	H

For example, a CD4046 "A"-Series type
in a dual-in-line plastic package will be
identified as the CD4046AE.

File No. 637 ———————————————————————————— **CD4046A Types**

DIMENSIONAL OUTLINES

(K) Suffix
JEDEC MO-004-AG
16-Lead Ceramic Flat Pack

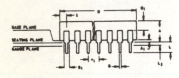

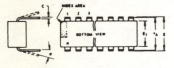

SYMBOL	INCHES		NOTE	MILLIMETERS	
	MIN.	MAX.		MIN.	MAX.
A	0.008	0.100		0.21	2.54
B	0.015	0.019	1	0.381	0.482
C	0.003	0.006	1	0.077	0.152
e	0.050 TP		2	1.27 TP	
E	0.200	0.300		5.1	7.6
H	0.600	1.000		15.3	25.4
L	0.150	0.350		3.9	8.8
N	16		3	16	
Q	0.005	0.050		0.13	1.27
S	0.000	0.025		0.00	0.63
Z	0.300		4	7.62	
Z_1	0.400		4	10.16	

92CS-I727IRI

NOTES:
1. Refer to Rules for Dimensioning (JEDEC Publication No. 13) for Axial Lead Product Outlines
2. Leads within .005'' (.12 mm) radius of True Position (TP) at maximum material condition.
3. N is the maximum quantity of lead positions.
4. Z and Z_1 determine a zone within which all body and lead irregularities lie.

(E), (F), and (Y) Suffix
JEDEC MO-001-AC 16-Lead Dual-In-Line
Plastic or Frit-Seal Ceramic Package

SYMBOL	INCHES		NOTE	MILLIMETERS	
	MIN.	MAX.		MIN.	MAX.
A	0.155	0.200		3.94	5.08
A_1	0.020	0.050		0.51	1.27
B	0.014	0.020		0.356	0.508
B_1	0.035	0.065		0.89	1.65
C	0.008	0.012	1	0.204	0.304
D	0.745	0.785		18.93	19.93
E	0.300	0.325		7.62	8.25
E_1	0.240	0.260		6.10	6.60
e_1	0.100 TP		2	2.54 TP	
e_A	0.300 TP		2, 3	7.62 TP	
L	0.125	0.150		3.18	3.81
L_2	0.000	0.030		0.000	0.76
a	0°	15°	4	0°	15°
N	16		5	16	
N_1	0		6	0	
Q_1	0.040	0.075		1.02	1.90
S	0.015	0.060		0.39	1.52

92CM-15967R3

NOTES:
Refer to Rules for Dimensioning (JEDEC Publication No. 13) for Axial Lead Product Outlines.
1. When the device is supplied solder dipped, the maximum lead thickness (narrow portion) will not exceed 0.013'' (0.33 mm).
2. Leads within 0.005'' (0.12 mm) radius of True Position (TP) at gauge plane with maximum material condition and unit installed.
3. e_A applies in zone L_2 when unit installed.
4. a applies to spread leads prior to installation.
5. N is the maximum quantity of lead positions.
6. N_1 is the quantity of allowable missing leads.

(D) Suffix
JEDEC MO-001-AE
16-Lead Dual-In-Line White-Ceramic Package

SYMBOL	INCHES		NOTE	MILLIMETERS	
	MIN.	MAX.		MIN.	MAX.
A	0.120	0.160		3.05	4.06
A_1	0.020	0.065		0.51	1.65
B	0.014	0.020		0.356	0.508
B_1	0.035	0.065		0.89	1.65
C	0.008	0.012	1	0.204	0.304
D	0.745	0.785		18.93	19.93
E	0.300	0.325		7.62	8.25
E_1	0.240	0.260		6.10	6.60
e_1	0.100 TP		2	2.54 TP	
e_A	0.300 TP		2, 3	7.62 TP	
L	0.125	0.150		3.18	3.81
L_2	0.000	0.030		0.000	0.76
a	0°	15°	4	0°	15°
N	16		5	16	
N_1	0		6	0	
Q_1	0.050	0.085		1.27	2.15
S	0.015	0.060		0.39	1.52

92SS-4286R4

When incorporating RCA Solid State Devices in equipment, it is recommended that the designer refer to "Operating Considerations for RCA Solid State Devices", Form No. 1CE-402, available on request from RCA Solid State Division, Box 3200, Somerville, N. J. 08876.

APPENDIX 8. CM860 TELEDYNE

ULTRA LOW NOISE
SILICON EPITAXIAL JUNCTION
N-CHANNEL FIELD EFFECT TRANSISTOR

2N6550
CM860

GEOMETRY 424

The 2N6550/CM860 is a high, g_m/I_D low noise junction F.E.T. for low level amplifier use. The min. g_m of 25,000 μmho assures a voltage gain of 25 min. with a 1K drain load. As a source follower, it has typical output impedance of 25 ohms. The 10mA operating point is easily held due to its low pinch-off voltage and is very close to its zero T.C. point for temperature stable operation.

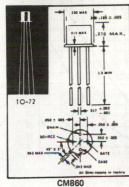

CM860

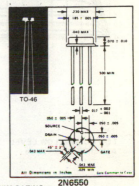

2N6550

The CM860 is in the four lead, TO-72 package which isolates all elements from the case, reducing stray capacitance and allowing the engineer greater design freedom.

ELECTRICAL DATA ABSOLUTE MAXIMUM RATING

PARAMETER	SYMBOL	2N6550	UNITS
Drain to Source Voltage	BV_{DSO}	20	Volts
Drain to Gate Voltage	BV_{DGO}	20	Volts
Gate to Source Voltage	BV_{GSO}	-20	Volts
D.C. Forward Gate Current	I_{GF}	50	mA
Junction Temp. (Operating & Storage)	T_J	-65°C to +200°C	
Power Dissipation (Free Air)	P_D	400 mW	
Lead Temp. (@ 1/16'' ± 1/32'' from case)	T_L	240°C for 10 sec.	
Derating Factor (Free Air)	D_F	2.3 mW/°C	

ELECTRICAL CHARACTERISTICS: T_A = 25°C (UNLESS OTHERWISE STATED)

PARAMETER	SYMBOL	CONDITION	2N6550 Min.	Typ.	Max.	UNITS
Gate Leakage Current	I_{GSS}	V_{GS} = -10V, V_{DS} = 0		0.1	3.0	nA
Gate Leakage Current	I_{GSS}	V_{GS} = -10V, V_{DS} = 0, T_A = 85°C		5	100	nA
Zero Gate Voltage Drain Current	I_{DSS}	V_{DS} = 10V, V_{GS} = 0	10	100		mA
Pinch-Off Voltage	V_{PO}	V_{DS} = 10V, I_D = 0.1mA	0.3	1.5	3.0	Volts
Transconductance	g_m	V_{DS} = 10V, I_D = 10mA, f = 1kHz	25	40		mmho
Input Capacitance	C_{iss}	V_{DS} = 10V, I_D = 10mA, f = 140kHz		30	35	pfd
Reverse Xfer Cap	C_{rss}	V_{DS} = 10V, f = 140kHz		17	20	pfd
Gate to Drain Capacitance	C_{GD}	V_{GD} = -10V, f = 140kHz		20		pfd
Output Admittance	Y_{OS}	V_{DS} = 10V, I_D = 10mA		50	100	μmho
Input Noise Voltage	e_n	V_{DS} = 5V, I_D = 10mA, f = 1kHz		1.4	2.0	$nV/Hz^{1/2}$
Input Noise Voltage	e_n	V_{DS} = 5V, I_D = 10mA, f = 10Hz		6.0	10	$nV/Hz^{1/2}$
Input Noise Voltage	e_n TOTAL	V_{DS} = 5V, I_D = 10mA, f = 10Hz to 20kHz		0.4	0.6	μVrms
Equivalent Open Ckt. Input Noise current	i_n	Rsource < 100K Ω f = 1 kHz		.01		$pA/Hz^{1/2}$

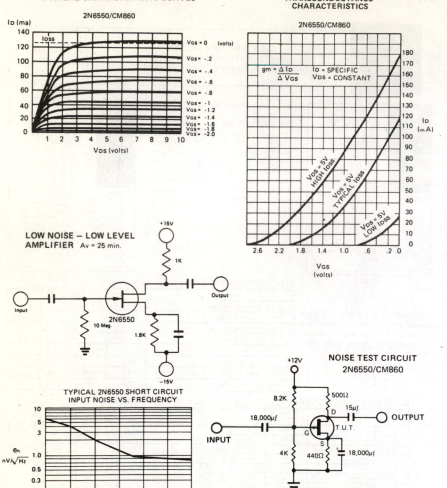

TYPICAL CHARACTERISTIC CURVES

2N6550/CM860

TRANSCONDUCTANCE CHARACTERISTICS

2N6550/CM860

$$gm = \frac{\Delta I_D}{\Delta V_{GS}} \qquad \begin{array}{l} I_D = \text{SPECIFIC} \\ V_{DS} = \text{CONSTANT} \end{array}$$

LOW NOISE – LOW LEVEL AMPLIFIER Av = 25 min.

TYPICAL 2N6550 SHORT CIRCUIT INPUT NOISE VS. FREQUENCY

NOISE TEST CIRCUIT
2N6550/CM860

APPENDIX 9. CP640 TELEDYNE CRYSTALLONICS

	CP640
BROADBAND RF FET	CP664
SILICON EPITAXIAL JUNCTION	CP665
N-CHANNEL FIELD EFFECT TRANSISTOR	CP666

HIGH DYNAMIC RANGE HF AND VHF AMPLIFIER
FOR USE IN COMMON GATE CONFIGURATION
- USABLE TO OVER 300 MHz
- 50 Ohm VSWR < 1.5:1 0.5-50 MHz (FIG. 1)
- LOW NOISE FIGURE — 2.2 dB TYPICAL @ 50 MHz
- INPUT Z CONSTANT 0.5-50 MHz
- HIGH IM INTERCEPT POINT — > + 40 dBm
- HIGH TRANSCONDUCTANCE — 100,000 μmhos (TYP.)
- 1 dB COMPRESSION POINT > + 20 dBm
- DYNAMIC RANGE >140 dB (TO 1 dB COMPRESSION)
- HIGH VOLTAGE—TO 50 V.

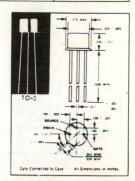

TYPICAL TWO TONE 3rd ORDER IM
PRODUCTS — CIRCUIT FIGURE 1

Tones at 3MHz/5MHz

Signal Level EMF	3rd Order Product
1 Volt	44 dB
0.3 Volt	75 dB
0.25 Volt (0dBM)	80 dB

ELECTRICAL DATA   ABSOLUTE MAXIMUM RATINGS

PARAMETER	SYMBOL	CP 640	CP 664	CP 665	CP 666	UNITS
Drain to Source Voltage	BVDSO	20	30	40	50	Volts
Drain to Gate Voltage	BVDGO	20	30	40	50	Volts
Gate to Source Voltage	BVGSO	− 15	− 20	− 20	− 20	Volts
Peak Drain Current	ID	1.2	1.2	1.2	1.2	Amps
Power Dissipation 25 °C CASE	PD	8.0	8.0	8.0	8.0	Watts
Derating Factor (slope)	DF	22	22	22	22	°C/W
Junction Temp (Oper. & Store)	TJ	− 55 °C to + 200 °C				

ELECTRICAL CHARACTERISTICS: T$_{CASE}$ = 25 °C (UNLESS OTHERWISE STATED)

PARAMETERS	CONDITIONS		SYMBOL	Min.	Typ.	Max.	UNITS
Gate Leakage Current	V_{GS} = 15V, V_{DS} = 0	25 °C	I$_{GSS}$		5	100	nA
		150 °C	I$_{GSS}$			10	μA
Operating Transconductance	V_{DS} = 15V, I$_{DS}$ = 40 mA		gf$_0$	40	60	80	mmho
Zero Bias Transconductance	V_{DS} = 15V, V_{GS} = 0(1)		gf$_0$	75	100	200	mmho
Gate-Source Cut-Off Voltage	V_{DS} = 5V, I$_{DS}$ = 1.0 mA		V_{GS} (off)	2	5	10	Volts
Zero Bias Drain Current	V_{DS} = 15V, V_{GS} = 0 (1)		I$_{DSS}$	100	400	800	mA
Gate to Source Cap.	V_{GS} = − 20V		C$_{GS}$		15	20	pf
Gate to Drain Cap.	V_{GD} = − 20V		C$_{GD}$		15	20	pf
Power Gain	I$_{DS}$ = 40mA, f = 50MHz, Fig. 1		Gpg	8	8.5	9.5	dB
Noise Figure	I$_{DS}$ = 40mA, f = 30MHz, Fig. 1		N.F.		2.2	3.0	dB
Voltage Standing Wave Ratio	f = 0.5-50MHz, 50 Ω Source, Fig. 1		VSWR			1.5:1	
Common Gate Input Conductance	f = 0.5-50MHz, V_{DS} = 15, I$_D$ = 40mA		gigs		60		mmho
Common Gate Output Conductance	f = − 50MHz, V_{DS} = 15, I$_D$ = 40mA		gogs		0.4		mmho

¹Pulse Measurement 1% Duty Cycle 10 mS Max.

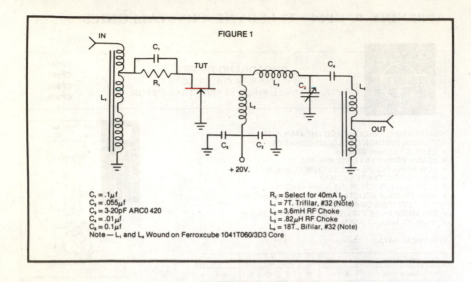

FIGURE 1

$C_1 = .1\mu f$
$C_2 = .055\mu f$
$C_3 = 3-20pF$ ARC0 420
$C_4 = .01\mu f$
$C_5 = 0.1\mu f$
Note — L_1 and L_4 Wound on Ferroxcube 1041T060/3D3 Core

$R_1 = $ Select for 40mA I_D
$L_1 = $ 7T. Trifilar, #32 (Note)
$L_2 = $ 3.6mH RF Choke
$L_3 = .82\mu H$ RF Choke
$L_4 = $ 18T., Bifilar, #32 (Note)

TYPICAL INTERCEPT AND COMPRESSION POINT

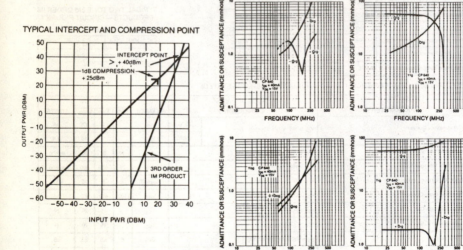

POWRFET™
SILICON EPITAXIAL JUNCTION
N-CHANNEL FIELD EFFECT TRANSISTOR

CP643

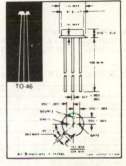

TO-46

- FOR HIGH DYNAMIC RANGE R.F. AMPLIFIERS
- SPECIFIED FOR H.F. BAND — USEABLE THRU 500 MHz
- LOW NOISE FIGURE DIRECT FROM 50 Ohm LINE[2]

ELECTRICAL DATA   ABSOLUTE MAXIMUM RATINGS

PARAMETER	SYMBOL		UNITS
Drain to Source Voltage	BVDSO	30	Volts
Drain to Gate Voltage	BVDGO	30	Volts
Gate to Source Voltage	BVGSO	−15	Volts
Peak Drain Current	ID	0.3	Amps
Power Dissipation 25°C CASE	PD	2.0	Watts
Derating Factor (slope)	DF	87	°C/W
Junction Temp. (Oper. & Store)	TJ	−55°C to +200°C	

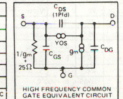

HIGH FREQUENCY COMMON
GATE EQUIVALENT CIRCUIT

**TYPICAL TWO TONE 3rd
ORDER IM PRODUCTS**

Tones at 3MHz/5MHz

Signal Level EMF	Typ.3rd Order Product
0.25 Volt (0dBM)	- 58dB

ELECTRICAL CHARACTERISTICS: T_CASE = 25°C (UNLESS OTHERWISE STATED)

PARAMETERS AND CONDITIONS	SYMBOL	CP 643			UNITS
		Min.	Typ.	Max.	
Gate Leakage Current $V_{GS} = -15V, V_{DS} = 0$	IGSS	–	1.0	10	nA
Gate Leakage Current $V_{GS} = -15V, V_{DS} = 0, T_C = 125°C$	IGSS	–	–	10	μA
Transconductance $V_{DS} = 15V, I_{DS} = 25$ mA	gm	20	25	30	mMhos
Pinch-Off Voltage $V_{DS} = 5V, I_{DS} = 1.0$ mA	VPO	2.0	4.0	7.0	Volts
Gain in Ckt. of TMF18 $I_{DS} = 25$ mA, f = 1 to 100 MHz.	A	8.0	9.0	10.0	dB
Gate to Source Cap $V_{GS} = -20V$	CGS	–	5	6	pfd
Gate to Drain Cap $V_{GD} = -20V$	CGD	–	5	6	pfd
Drain Current[1] $V_{DS} = 15V, V_{GS} = 0$	IDSS	50	100	250	mAmps
TMF18[3] $I_{DS} = 25$ mA, f = 1 MHz.	N.F.	–	4.0	5.0	dB

1 Pulse Measurement 1% Duty Cycle 10 mS Max.
2 The noise figure will be improved at the cost of gain when used in a 75Ω line with a 2:1 output winding ratio or in a 50Ω line with an input step up transformer.
3 The gain may be raised at a sacrifice in bandwidth by increasing the output transformer ratio.

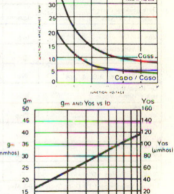

JUNCTION CAPACITANCE VS VOLTAGE

POWRFET™
SILICON EPITAXIAL JUNCTION
N-CHANNEL FIELD EFFECT TRANSISTORS

CP650
CP651

- LOW C_{GD} — 20 pfd TYPICAL
- HIGH I_{DSS} — 0.5 Amp TYPICAL
- HIGH g_m — 150,000 μmhos TYPICAL

ELECTRICAL DATA   ABSOLUTE MAXIMUM RATINGS

PARAMETER	SYMBOL	CP650	CP651	UNITS
Drain to Source Voltage	BV_{DSO}	25	20	Volts
Drain to Gate Voltage	BV_{DGO}	25	20	Volts
Gate to Source Voltage	BV_{GSO}	-25	-20	Volts
Peak Drain Current	I_D	1.2	0.6	Amps
Power Dissipation 25°C Case	P_D	8.0	8.0	Watts
Derating Factor (slope)	D_F	22	22	°C/W
Junction Temp. (Oper. & Store)	T_J	-65°C to +200°C		

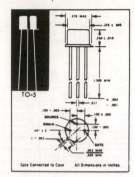

TO-5

Gate Connected to Case   All Dimensions in Inches

ELECTRICAL CHARACTERISTICS: T_{CASE} = 25°C (UNLESS OTHERWISE STATED)

PARAMETERS AND CONDITIONS	SYMBOL	AMPLIFIERS						UNITS
		CP650			CP651			
		Min.	Typ.	Max.	Min.	Typ.	Max.	
Gate Leakage Current V_{GS} = -15V, V_{DS} = 0	I_{GSS}	–	5.0	100	–	5.0	100	nA
Gate Leakage Current V_{GS} = -15V, V_{DS} = 0, TC = 100°C	I_{GSS}	–	–	10	–	–	10	μA
Transconductance[1] V_{DS} = 15V, V_{GS} = 0	g_m	0.1	0.15	0.25	0.075	0.1	0.2	mhos
Pinch-Off Voltage V_{DS} = 5V, I_{DS} = 1.0mA/3nA*	V_{PO}	2.0	5.0	10	2.0	5.0	10	Volts
On Resistance I_{DS} = 10mA, V_{GS} = 0	R_{DS}	–	4.0	–	–	7.0	–	Ohms
Gate to Source Cap. V_{GS} =-20V	C_{GS}	–	20	25	–	20	25	pfd
Gate to Drain Cap. V_{GD} =-20V	C_{GD}	–	20	25	–	20	25	pfd
Drain Current[1] V_{DS} = 15V, V_{GS} = 0	I_{DSS}	0.3	0.6	1.2	0.1	0.3	0.5	Amps
Gain-Bandwidth Product V_{DS} = 15V, V_{GS} = 0	F_t	–	1.0	–	–	1.0	–	GHz

[1] Pulse Measurement 1% Duty Cycle 10 mS Max.

CP640 , CP650 , CP651

TYPICAL CHARACTERISTICS

ON RESISTANCE VS. TEMPERATURE

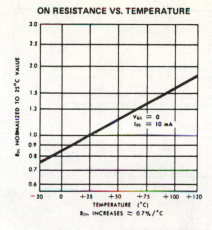

$V_{GS} = 0$
$I_{DS} = 10$ mA

R_{ON} NORMALIZED TO 25°C VALUE

TEMPERATURE (°C)

R_{ON} INCREASES $\approx 0.7\%/°C$

GATE LEAKAGE CURRENT VS. TEMPERATURE

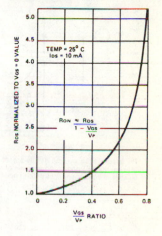

I_{GSS} NORMALIZED TO 25°C VALUE

TEMPERATURE (°C)

I_{GSS} DOUBLES EACH 10°C

JUNCTION CAPACITANCE VS. VOLTAGE

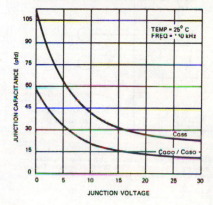

JUNCTION CAPACITANCE (pfd)

TEMP = 25° C
FREQ = 100 kHz

C_{GSS}

C_{GDO} / C_{GSO}

JUNCTION VOLTAGE

ON RESISTANCE VS. GATE VOLTAGE

R_{DS} NORMALIZED TO $V_{GS} = 0$ VALUE

TEMP = 25° C
I_{DS} = 10 mA

$$R_{ON} \approx \frac{R_{DS}}{1 - \frac{V_{GS}}{V_P}}$$

$\frac{V_{GS}}{V_P}$ RATIO

TECHNICAL MEMO

This RF broadband amplifier uses a constant current source in place of a source biasing resistor. The current source automatically sets the operating point regardless of I_{DSS} or V_p. The alternatives are selection of the source resistor or use of a high negative supply voltage ($\gg V_p$) and a high source resistor value.

The 2N2222 may be replaced by any NPN transistor with a reasonable β at 50mA. The base voltage divider includes a diode to temperature compensate for V_{BE} and sets up about 1V at the emitter. This voltage must be lower than the minimum V_{GS} at the specified I_{DS}.

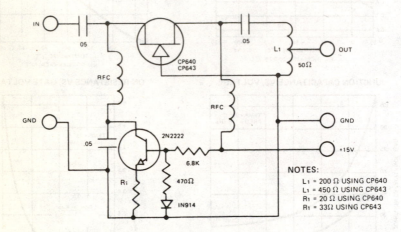

NOTES:

L_1 = 200 Ω USING CP640
L_1 = 450 Ω USING CP643
R_1 = 20 Ω USING CP640
R_1 = 33Ω USING CP643

The upper frequency limit of this broadband circuit can be extended by the addition of a peaking network between the FET and L_1. An example of such a network appears in the circuit on the CP640 data sheet. The broadband input impedance of this circuit is typically 20 ohms for the CP640 and 40 ohms for the CP643.

APPENDIX 10. LM194 NATIONAL SEMICONDUCTOR

National Semiconductor

LM194/LM394 Supermatch Pair

General Description

The LM194 and LM394 are junction isolated ultra well-matched monolithic NPN transistor pairs with an order of magnitude improvement in matching over conventional transistor pairs. This was accomplished by advanced linear processing and a unique new device structure.

Electrical characteristics of these devices such as drift versus initial offset voltage, noise, and the exponential relationship of base-emitter voltage to collector current closely approach those of a theoretical transistor. Extrinsic emitter and base resistances are much lower than presently available pairs, either monolithic or discrete, giving extremely low noise and theoretical operation over a wide current range. Most parameters are guaranteed over a current range of 1µA to 1 mA and 0V up to 40V collector-base voltage, ensuring superior performance in nearly all applications.

To guarantee long-term stability of matching parameters, internal clamp diodes have been added across the emitter-base junction of each transistor. These prevent degradation due to reverse biased emitter current—the most common cause of field failures in matched devices. The parasitic isolation junction formed by the diodes also clamps the substrate region to the most negative emitter to ensure complete isolation between devices.

The LM194 and LM394 will provide a considerable improvement in performance in most applications requiring a closely matched transistor pair. In many cases, trimming can be eliminated entirely, improving reliability and decreasing costs. Additionally, the low noise and high gain make this device attractive even where matching is not critical.

The LM194 and LM394/LM394B/LM394C are available in an isolated header 6-lead TO-5 metal can package. The LM394/LM394B/LM394C are also offered in an 8-lead DIP. The LM194 is identical to the LM394 except for tighter electrical specifications and wider temperature range.

Features

- Emitter-base voltage matched to 50µV
- Offset voltage drift less than 0.1µV/°C
- Current gain (h_{FE}) matched to 2%
- Common-mode rejection ratio greater than 120 dB
- Parameters guaranteed over 1µA to 1 mA collector current
- Extremely low noise
- Superior logging characteristics compared to conventional pairs
- Plug-in replacement for presently available devices

Typical Applications

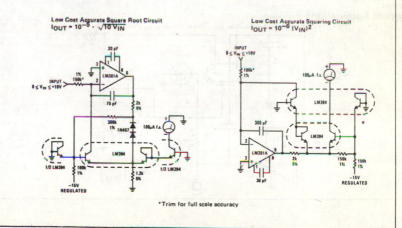

Low Cost Accurate Square Root Circuit
$I_{OUT} = 10^{-5} \cdot \sqrt{10\,V_{IN}}$

Low Cost Accurate Squaring Circuit
$I_{OUT} = 10^{-6}\,(V_{IN})^2$

*Trim for full scale accuracy

Absolute Maximum Ratings

Collector Current	20 mA	Collector-Collector Voltage	40V
Collector-Emitter Voltage	V_{MAX}	LM394C	20V
Collector-Emitter Voltage	40V	Base-Emitter Current	±10 mA
LM394C	20V	Power Dissipation	500 mW
Collector-Base Voltage	40V	Junction Temperature	
LM394C	20V	LM194	−55°C to +125°C
Collector-Substrate Voltage	40V	LM394/LM394B/LM394C	−25°C to +85°C
LM394C	20V	Storage Temperature Range	−65°C to +150°C
		Lead Temperature (Soldering, 10 seconds)	300°C

Electrical Characteristics $(T_J = 25°C)$

PARAMETER	CONDITIONS	LM194 MIN	LM194 TYP	LM194 MAX	LM394 MIN	LM394 TYP	LM394 MAX	LM394B/LM394C MIN	LM394B/LM394C TYP	LM394B/LM394C MAX	UNITS
Current Gain (h_{FE})	$V_{CB} = 0V$ to V_{MAX} (Note 1)										
	$I_C = 1$ mA	500	700		300	700		225	500		
	$I_C = 100\mu A$	400	550		250	550		200	400		
	$I_C = 10\mu A$	300	450		200	450		150	300		
	$I_C = 1\mu A$	200	300		150	300		100	200		
Current Gain Match (h_{FE} Match) $= \dfrac{100\, [\Delta I_B]\, [h_{FE(MINI)}]}{I_C}$	$V_{CB} = 0V$ to V_{MAX}										
	$I_C = 10\mu A$ to 1 mA		0.5	2		0.5	4		1.0	5	%
	$I_C = 1\mu A$		1.0			1.0			2.0		%
Emitter-Base Offset Voltage	$V_{CB} = 0$		25	50		25	150		50	200	μV
	$I_C = 1\mu A$ to 1 mA										
Change in Emitter-Base Offset Voltage vs Collector-Base Voltage (CMRR)	(Note 1) $I_C = 1\mu A$ to 1 mA, $V_{CB} = 0V$ to V_{MAX}		10	25 −		10	50		10	100	μV
Change in Emitter-Base Offset Voltage vs Collector Current	$V_{CB} = 0V$, $I_C = 1\mu A$ to 0.3 mA		5	25		5	50		5	50	μV
Emitter-Base Offset Voltage Temperature Drift			0.08	0.3		0.08	1.0		0.2	1.5	μV/°C
			0.03	0.1		0.03	0.3		0.03	0.5	μV/°C
Logging Conformity	$I_C = 3$ nA to 300μA, $V_{CB} = 0$, (Note 3)		150			150			150		μV
Collector-Base Leakage	$V_{CB} = V_{MAX}$		0.05	0.25		0.05	0.5		0.05	0.5	nA
Collector-Collector Leakage	$V_{CC} = V_{MAX}$		0.1	2.0		0.1	5.0		0.1	5.0	nA
Input Voltage Noise	$I_C = 100\mu A$, $V_{CB} = 0V$, $f = 100$ Hz to 100 kHz		1.8			1.8			1.8		nV/$\sqrt{Hz}$
Collector to Emitter Saturation Voltage	$I_C = 1$ mA, $I_B = 10\mu A$		0.2			0.2			0.2		V
	$I_C = 1$ mA, $I_B = 100\mu A$		0.1			0.1			0.1		V

Note 1: Collector-base voltage is swept from 0 to V_{MAX} at a collector current of $1\mu A$, $10\mu A$, $100\mu A$, and 1 mA.

Note 2: Offset voltage drift with $V_{OS} = 0$ at $T_A = 25°C$ is valid only when the ratio of I_{C1} to I_{C2} is adjusted to give the initial zero offset. This ratio must be held to within 0.003% over the entire temperature range. Measurements taken at +25°C and temperature extremes.

Note 3: Logging conformity is measured by computing the best fit to a true exponential and expressing the error as a base-emitter voltage deviation.

Typical Applications (Continued)

Fast, Accurate Logging Amplifier, $V_{IN} = 10V$ to 0.1 mV or $I_{IN} = 1$ mA to 10 nA

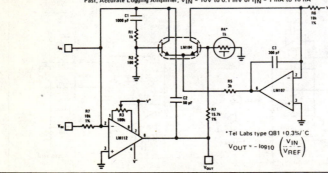

*Tel Labs type Q81 ±0.3%/°C

$$V_{OUT} = -\log_{10}\left(\frac{V_{IN}}{V_{REF}}\right)$$

APPENDIX 11. TEXAS INSTRUMENTS DIGITAL PHASE-LOCKED-LOOP FILTERS

TTL
LSI

TYPES SN54LS297, SN74LS297
DIGITAL PHASE-LOCKED-LOOP FILTERS

SN54LS297 . . . J OR W PACKAGE
SN74LS297 . . . J OR N PACKAGE

- **Digital Design Avoids Analog Compensation Errors**

- **Easily Cascadable for Higher Order Loops**

- **Useful Frequency from DC to:**
  50 MHz Typical (K Clock)
  35 MHz Typical (I/D Clock)

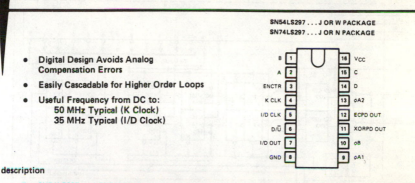

B	1	16	V_CC
A	2	15	C
ENCTR	3	14	D
K CLK	4	13	φA2
I/D CLK	5	12	ECPD OUT
D/Ū	6	11	XORPD OUT
I/D OUT	7	10	φB
GND	8	9	φA1

description

The SN54LS297 and SN74LS297 devices are designed to provide a simple, cost-effective solution to high-accuracy, digital, phase-locked-loop applications. These devices contain all the necessary circuits, with the exception of the divide-by-N counter, to build first order phase-locked loops as described in Figure 1 in the operations section.

Both exclusive-OR (XORPD) and edge-controlled (ECPD) phase detectors are provided for maximum flexibility.

Proper partitioning of the loop function, with many of the building blocks external to the package, makes it easy for the designer to incorporate ripple cancellation or to cascade to higher order phase-locked loops.

The length of the up/down K counter is digitally programmable according to the K counter function table. With A, B, C, and D all low, the K counter is disabled. With A high and B, C, and D low, the K counter is only three stages long, which widens the bandwidth or capture range and shortens the lock time of the loop. When A, B, C, and D are all programmed high, the K counter becomes seventeen stages long, which narrows the bandwidth or capture range and lengthens the lock time. Real-time control of loop bandwidth by manipulating the A through D inputs can maximize the overall performance of the digital phase-locked loop.

The 'LS297 can perform the classic first-order phase-locked loop function without using analog components. The accuracy of the digital phase-locked loop (DPLL) is not affected by V_{CC} and temperature variations, but depends solely on accuracies of the K clock, I/D clock, and loop propagation delays.

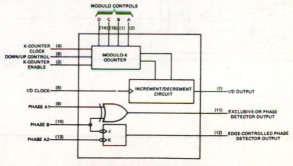

FIGURE 1—SIMPLIFIED BLOCK DIAGRAM

TYPES SN54LS297, SN74LS297
DIGITAL PHASE-LOCKED-LOOP FILTERS

functional block diagram

TYPES SN54LS297, SN74LS297
DIGITAL PHASE-LOCKED-LOOP FILTERS

K COUNTER FUNCTION TABLE
(DIGITAL CONTROL)

D	C	B	A	MODULO (K)
L	L	L	L	Inhibited
L	L	L	H	2^3
L	L	H	L	2^4
L	L	H	H	2^5
L	H	L	L	2^6
L	H	L	H	2^7
L	H	H	L	2^8
L	H	H	H	2^9
H	L	L	L	2^{10}
H	L	L	H	2^{11}
H	L	H	L	2^{12}
H	L	H	H	2^{13}
H	H	L	L	2^{14}
H	H	L	H	2^{15}
H	H	H	L	2^{16}
H	H	H	H	2^{17}

FUNCTION TABLE
EXCLUSIVE-OR PHASE DETECTOR

$\phi A1$	ϕB	XORPD OUT
L	L	L
L	H	H
H	L	H
H	H	L

FUNCTION TABLE
EDGE-CONTROLLED PHASE DETECTOR

$\phi A2$	ϕB	ECPD OUT
H or L	↓	H
↓	H or L	L
H or L	↑	No change
↑	H or L	No change

H = steady-state high level
L = steady-state low level
↓ = transition from high to low
↑ = transition from low to high

schematics of inputs and outputs

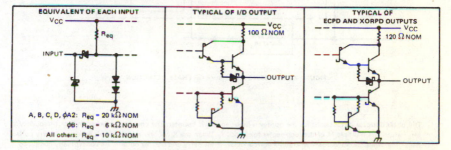

EQUIVALENT OF EACH INPUT

A, B, C, D, $\phi A2$: R_{eq} = 20 kΩ NOM
ϕB: R_{eq} = 6 kΩ NOM
All others: R_{eq} = 10 kΩ NOM

TYPICAL OF I/D OUTPUT

V_{CC} 100 Ω NOM

TYPICAL OF
ECPD AND XORPD OUTPUTS

V_{CC} 120 Ω NOM

operation

The phase detector generates an error signal waveform that, at zero phase error, is a 50% duty cycle square wave. At the limits of linear operation, the phase detector output will be either high or low all of the time, depending on the direction of the phase error ($\phi_{in} - \phi_{out}$). Within these limits, the phase detector output varies linearly with the input phase error according to the gain k_d, which is expressed in terms of phase detector output per cycle of phase error. The phase detector output can be defined to vary between ±1 according to the relation:

$$\text{PD Output} = \frac{\% \text{ high} - \% \text{ low}}{100} \qquad (1)$$

The output of the phase detector will be $k_d \phi_e$, where the phase error $\phi_e = \phi_{in} - \phi_{out}$.

Exclusive-OR phase detectors (XORPD) and edge-controlled phase detectors (ECPD) are commonly used digital types. The ECPD is more complex than the XORPD logic function, but can be described generally as a circuit that changes states on one of the transitions of its inputs. k_d for an XORPD is 4 because its output remains high (PD output = 1) for a phase error of 1/4 cycle. Similarly, k_d for the ECPD is 2 since its output remains high for a phase error of 1/2 cycle. The type of phase detector will determine the zero-phase-error point, i.e., the phase separation of the phase detector inputs for ϕ_e defined to be zero. For the basic DPLL system of Figure 2, ϕ_e = 0 when the phase detector output is a square wave. The XORPD inputs are 1/4 cycle out of phase for zero phase error. For the ECPD, ϕ_e = 0 when the inputs are 1/2 cycle out of phase.

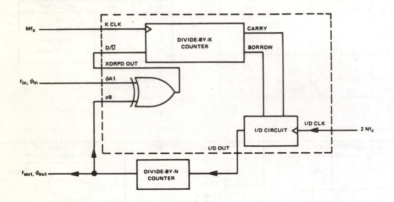

FIGURE 2—DPLL USING EXCLUSIVE-OR PHASE DETECTION

The phase detector output controls the up/down input to the K counter. The counter is clocked by input frequency Mf_c, which is a multiple M of the loop center frequency f_c. When the K counter recycles up, it generates a carry pulse. Recycling while counting down generates a borrow pulse. If the carry and borrow outputs are conceptually combined into one output that is positive for a carry and negative for a borrow, and if the K counter is considered as a frequency divider with the ratio Mf_c/K, the output of the K counter will equal the input frequency multiplied by the division ratio. Thus the output from the K counter is $(k_d \phi_e Mf_c)/K$.

The carry and borrow pulses go to the increment/decrement (I/D) circuit, which, in the absence of any carry or borrow pulse, has an output that is 1/2 of the input clock I/D CLK. The input clock is just a multiple, 2N, of the loop center frequency. In response to a carry or borrow pulse, the I/D circuit will either add or delete a pulse at I/D OUT. Thus the output of the I/D circuit will be $Nf_c + (k_d \phi_e Mf_c)/2K$.

The output of the N counter (or the output of the phase-locked loop) is thus:

$$f_o = f_c + (k_d \phi_e Mf_c)/2KN$$

If this result is compared to the equation for a first-order analog phase-locked loop, the digital equivalent of the gain of the VCO is just $Mf_c/2KN$ or f_c/K for M = 2N.

Thus the simple first-order phase-locked loop with an adjustable K counter is the equivalent of an analog phase-locked loop with a programmable VCO gain.

TYPES SN54LS297, SN74LS297
DIGITAL PHASE-LOCKED-LOOP FILTERS

electrical characteristics over recommended operating free-air temperature range (unless otherwise noted)

PARAMETER		TEST CONDITIONS[†]		SN54LS297 MIN	TYP[‡]	MAX	SN74LS297 MIN	TYP[‡]	MAX	UNIT
V_{IH}	High-level input voltage			2			2			V
V_{IL}	Low-level input voltage					0.7			0.8	V
V_{IK}	Input clamp voltage	V_{CC} = MIN,	I_I = −18 mA			−1.5			−1.5	V
V_{OH} High-level output voltage	I/D OUT	V_{CC} = MIN, V_{IH} = 2 V,	I_{OH} = MAX	2.4			2.4			V
	Others	V_{IL} = V_{IL} max	I_{OH} = MAX	2.5			2.7			
V_{OL} Low-level output voltage	I/D OUT	V_{CC} = MIN, V_{IH} = 2 V,	I_{OL} = 12 mA		0.25	0.4		0.25	0.4	V
			I_{OL} = 24 mA					0.35	0.5	
	Others	V_{IL} = V_{IL} max	I_{OL} = 4 mA		0.25	0.4		0.25	0.4	
			I_{OL} = 8 mA					0.35	0.5	
I_I	Input current at maximum input voltage	V_{CC} = MAX,	V_I = 7 V			0.1			0.1	mA
I_{IH} High-level input current	U/D̄, EN, φA1	V_{CC} = MAX,	V_I = 2.7 V			40			40	μA
	φB					60			60	
	All others					20			20	
I_L Low-level input current	A,B,C,D,φA1	V_{CC} = MAX,	V_I = 0.4 V			−0.4			−0.4	mA
	φB	V_I = 0.4 V				−1.2			−1.2	
	All others					−0.8			−0.8	
I_{OS} Short-circuit output current §	I/D OUT	V_{CC} = MAX		−30		−130	−30		−130	mA
	Others			−20		−100	−20		−100	
I_{CC}	Supply current	V_{CC} = MAX, All inputs grounded, All outputs open			75	120		75	120	mA

†For conditions shown as MIN or MAX, use the appropriate value specified under recommended operating conditions.
‡All typical values are of V_{CC} = 5 V, T_A = 25°C.
§ Not more than one output should be shorted at a time and the duration of the short-circuit should not exceed one second.

switching characteristics, V_{CC} = 5 V, T_A = 25°C

PARAMETER[¶]	FROM (INPUT)		TO (OUTPUT)	TEST CONDITIONS	MIN	TYP	MAX	UNIT
f_{max}	KCLK		I/D OUT	R_L = 667 Ω, C_L = 45 pF, See Note 2	30	50		MHz
	I/D CLK		I/D OUT		15	35		
t_{PLH}	I/D CLK ↑		I/D OUT			15	25	ns
t_{PHL}	I/D CLK ↑		I/D OUT			22	35	ns
t_{PLH}	φA1 or φB	Other input low	XOR OUT	R_L = 2 kΩ, C_L = 45 pF, See Note 2		10	15	ns
	φA1 or φB	Other input high	XOR OUT			17	25	
t_{PHL}	φA1 or φB	Other input low	XOR OUT			15	25	ns
	φA1 or φB	Other input high	XOR OUT			17	25	
t_{PLH}	φB ↓		ECPD OUT			20	30	ns
t_{PHL}	φA2↓		ECPD OUT			20	30	ns

¶t_{PLH} = propagation delay time, low-to-high level output
t_{PHL} = propagation delay time, high-to-low level output
NOTE 2: Load circuit and voltage waveforms are shown on page 3-11 of *The TTL Data Book for Design Engineers*, Second Edition, LCC4112.

INDEX